Contributions from Science Education Research

Volume 9

Contributions from Science Education Research is the international, multidisciplinary book series of the European Science Education Research Association (ESERA). The aim of the series is to synthesize, for the benefit of the scholarly community, the findings of high quality, theoretically-framed research in the domain of science education as well as comprehensive explorations of specific methodological strands in science education research. The series aims to publish books that are innovative in attempting to forge new ways of representing emergent knowledge in the field. The series includes edited collections of chapters, monographs and handbooks that are evaluated on the basis of originality, scientific rigor and significance for science education research. The book series is intended to focus mainly on work carried out in Europe. However, contributions from researchers affiliated with non-European institutions and non-members of the European Science Education Research Association are welcomed. The series is designed to appeal to a wide audience of researchers and post-graduate students in science education. Book proposals for this series may be submitted to the Publishing Editor: Claudia Acuna E-mail: Claudia.Acuna@springer.com

More information about this series at http://www.springer.com/series/11180

Olivia Levrini • Giulia Tasquier
Tamer G. Amin • Laura Branchetti
Mariana Levin
Editors

Engaging with Contemporary Challenges through Science Education Research

Selected papers from the ESERA 2019 Conference

Springer

Editors
Olivia Levrini
Department of Physics and Astronomy
"A. Righi"
Alma Mater Studiorun - University
of Bologna
Bologna, Italy

Giulia Tasquier
Department of Physics and Astronomy
"A. Righi"
Alma Mater Studiorun - University
of Bologna
Bologna, Italy

Tamer G. Amin
Department of Education
American University of Beirut
Beirut, Lebanon

Laura Branchetti
Department of Mathematics
"Federigo Enriques"
University of Milan
Milan, Italy

Mariana Levin
Department of Mathematics
Western Michigan University
Kalamazoo, Michigan, USA

ISSN 2213-3623 ISSN 2213-3631 (electronic)
Contributions from Science Education Research
ISBN 978-3-030-74489-2 ISBN 978-3-030-74490-8 (eBook)
https://doi.org/10.1007/978-3-030-74490-8

This Springer imprint is published by the registered company Springer Nature Switzerland AG
The registered company address is: Gewerbestrasse 11, 6330 Cham, Switzerland

In memory of Nella (Lella) Grimellini Tomasini.

Nella (Lella) Grimellini Tomasini passed away on March 27, 2020.

We wish to dedicate this book to her.

Lella founded the research group in Physics Education in Bologna in the 1960s and took an active part in the establishment of ESERA in Leeds, 1995. We decided to organize the ESERA conference in Bologna mainly because of her special feelings of affection toward ESERA.

She made original contributions in research on conceptual change, teacher education, and the role of the laboratory in the teaching and learning of physics as a culture.

We all remember her for her lively and brilliant intelligence, her iron determination and attention to detail, her elegant passion for knowledge, and for "the pleasure of understanding." Her special way of "looking at science to see human thought reflected in it" is etched into how we continue to carry out our research and nurture our intellectual and human collaborations.

We will miss her.

Introduction

This edited volume is composed of selected papers that were presented at the 13th European Science Education Research Association (ESERA) Conference, held in Bologna, Italy, from the 26th to the 30th of August 2019. The ESERA 2019 Conference theme was ***The beauty and pleasure of understanding: Engaging with contemporary challenges through science education***.

The organization of the ESERA 2019 conference was undertaken by Olivia Levrini (Conference President) and Giulia Tasquier (Conference Manager) in collaboration with the research group in Physics Education and History of Physics at ALMA MATER STUDIORUM – University of Bologna, and with the support of the Steering Committee, the Local Organising Committee, the Scientific Committee and the ESERA Executive Board. Technical and logistical support was provided by EGA Worldwide.

ESERA is an international organization for science education researchers and science educators, and it aims to: (i) enhance the range and quality of research and research training in science education; (ii) provide a forum for collaboration in science education research; (iii) represent the professional interests of science education researchers in Europe; (iv) seek to relate research to the policy and practice of science education in Europe; and (v) foster links between science education researchers in Europe and elsewhere in the world (www.esera.org).

The ESERA community consists of professionals with diverse disciplinary backgrounds, ranging from the natural to the social sciences. Such diversity provides a broad range of perspectives on research, practice and policy in science education and is well reflected in this volume. The biennial ESERA conference is the main forum for direct scientific discourse within the community, for exchange of insightful practices, and for extending networks among researchers and educators. The contributions in this volume showcase current orientations of research in science education.

Overall, this book will be of interest to an international audience of science teachers, teacher educators and science education researchers who have a commitment to evidence-based and innovative science teaching and learning.

Behind the Scientific Organization of the ESERA 2019 Conference

As we worked to create the scientific program of the conference, we (the Scientific Committee for ESERA 2019) tried to imagine the story that we hoped would emerge from the dynamics of the conference – that would be "in the air" and would stimulate formal and informal discussions among attendees. The story that emerged was pre-pandemic, but already it was strongly influenced by dramatic changes that were occurring in our respective societies. We were (and continue to be) living in what the sociologist Hartmut Rosa calls "the society of acceleration" (Rosa 2013), a society accelerated by the impressive velocity of scientific and technological (S&T) development. And we were, and are, living in a moment of deep social, political, and environmental global change, exacerbated by the COVID-19 crisis.

Many demanding contemporary challenges, that involve science education, deeply affect the present and the future of the younger generation and of the planet: climate change, multiculturalism, the flourishing of new interdisciplinary domains (like cognitive neuroscience, artificial intelligence, digital humanities to name a few), as well as issues stemming from living in the digital and post-truth era. With this backdrop, the questions that arose during the process of organizing the conference were: how can we, as researchers in science education, contribute to equipping the younger generation with what they need to cope with contemporary challenges like these? In particular, what contribution can a conference like this one make?

In this frenetic and fast-changing society, we imagined a conference where it would be possible to take time to deeply reflect on what was happening in the present, while also taking time to push our imagination forward – to think about possible, alternative, desirable future scenarios for science education and for the relation between science and society. Specifically, to enhance these discussions, we started from the belief that, being science educators, generating understanding is for us the preferred way to address these challenges, recognizing also that these challenges are so deep and novel that addressing them via science education necessitates collectively searching for new narratives, languages and forms of beauty that capture our attention and trigger new ways of thinking.

Accordingly, we decided upon the theme of the conference, *The beauty and pleasure of understanding: engaging with contemporary challenges through science education,* and tried to create moments and contexts that would nurture a deeply reflective attitude on the present and on current science education research and, at the same time, inspire a provocative and visionary attitude toward the future.

Indeed, both the theme for the conference and the invited speakers and symposia were chosen to be "foundational" (for the purpose of orienting the community towards reflection on current science education research and practice) and/or "visionary" (able to open new, positive, and active windows towards the future). Specifically, the opening speech by Igal Galili was about "The beauty and pleasure of understanding" detailing the importance of aesthetic engagement in science education, moving beyond "understanding" in its more narrow sense, as well as pointing out the special history of the city of Bologna in debates on beauty and understanding. The first plenary lecture was given by noted Icelandic writer, Andri Snær Magnasson, who alerted us to the ways climate change challenges all our available forms of describing a phenomenon, from numerical representation to the myths of a society. This theme is engaged with in the first chapters of the book. The other three plenary talks and one panel by prominent researchers led to chapters that appear in different sections of the volume and were titled (ii) Where are we? Syntheses and Synergies in Science education research and practice (Bruce Sherin); (iv) Embodied cognition: From Neuroscience to Science Education (Corrado Sinigaglia & Tamer Amin); (v) Socioscientific-issues: Searching for new perspectives (Maria Evagorou and Jan Alexis Nielsen); (vi) Science Education in Multicultural and Multilingual Contexts (Mariona Espinet, Saouma BouJaoude, Sonya N. Martin, Audrey Msimanga and Alberto J. Rodríguez).

A total of 1792 single and multi-paper proposals were submitted to the ESERA 2019 conference in early 2019. Of the 1061 proposals submitted for single oral presentations, 824 were presented at the conference. A total of 410 proposals were presented as interactive posters and this included contributions from 91 young researchers who had attended the ESERA summer schools (in 2017, 2018, 2019). In total, of the 73 submitted, 63 symposia (each with four papers) were presented at the conference, of which 16 were invited symposia. Each symposium was organized by a coordinator around a specific topic and each of the papers addressed the topic from different perspectives by authors from different countries. Moreover, 15 sessions were presented in the format of an ICT demonstration, hands-on workshop or as a World Café. The conference week was thus richly scheduled with single oral presentations, symposia, interactive posters, ICT demonstrations and workshops divided into 18 different strands based on their topic (see www.esera2019.org).

After the conference, all presenters were invited to submit revised and extended papers on their conference presentation to the electronic proceedings of the ESERA 2019 Conference, which is available at https://www.esera.org/publications/esera-conference-proceedings/esera-2019 (Levrini, O. & Tasquier, G. (Eds.) (2020). Electronic Proceedings of the ESERA 2019 Conference. The Beauty and Pleasure of Understanding: Engaging with Contemporary Challenges through Science Education, Bologna: ALMA MATER STUDIORUM – University of Bologna. ISBN 978-88-945874-0-1).

The ESERA 2019 Conference was attended by 1609 science education researchers from 58 countries around the world and thus the conference was indeed a very international meeting. While presenting one's own research and engaging with others in discussion were among the most important aspects of the conference, having an opportunity to meet other science education researchers was just as valuable. The discussions at conference sessions provided opportunities for researchers and practitioners to exchange their experiences and approaches. The countless encounters with other researchers throughout the week enabled the participants to strengthen their existing networks, make new acquaintances and sow seeds for future cooperation.

Overview of the Organization of the Volume

This volume includes science education research presented at the ESERA 2019 conference identified by the strand chairs and the scientific committee as particularly interesting and representative of current work in the field. The topics discussed will generate interest and spark debate within the community of science education researchers and science educators. The editorial team is very grateful for all the work carried out by the international panel of strand chairs and reviewers who made it possible to include these selected papers in this compilation. Following the conference, the strand chairs recommended interesting conference contributions as possible papers for this book by following common criteria. The selection made by strand chairs was examined by the scientific committee and a selection of recommended authors were invited to submit full manuscripts. The papers underwent a rigorous scientific review process involving at least two reviewers per paper and the scientific committee. As the final product of the review process, this volume is composed of 25 chapters, organised in four sections: (1) Meeting societal challenges, (2) Expanding the evidence base, (3) Developing innovative theoretical perspectives and methodologies, and (4) Designing research-based instruction.

In the first section, we included chapters that examined how science education research could help meet contemporary societal challenges. These chapters engaged with broad policy issues, examined novel curricular approaches to meet societal concerns and reported on studies of science learning and instruction that focused on how learners could be prepared for meeting the pressing challenges of our time.

In the second section, we brought together chapters that focused on expanding the empirical base of current science education research. Any empirical field relies on its evidence base to validate its claims. The field of science education relies on evidence-based claims to ground its practical curricular and instructional recommendations. Indeed, the community of science educators is acutely aware that theoretical frameworks are complex knowledge structures that must be supported by a collective body of evidence. Thus, expanding the evidence base cumulatively via theoretically framed and methodologically rigorous investigations is crucial. This section includes chapters that report on empirical studies with clear theoretical framing and carefully designed quantitative and/or qualitative methods.

The third section is organized around a shift in focus toward the development of innovative theoretical perspectives and methodologies. To meet societal challenges that are increasingly complex, our theoretical understanding of science learning and instruction needs to match this complexity. This demands exploring new theoretical perspectives and crafting novel methods as appropriate. This section includes chapters that focus primarily on developing new, and often interdisciplinary, theoretical foundations and enriching the methodological tools available to the science education research community.

The last section concerns designing research-based instruction. Among the main goals of science education research, one of the most important is to contribute to improving teaching practice and make research results operational, impacting education in formal and informal contexts. These general objectives become even more challenging to pursue if science education is expected to be effective in dealing with contemporary challenges. This demands designing research-based teaching materials, paths, and programs and to test them in real contexts. This section includes chapters that focus primarily on innovative instructional design or on programs to infuse formal and informal teaching with novel pedagogical principles or methods.

Highlights of the Chapters

In what follows, we will highlight the main themes addressed in the four sections by reporting how each individual chapter within each section contributes to the larger narrative of the volume, specifically, and the conference, more broadly.

Section 1: Meeting Societal Challenges

The first section launches the theme of societal challenges with the inspirational chapter by **Andri Snær Magnasson**. This chapter is not a research paper but the section could not be opened in a more appropriate way. Magnasson's particularly effective prose concerning the representational challenges we face, specifically challenges posed by the "flatness" of numerical representations, for understanding and communicating about the nature of climate change. The section continues with three chapters focusing on how science education research can help meet contemporary societal challenges of various kinds. The challenges include: (i) global environmental and health issues, (ii) the need to re-think knowledge organization by stressing interdisciplinarity for dealing with Responsible Research and Innovation, and (iii) multiculturalism and multilingualism. The last three chapters of the section engage with policy issues, examine novel curricular approaches to meet societal concerns and report on studies of science learning and instruction that focus on how learners can be prepared for meeting the pressing challenges of our time.

More specifically, after the inspirational chapter by Andri Snær Magnasson, the first research paper is by **Zeyer and colleagues** and discusses two of the first priorities citizens expect science education should address: environment and health. In dealing with these challenges, this chapter touches on the importance of educating toward an understanding of complexity, increasing systemic views and the development of comprehensive approaches to deal with these huge and problematic issues. The themes of complexity and systemic thinking are picked up in several of the pieces throughout the section and volume.

Another running theme within the volume is the issue of interdisciplinary teaching. This theme is the core of the chapter by **Fazio and colleagues**. Specifically, their paper argues that we need a perspectival change within research on Inquiry-Based Science Education (IBSE) since societal challenges and the Responsible

Research and Innovation (RRI)[1] framework require an interdisciplinary approach, if they are to be tackled in an original and suitably complex way. In their argument, the authors also stress the importance of attending to the institutional issue of improving teacher training in order to harmonize the interdisciplinary approaches across different school levels.

Finally, the fourth chapter focuses on the increased multiculturalism and multilingualism of our societies in Europe and the world. The chapter represents the panel discussion that **Espinet and colleagues** offered to the conference. With representation across a wide variety of countries, the authors illustrate an interesting range of challenges that science education research is addressing. The main challenge discussed in the chapter deals with the design and implementation of instructional approaches that make sense in these new cultural contexts. At the same time, Espinet et al. raise awareness within the research community of the forms of knowledge produced in the field of science education as we study multicultural and multilingual contexts.

After the discussion of these three main challenges, the section turns to three pieces relating to crucial policy issues. In the chapter by **Duschl and colleagues**, recommendations from four 21st Century Education Policy reports were discussed by a panel of international leaders in the science education community. The examined frameworks address curriculum, international assessments, instructional policies, and teachers' practices. The panelists specifically speculate on how models of education need to change in order to prepare students and citizens for life with uncertain global conditions and for workforce dynamics that are rapidly changing.

Then, the chapter by **Osborne and colleagues** offers a macro look at PISA data and OECD analyses to point out research orientations needed to address changes in school systems. This chapter reports a collection of four studies. Two of them represent a second layer of analysis of OECD-PISA data to discuss and check results coming from analysis carried out by OECD and from which substantive claims are made about the strengths and weaknesses of certain forms of teaching, like inquiry-based instruction. These studies argue that it is important for the research community to conduct secondary analyses of the data. The other two studies of the collection make a case for the need to avoid hyper-simplified conclusions from the data since several dimensions of complexity underlying student performance can be unpacked.

The section closes with a piece by **Bruun and colleagues**, that, in contrast to the previous chapters, offers both a more micro view of curriculum change and also tools that science education research can offer in the service of studying curriculum change. The authors focus on a specific course in Denmark called "*Basic Science Course*" in which the ministry of education has regulated that topics pertaining to scientific literacy, inquiry-based science teaching, Bildung, and interdisciplinarity should be emphasized. By applying a combination of qualitative and quantitative

---

[1] von Schomberg, R. (2013). A vision of responsible research and innovation. In R. Owen, J. Bessant and M. Heintz (eds), Responsible Innovation (pp. 51–74.). Chichester: John Wiley & Sons.

methods, the authors track the type of innovation incorporated in official curriculum texts and the kind of policy change they implicitly and explicitly introduce over time.

Section 2: Expanding the Evidence Base

As we have seen, section one of this book reflects the science education research community's call to action and its proclamation that it must play a role in meeting societal challenges. But if effective action is to be taken and if any challenges are to be met, this community's distinctive contribution will be to provide the needed evidence base to validate our understanding of science teaching and learning and ultimately support practical recommendations. This section includes a number of chapters that report on empirical studies examining the development of student conceptual understanding in a number of domains, learners' epistemological cognition, learners' developing self-concepts (as these relate to learning in specific domains) and learner identity. The section concludes with two chapters focusing on teachers, examining pre-service teachers' sense of "psychological distance" with respect to socio-scientific issues in the domains of health and the environment and in-service teachers' sense of self-efficacy as this relates to education for sustainable development.

The first two chapters address an important theme that has increasingly engaged the science education research community in research years: long term developmental patterns in students' understanding in a domain, often referred to as "learning progressions." This work has been important in synthesizing the large body of work on learner conceptions, putting it to the service of curriculum design and assessment. **Bernholt and Höft**'s longitudinal study examines students' developing understanding of core concepts in chemistry spanning the grade brackets 5–8 and 9–12. They make an important methodological contribution, showing how different approaches to analysing students' responses to test items lead to diverging conclusions regarding developmental patterns across grades. This work is emblematic of how carefully designed quantitative methods make indispensable contributions to our understanding of learning. Similarly, **Scheuch and colleagues** report on a longitudinal study of the development of conceptual understanding – in this case, students developing understanding of variation and change in evolutionary theory. They present case studies of three students mapping their developing understanding in this domain over the grades 8–12. This work documents students' evolving conceptions of variation and change over this period. Crucially, while improved scientific understanding can be seen over this period, non-canonical teleological, essentialist and anthropomorphic forms of reasoning persist. This work is a good illustration of how entrenched naive reasoning patterns limit student learning if not explicitly addressed in the curriculum.

The third chapter in this section, by **Tena and Couso**, examines the impact of a teaching intervention on children's conceptions of clean and polluted air, a central environmental problem in cities. This area of student understanding has not been well studied, but is very important if we are to successfully promote scientific literacy and responsible, environmentally-oriented citizenship. In this study, elementary/primary students participated in a modeling activity sequence. The results show that while most children were capable of thinking about air as a discrete substance without macroscopic differences when it is polluted, they faced difficulties in interpreting the nature of the different "particles" they identified in both clean and polluted air. These results support the view that elementary/primary school science curricula should emphasize macro and meso scale perspectives as a precursor to the later introduction of the atomic-molecular and subatomic scale.

It is now well understood that learners' epistemological understanding of the nature of scientific knowledge and knowledge change contributes to their developing conceptual understanding in science domains. Moreover, it is an important goal of science education in its own right as a central component of scientific literacy. **Kim and Alonzo**'s chapter in this section, investigates undergraduate students' evaluation of the trustworthiness of knowledge claims when considering socio-scientific issues. They show that Duncan et al.'s (2018) Grasp of Evidence framework is able to distinguish between expert and lay understanding of how to use evidence to evaluate claims. They extend this with a grounded theory analysis, identifying novel epistemic concepts not previously identified in the literature.

Learners' understanding and beliefs about themselves as learners have an important impact on their learning. Moreover, these vary considerably across learners, with gender and sociocultural variables influencing these understandings and beliefs. Understanding these influences is important if we are to ensure equity in science classrooms. **Rüschenpöhler and Markic** investigate learners' self-concept in the context of chemistry education in secondary schools in Germany. Participants included 585 students, belonging to migrant communities (mainly from Turkey). The study examined the relationships between learners' chemistry self-concept and a number of variables including gender, cultural background, learning goal orientations in chemistry, and the learners' perceptions of linguistic abilities and their social context.

The chapter by **Cavalcante and Gonsalves** broadens the perspective on science learning even further, considering how university undergraduates' science identity takes shape. Using narratives collected from three students majoring in science and participating as "local experts" in a science outreach program, the authors characterize aspects of these students' developing science identities. Central to this characterization was the notion of "science capital" understood as the scientific knowledge, understanding and social connections one has in the science community. They argue that accumulating science capital of multiple forms through early experiences and schooling leads to strong science identities. This work is important for illuminating how strong science identities are formed and to help teachers see a wider range of ways of relating to their students beyond the traditional lecture.

The last two chapters in this second section shift the focus to teachers. The first of these, by **Büssing, Dupont and Menzel**, presents the results of a survey carried out with a sample of 189 pre-service biology teachers at four different German universities. The survey was designed to explore teachers' "psychological distance" in relation to socioscientific issues dealing with the environment and health. The results capture the differences in psychological distance with regards to the SSI issues of climate change, returning wolves, and pre-implantation genetic diagnosis. The second, by **Mogias, Malandrakis, Papadopoulou and Gavrilakis**, uses a quantitative methodology to investigate in-service teachers' self-efficacy regarding education for sustainable development. This study increases our understanding of the factors affecting teachers' choices and attitudes toward how science education can contribute to a very important societal challenge.

Section 3: Developing Innovative Theoretical Perspectives and Methodologies

In order to bridge from where we are to where we must be in terms of meeting contemporary challenges, new theoretical perspectives and methodologies are needed. The first two pieces in this section, by Sherin and by diSessa and Levin, take up the issue of building theory in science education.

Sherin's piece focuses on the landscape of research on conceptual change, including taking stock of the challenges and possibilities of finding theoretical common ground in a field that has been so fraught with controversy and seemingly inconsistent findings, especially as these relate to the degree of coherence in pre-instruction conceptual understanding. Interestingly, Sherin downplays the significance of the often heated empirical debates in the field claiming that differences in results are inextricably linked to the way researchers have decided to ask questions and investigate them. Sherin argues that it is important to recognize that the field is broad and encompasses a wide range of contexts, domains, and research foci and apparently divergent findings need to be interpreted in this light. It may not be reasonable to expect uniform conceptual change processes across such a diverse landscape. However, in seeking out those places where there may be points of convergence, Sherin suggests that it would be helpful to adopt some minimal consensus language. He proposes that the constructs "elements," "ensembles," and "dynamic mental constructs" could serve this function.

In the second chapter of the section, **diSessa and Levin** reflect on the processes of building theories of learning. They draw lessons for theory building from a cross-case analysis of three quite different theories that were at different stages of development, but all of which came out of dialogue with a common orienting framework, Knowledge in Pieces (diSessa 1993, 2018). In some sense, diSessa and Levin's cross case analysis can be seen as an example of Sherin's call to action with respect to theory building emphasizing the power of a common generic language for thinking about learning that then gets specified and elaborated in particular contexts such as the nature and form of intuitive knowledge, the form of expert understanding, and processes of problem solving.

In the third chapter in the section, **Amin** explores what science education and cognitive neuroscience have to offer each other, a clear reflection of the theme of interdisciplinarity that runs throughout the volume. Amin begins by reviewing research that examines the neural underpinnings of conceptual representations and the processes of conceptual change in science learning. Amin points out how this work has not engaged with findings from the learning sciences and science education that suggest that intuitive knowledge – such as that based on sensorimotor experience – is put to use in various aspects of scientific understanding and reasoning. He hypothesizes that the same construct – image schema – is appealed to in research on science learning and in cognitive neuroscience to understand how higher level cognition is grounded in sensorimotor experience. Resonant with Sherin's call for seeking consensus constructs, Amin argues that image schemata could serve as a natural interdisciplinary bridge between research in science education and cognitive neuroscience and point the way for a productive program of research in educational neuroscience.

New methodologies are also needed. In their chapter, **Saucedo and Pietrocola** describe an innovative qualitative method they call Emotions Microsociology that can support the investigation of a challenging phenomenon, young children's emotional engagement with science. Understanding this early engagement is very important as it sets the scene for children's subsequent trajectory as science learners. Saucedo and Pietrocola's chapter illustrates the application of this method to capture a group of young children's heightened emotional engagement with a science demonstration. They show how the analytical technique illuminates children's interaction with each other, with their teacher and with the demonstration itself. Their methodology illustrates the importance of interdisciplinarity in expanding our methodological toolboxes as science education researchers.

The last chapter in this section, by **Kapon and Erduran**, analyzes curricular interdisciplinarity in STEM across three different projects. The authors offer a theoretical reflection on the different approaches adopted to crossing boundaries between science, math, and technology. The relationship between STEM interdisciplinary approaches and disciplinary teaching is explored in three different cases, with the common theoretical lens of the boundary crossing mechanisms proposed by Akkerman and Bakker (2011). They show that the boundaries between the disciplines can be crossed in several ways, with different goals and strategies. The analysis is theoretically innovative since there are many interpretations of STEM and interdisciplinarity and the authors introduced a metalevel of analysis that allowed them to compare approaches framed within different theoretical frameworks.

Section 4: Designing Research-Based Instruction

If science education is to meet contemporary challenges there will be a need for rigorous empirical research framed within existing theoretical frameworks and for innovations in both theory and research methodology. But all this research must serve the goal of developing effective innovative teaching tools and instructional approaches and these must, in turn, be subjected to empirical evaluation in real contexts. In the fourth section of the book, we include chapters that report on innovations in instructional tools and strategies and evaluate their effectiveness. These tools and strategies are shown to support the development in students of the kinds of scientific understanding, thinking skills and dispositions that they will need to meet contemporary challenges.

For example, in the first chapter in this section, **Nielsen and Brandt** report on their European project, ARsci, conducted in lower secondary science classrooms in Denmark, Norway, and Spain, which used a design-based research approach. They show how understandings related to the environment and ecology, as well as systems thinking and meta-modeling competencies can be developed in an Augmented Reality (AR) learning environment. The augmented reality environment allowed learners to take up the role of producers and engage in collaborative modeling activities, allowing them to have embodied experiences that ground and make accessible ideas that were otherwise abstract and remote.

In their chapter, **Tytler, White and Mulligan** focus on the early development of the skills of constructing, evaluating, and coordinating multiple representations, which are all central to scientific and mathematical thinking. They evaluate a lesson sequence in astronomy offered to 150 grade 1 students (6 year-olds) in two schools. In this lesson sequence, the children constructed, evaluated, and coordinated spatial representations of the movements of the Earth and Sun to make sense of shadows changing and moving throughout the day and to explain cycles of day and night. The results show the power of an interdisciplinary, guided inquiry pedagogy applying the principles of the Representation-Construction-Approach.

Next, **Buongiorno and colleagues** examine the effects of an active learning approach to teaching physics at the university level. Active learning is relatively rare at this level, but science education research is increasingly exploring ways to move away from the traditional lecture-based pedagogy that dominates science teaching at the tertiary level. The project reported here provides evidence that an innovative active learning approach can be applied across countries and contexts and that it is possible to successfully integrate conceptual understanding, problem solving and lab work in university physics instruction.

The first three chapters in this section show how innovative tools and instructional strategies can develop scientific understanding in abstract domains and help learners of various ages engage in scientific epistemic practices such as modeling, systemic thinking, constructing, and evaluating representations and laboratory investigations. But developing positive dispositions to these practices are also important. In the fourth and last chapter in this section, **Vilhunen and colleagues** examined how various instructional activities carried out within a project-based science learning unit predicted the different kinds of epistemic emotions experienced by upper secondary school students in Finland. This is methodologically challenging research requiring diverse and carefully applied methods. The authors used experience sampling, video observations and stimulated recall to investigate the participants' epistemic emotions during the implementation of the project-based learning unit. Using multi-level regression analyses, they found that initial project orienting activities were associated with positive epistemic emotions such as excitement and curiosity, whereas skills and content tasks were more associated with negative emotions such as confusion, anxiety, and frustration. As Vilhunen and colleagues point out, this research can help teachers become more aware of the emotional implications of the different design features of learning environments.

Concluding Remarks

Together, we feel that the chapters included in this volume illustrate well how the science education research community is responding to contemporary challenges. Researchers are working on many fronts: they are re-examining and evaluating current curricula, assessment, and policy of relevance to current challenges; they are conducting theory-driven empirical studies to add to our knowledge base; they are proposing novel theoretical frameworks and methodological approaches to capture the complexities of learning and instruction, that may often need to cut across disciplines; and they are designing and evaluating new educational tools and strategies. While this large body of work is multifaceted and diverse, we hope to have offered the reader a well-organized and clear view of current research in science education and we hope that you share our pleasure in the understanding that emerges.

We wish to end with a sincere thank you to the ESERA Board for the opportunity, confidence and support they provided to us in the organization of such a stimulating conference in Bologna.

Referecnces

Akkerman, S. F., & Bakker, A. (2011). Boundary crossing and boundary objects. *Review of Educational Research, 81*(2), 132–169. https://doi.org/10.3102/0034654311404435

diSessa, A. A. (1993). Toward an Epistemology of Physics. *Cognition and Instruction, 10*(2 & 3), 105–225.

diSessa, A. A. (2018). A friendly introduction to "knowledge in pieces": Modeling types of knowledge and their roles in learning. In *Invited lectures from the 13th international congress on mathematical education* (pp. 65–84). Springer International Publishing.

Duncan, R. G., Chinn, C. A., & Barzilai, S. (2018). Grasp of evidence: Problematizing and expanding the next generation science standards' conceptualization of evidence. *Journal of Research in Science Teaching, 55*(7), 907–937. https://doi.org/10.1002/tea.21468

Rosa, H. (2013). *Social acceleration: A new theory of modernity (new directions in critical theory).* Columbia University Press.

Alma Mater Studiorum – University of Bologna
Bologna, Italy
Olivia Levrini

Alma Mater Studiorum – University of Bologna,
Bologna, Italy
Giulia Tasquier

American University of Beirut
Beirut, Lebanon
Tamer G. Amin

University of Milan
Milan, Italy
Laura Branchetti

Western Michigan University
Kalamazoo, MI, USA
Mariana Levin

Contents

Editors and Contributors

About the Editors

Olivia Levrini is Associate Professor in Physics Education and History of Physics at the Department of Physics and Astronomy of the University of Bologna, Italy. Her current research work concerns interdisciplinarity in STEM education, cognition and conceptual change, identity and processes of appropriation, instruction design on future-oriented STEM issues (climate change, artificial intelligence, quantum computing), educational reconstruction of advanced current topics in physics (thermodynamics, relativity, quantum physics). She currently coordinates the European Erasmus + Project called IDENTITIES (www.identitiesproject.eu) and the Horizon 2020 project titled FEDORA – Future-oriented Science EDucation to enhance Responsibility and engagement in the society of Acceleration and uncertainty (www.fedora-project.eu).

Giulia Tasquier is Junior Assistant Professor in Physics Education at the Department of Physics and Astronomy of the University of Bologna, Italy. Her research interests include design and implementation of innovative teaching materials on modern physics and SSI; the correlation between knowledge and behaviour in dealing with climate change; the role of epistemological knowledge on models and modelling in teaching/learning physics; qualitative methods of data analysis; development of strategies, tools, and activities for transforming scientific knowledge into transversal skills about the future. She is the coordinator of the national network of the Horizon 2020 project called SEAS (https://www.seas.uio.no).

Tamer G. Amin is currently Associate Professor of Science Education in the Department of Education at the American University of Beirut, Lebanon. His research focuses on conceptual change in science learning. He has been examining how the cognitive linguistic theory of conceptual metaphor can help uncover image schematic structures implicit in the language of science and how these can support and sometimes hinder learning scientific concepts. In a parallel line of research, he is investigating the challenges of teaching and learning science in the multilingual contexts of the Arab world and how these challenges might be overcome.

Laura Branchetti is tenure-track researcher of Mathematics Education at the Department of Mathematics "Federigo Enriques" at the University of Milan, Italy. She is a member of an interdisciplinary research group in physics, mathematics and computer science education in Bologna and she has been involved in European projects about STEM education and interdisciplinarity in preservice teacher training. Her main research interests concern mathematics education and interdisciplinarity in secondary school and at the transition from secondary to tertiary education; in particular she carried out research in the teaching of calculus and analysis and of the interplay between mathematics and physics.

Mariana Levin is Associate Professor of Mathematics Education in the Department of Mathematics at Western Michigan University. Her research focuses on understanding the role of knowledge and epistemic affect in moment-by-moment processes of reasoning and sense-making. Her edited volume "Knowledge and Interaction: A Synthetic Agenda for the Learning Sciences" (with A. A. diSessa and N.J.S. Brown) explores this line of work, connecting insights from diverse research traditions on learning processes as they unfold in real-time in real-world contexts. Her current work explores the development of mathematical agency and autonomy in undergraduate students' experiences in proof-intensive mathematics courses.

Contributors

Alicia C. Alonzo is Associate Professor of science education in the Department of Teacher Education at Michigan State University in the United States. She has a background in physics (with a Ph.D. in applied physics from the California Institute of Technology). Her current research broadly concerns how school science intersects with the ways that students encounter science in their everyday lives. She is interested in how school science can both leverage the ideas and experiences that students bring to the classroom and prepare students to make science-related decisions that are personally and civically meaningful.

Nuria Álvaro is a biology and geology teacher in compulsory secondary education and high school and a doctoral student in the Department of didactics of experimental and social sciences in the teacher training faculty of the University of Valencia (Spain). She is a specialist in environmental health and environmental science education. She has also authored teaching materials for the Valencia Cultural Territorial Cathedra.

Heidemarie Amon has obtained her Master in Education at the University of Vienna and has been working as a teacher of biology in Vienna since 1987. She has also cooperated with the Austrian Educational Competence Centre for Biology at the University of Vienna since 2008. Her interests include the development of competency-based learning tasks for students and biology teacher education (pre-service and in-service). In 2010, together with Dr. Patricia Hoffmeister (Jelemenská), she started to develop teaching and learning sequences on different levels of lower and upper secondary school to work with students´ conceptions about evolution.

Julia Arnold studied Biology and English for Secondary Education. After her Ph.D. in Biology Education at Kassel University, she has been a post-doctoral researcher at RWTH Aachen University and at the Leibniz Institute for Science and Mathematics Education in Kiel, Germany. Now she is head of the research group of Biology Education at the Centre for Science and Technology Education at the School of Education FHNW in Basel, Switzerland. Her research interests are in health and environmental education, inquiry learning and teachers' TPACK. In these fields she mainly focuses on the interplay between knowledge, motivational factors and actions.

Davide Azzolini is a Research Fellow at the Research Institute for the Evaluation of Public Policies of the Bruno Kessler Foundation (Italy) and Affiliated Scholar with the Urban Institute (Washington, DC, USA). He is interested in evidence-based policy and program evaluation related to student achievement, children of immigrants' education, student aid programs and the role of digital technology in learning and training. He was a Visiting Scholar at the Urban Institute and at the Office of Population Research, Princeton University.

Deidre Bauer is a research associate in biology education at Leibniz University Hannover, Germany and a teacher for biology and Latin language in a grammar school in Lower Saxony, Germany. Her research interests are knowledge and attitudes in Education for Sustainable Development (ESD).

Nicola Bazoli is a Junior Research at the Research Institute for the Evaluation of Public Policies of the Bruno Kessler Foundation (Italy). He has achieved a Ph.D. in Statistical Science at the University of Bologna. His research interests include the analysis of students' skills, with a psychometric approach, and the analysis of social inequalities in educational achievement.

Sascha Bernholt works as a research scientist at the Leibniz Institute for Science and Mathematics Education (IPN) at the University of Kiel. He studied chemistry and mathematics in a degree of education program for teaching at secondary level at the University of Oldenburg, Germany. He has a Ph.D. in chemistry education from the University of Oldenburg, Germany, and he finished his habilitation in 2020 at the University of Kiel. His research interests are largely related to students' learning and conceptual understanding and how these develop over time.

Saouma BouJaoude is presently Professor of Science Education and Director of the Center for Teaching and Learning at the American University of Beirut. His research interests include evolution education, teaching science in multilingual settings, curriculum and teaching methods, and the nature of science. BouJaoude has published in international journals such as the *Journal of Research in Science Teaching*, *Science Education*, *International Journal of Science Education*, *Journal of Science Teacher Education*, *The Science Teacher*, *Science & Education*, *Research in Science Education*, and *School Science Review*. Additionally, he has presented his research at local, regional, and international education conferences.

Amélia Branco is Assistant Professor at ISEG (ULisboa), in the Department of Social Sciences. She holds a Ph.D. in Economic and Social History. Researcher at Economic and Social History Centre (GHES/CSG – ISEG) that integrates the consortium Research in Social Sciences & Management, where she makes part of the research group Sustainability and Policy. Her main research interests are forestry value chain along the nineteenth and twentieth century, where she has several publications. Currently, she participates in two research projects: Capital Markets in Portugal (nineteenth century until WWI) and SustInAfrica – Sustainable intensification of food production through resilient farming systems in West & North Africa.

Harald Brandt is Associate Professor at VIA University College. He is teaching and supervising student teachers specialized in Physics & Chemistry and researcher at the Research Centre for Pedagogy and Education. He has been involved in various research and development projects focused in particular on technology enhanced learning in science and models and modelling. Currently, he is the Danish project leader in the EU-ERASMUS project "Ocean Connection".

Jesper Bruun is Assistant Professor at the Department of Science Education at the University of Copenhagen. He uses network analysis as a methodological tool in physics and science education research and views educational systems at all levels as complex systems.

Daniele Buongiorno is a post doc researcher in Physics Education in the Department of Mathematics, Informatics and Physics at the University of Udine in Italy. He teaches Physics for Biotechnology Students in the University of Udine and Math and Phys in Secondary School. In 2018 he received the award for the Ph.D. study on Optical Spectroscopy education in GIREP Congress. 12 papers in peer review publications characterize his research work.

Alexander Georg Büssing worked as a research assistant in the Department of Didactics of Biology at Osnabrück University (Germany) and wrote his doctoral thesis about pre-service teachers' emotions for teaching about environmental socio-scientific issues. Currently, he works at the Institute for Science Education at Leibniz University Hannover (Germany) as a postdoctoral researcher and fellow of the German Telekom Foundation in the working group for biology education. His research concentrates on the effects of emotions in digital learning environments (e.g. virtual reality and social media).

Alexandre Cavalcante is an Assistant Professor in the Department of Curriculum, Teaching and Learning at OISE/University of Toronto, Canada. His research interests are geared toward STEM education and how it can respond to contemporary socioeconomic changes in the world.

Mojca Čepič Ph.D., is Full Professor of General Physics and Physics Education. Her research is focused on two fields, on theoretical soft matter physics, more precisely on polar smectic liquid crystals, and on physics education. She has led several research projects on the introduction of new fundamental physics findings into all levels of education and their application to identifying gifted students. For several years she was Head of the Department of Physics and Technology. Currently she is Head of the Institute for Science and Arts at the Faculty of Education of the University of Ljubljana, and she acts as a chief editor of European Journal of Physics.

Digna Couso is reader at the Science Education unit at the faculty of Education and director of the research center CRECIM at Universitat Autònoma de Barcelona, Spain. Her research interests are models and modelling for the teaching of science, participatory design-research approaches for teacher education within the context and competences-based framework for STEM, with special interest in ICTs, and both equity and gender balance in STEM education. She has led and participated in a wide range of national and European projects. She is referee of some prestigious journals in field and she is also part of the expert group of PISA 2024.

Iztok Devetak Ph.D. is a Full Professor of Chemical Education at University of Ljubljana, Faculty of Education, Slovenia. His research focuses on students' learning chemistry (the influence of chemistry in context and active learning approaches), using eye-tracking in explaining chemistry learning, environmental chemistry education, developing teachers' health-managing competences etc. He was the national coordinator of PROFILES project (7th Framework Program) for 4.5 years and ERASMUS+ project Non-formal Education in Science for Students' Diversity (DiSSI). He is a Chair of Chemical Education Division in Slovenian Chemical Society and Vice-chair for Eastern Europe of European Chemical Society Division of Chemical Education.

Andrea A. diSessa holds degrees in physics from MIT (Ph.D.) and Princeton (AB). He is a member of the National Academy of Education, a Fellow of the American Educational Research Association, and Corey Professor Emeritus at UC Berkeley. His research centers on the role of intuitive knowledge in learning scientific concepts, and computational literacies. He is the prime designer of Boxer, a medium to support computational literacy. diSessa has authored over 100 articles and chapters and authored or edited seven volumes, including *Changing Minds:*

Computers, Learning and Literacy, and *Turtle Geometry: The Computer as a Medium for Exploring Mathematics*.

Sara Dozier is a Ph.D. candidate in Science Education in the Graduate School of Education, Stanford University, USA. Her current research interests focus on both classroom and large-scale assessment practices and how assessment is related to issues of educational equity. She was a high school science teacher and county-level administrator in the California public school system. Her biological research focused on the molecular biology of the hypoxic response in murine models.

Mylène Duclos is a neuro-psychologist, doctoral student in Science Education at the Education Laboratory of the Ecole Normale Supérieure de Lyon in France. She is preparing a Ph.D. on the investigation of the links between the understanding and resolution of tasks related to scientific culture by students and their socio-economic-cultural status.

Jacqueline Dupont works as a research assistant in the Department of Didactics of Biology at Osnabrück University (Germany). In her doctoral thesis she is working on the sustainable consumption behaviour of children and adolescents. One research focus is the acceptance of meat alternatives such as insects and cultured meat.

Richard A. Duschl Executive Director, Caruth Institute for Engineering Education and Texas Instruments Distinguished Professor, Lyle School of Engineering, Southern Methodist University; Professor Emeritus, Penn State University. He has served as Editor, *Science Education*; President of US NARST (2009–2011); Director, Division for Research on Learning, NSF (2012–2015) and chaired the US National Research Council research synthesis report *Taking Science to School: Learning and Teaching Science in Grades K-8*. He received the JRST Award for best publication in 1989 and 2003. In 2014, he received the NARST Distinguished Career Research Award.

Sibel Erduran is a Professor of Science Education and Fellow of St Cross College at University of Oxford, United Kingdom. She is also Professor II at University of Oslo, Norway. She is the President of the *European Science Education Research Association*; Editor-in-Chief of *Science & Education* and an Editor for *International Journal of Science*. Her work experience includes positions in the USA, Ireland as well as the UK. Her research interests focus on the infusion of epistemic practices of science in science education. Her work on argumentation has received international recognition through awards from NARST and EASE.

Mariona Espinet is Professor in the Didactics of Science and Mathematics Department at the Autonomous University of Barcelona in Catalonia, Spain. She is the coordinator of two UAB research groups ACELEC (School Science Activity: Languages, tools and contexts) and Gresc@ (Education for sustainability, school and community) and member of the ESERA Board. Her research and innovation interests focus on education for sustainability, classroom discourse and critical literacy in multilingual science education contexts. She has been the coeditor in 2020 of a special issue in International Journal of Science Education about International perspectives on science education in multilingual contexts.

Robert Harry Evans is an Associate Professor in the Department of Science Didactics at the University of Copenhagen in Denmark. He teaches pedagogical courses for University science faculty, science Ph.D. students and undergraduates who wish to become Physics, Chemistry, Biology or Geography teachers. He supervises various levels of graduate student projects. His research interests include inquiry-based science teaching and learning and the role that self-efficacy has in facilitating inquiry and problem-based instruction. Professionally he is a member of the European Science Research Association Board with the Ph.D. summer school portfolio.

Sergej Faletič received his Ph.D. in Physics Education from the University of Ljubljana. For the past 15 years, he worked as a high school physics teacher and a researcher in the field of physics education research at the University of Ljubljana. His focus is on active engagement methods in physics education. His main work is on open-ended laboratory work and teaching and learning quantum mechanics at high school level.

Cláudia Faria is Assistant Professor at the Institute of Education, University of Lisbon. She has a background in Biology and two Ph.D., one Ph.D. in Biology – Ecology and Biosystematics and one Ph.D. in Science Education, both from the University of Lisbon. Her present research is related to the development of innovative science teaching and learning strategies in formal and non-formal contexts, such as museums, science centers, and research laboratories. She has published more than 80 scientific papers, books and book chapters. She coordinated three national research projects, related with the promotion of scientific literacy, and participated in several national and international projects.

Claudio Fazio is an Associate Professor in Physics Education at the University of Palermo, Italy. He teaches Physics Education for Master Students in Physics, History of Physics for Bachelor Students in Physics and General Physics and Physics Education for prospective primary teachers. He is member of the GIREP Board. His main research interests focus on the study of students' mental models and use of cognitive resources, on the study Pedagogical Content Knowledge development in prospective teachers, and on the development and use of quantitative and qualitative analysis methods in research, with particular reference to cluster analysis.

Odilla E. Finlayson is Associate Professor of Science Education in the School of Chemical Sciences, Dublin City University (DCU). She is one of the founding members of CASTeL (Centre for the Advancement of STEM Teaching and

Learning) at DCU and plays an active part in its management and development. She is involved in teaching chemistry to undergraduate students and pre-service teachers. Current research interests are in sustaining science across transitions and, in particular, in developing appropriate science (chemistry) curricula and assessment.

Valérie Fontanieu is a statistician at the French Institute of Education at the Ecole Normale Supérieure in Lyon (France). She is involved in several national and international research projects in science education.

Cory Forbes is Associate Professor of Science Education at the University of Nebraska-Lincoln and Director of the National Collaborative for Research on Food, Energy, & Water Education (NC-FEW), USA.

Gerald Fuchs grew up on a farm in Austria and studied biology and sports education at the University of Vienna where he also worked as a tutor for the course of outdoor education in biology teaching for several years, which meets his interest in outdoor learning. In his master thesis, written at the AECC Biology, he tried to find new aspects about the understanding of the term 'population' among students during the ongoing process of learning about evolution. Throughout his time at university, he has been restructuring the farm into an organic one and started breeding an old breed of sheep which was threatened with extinction.

Igal Galili holds a Ph.D. in theoretical physics from the Hebrew University of Jerusalem. His post doctorate work was in physics education at UC San Diego and UC Berkeley. Currently, he is Professor of Science Education in the Faculty of Mathematics and Natural Science at the Hebrew University of Jerusalem. His research has explored the physics knowledge of students structured in scheme-facets. He investigated representations of physical concepts and the structure of theory. He produced the paradigm of discipline-culture implied in curriculum. He argued for using artistic images to represent scientific ideas and science nature. He produced a textbook of cultural representation of light theory at schools.

Cecília Galvão is Full Professor at Institute of Education, University of Lisbon. She has a background in Biology and a Ph.D. in Education. She coordinates the Ph.D. Programme in Sustainability Science of the University of Lisbon, the Ph.D. in Education and the Joint Master Degree in Scientific Culture and Outreach of Sciences. She also coordinates the Master program on Biology and Geology teaching. She coordinates the Science Education Research Group and develops research in science education and teachers' professional development. She was the Portuguese coordinator of EU FP6 PARSEL and FP7 SAILS projects. She is member of COST Action CA15212 Citizen Science (2016–2020).

Valentín Gavidia is honorary professor at the University of Valencia, Doctor of Biological Sciences with an extraordinary doctorate award. He is teacher of Didactics of Biology and Health Education in the Department of Didactics of

Science, of which he has been the director. He has given conferences, workshops, postgraduate courses, directed doctoral theses and research projects both in Spain and abroad. He has received several local, national and international awards for educational innovation and research. His current lines of research are scientific and health education competencies for teachers. He is a friend of his friends and tries to understand those who are not, although sometimes he does not succeed.

Costas Gavrilakis is an Assistant Professor and Director of the Geographical and Environmental Education Laboratory in the Department of Primary Education at the University of Ioannina, Greece. His academic and research interests include several dimensions of the theory and practice of Environmental and Sustainability Education, such as student/teachers' ideas about environmental and sustainability issues, school-community collaboration, planning and evaluation of educational programs, and ICTs.

Allison J. Gonsalves is an Assistant Professor in the Department of Integrated Studies in Education at McGill University in Montréal, Canada. Her research interests are in the area of science identities, with a focus on gender and equity in higher education and out of school science learning contexts.

Luís F. Goulão is Ph.D. in Agronomic Engineering from the University of Lisbon. Assistant Professor at the School of Agriculture – University of Lisbon, focusing his research and teaching to link agriculture, food security and nutrition with Development outcomes. He is the coordinator of curricular units "Food Security and Food Policy", "Food Security and Global Dynamics" and "Tropical Farming and Agroforestry Systems" and member of the Scientific Committee of Doctoral Programmes "Development Studies" and "Sustainability Science". He is member of the Coordination Committee at the ISA's Research Centre "Linking Landscape, Environment, Agriculture and Food". Published more than 75 scientific and technical papers, book chapters, strategic briefs and more than 100 communications in congresses.

Jenaro Guisasola received his BS. in Physics and a MS. in Theoretical Physics, both from the University of Barcelona, as well as a Ph.D. in Applied Physics from the University of the Basque Country. He is Assistant Professor of Physics at the Applied Physics Department at the University of Basque Country. Since 2008 he teaches also Physics Education in the MA for Initial Training of Secondary Science Teachers. His research interest follows two related paths: (1) How Design Based Research can promote instructional models and enhance learning in science curriculum topics. With the support from several grants of Spanish and European projects, this research has given rise to new knowledge about the design of materials and teaching strategies. (2) The use of history and philosophy of science as tools to help organize teaching and learning in science curriculum. The agenda includes the understanding of development of scientific knowledge to apply it to science classrooms.

Christine Heidinger is a psychologist, long-term cooperation partner in Science Education (e.g., research assistant at the Austrian Educational Competence Centre for Biology, University of Vienna). Her expertise is in quantitative and qualitative research methods of the social sciences, philosophy of science and learning theories. Her research and development priorities in biology education are authentic learning environments for Inquiry Based Science Education, Nature of Science and education for sustainable development. As part of national and international third-party-funded projects (e.g. EU-project PARRISE), she developed and implemented various student-scientist-partnerships as well as learning environments (for students, pre- and in-service teachers) for Responsible Research and Innovation.

Paula Heron is a Professor of Physics at the University of Washington. She earned a Ph.D. in theoretical physics at the University of Western Ontario in 1995. Dr. Heron is co-Founder of the biannual "Foundations and Frontiers in Physics Education Research" conference series, Past Chair of the Executive Committee of the Topical Group on Physics Education Research of the American Physical Society and Associate Editor of *Physical Review – PER*. Dr. Heron is co-author of the report *Phys21: Preparing Physics Students for 21st Century Careers*. She is a Fellow of the APS and a co-recipient of the 2008 APS Education award.

Lars Höft studied chemistry and physics at the University of Kiel to become a secondary school teacher. After his first state examination, he started his research at the Leibniz Institute for Science and Mathematics Education (IPN) and successfully obtained his Ph.D. in 2020. His research interests focus on the interplay of students' conceptual understanding and their interests in specific school science activities, with a special emphasis on longitudinal development and interactions over the course of secondary school. At present, his is in secondary school teacher training at the Isarnwohld-Schule Gettorf, Germany.

Doris Jorde is Professor in Science Education at the University of Oslo, Norway. Her research activities have concentrated on bridging classroom research with development of curriculum materials and courses for teachers. She has participated in the development of science education policy documents in Europe and participated in multiple EU supported projects in Science Education. She served as President of ESERA (2003–2007). She has been vice rector at the University of Oslo, leader of the Norwegian Center for Science Education and is currently leader for the Norwegian Center of Excellence in Teacher Education.

Kalle Juuti is an Associate Professor of digital learning at schools in the Faculty of Educational Sciences, University of Helsinki. He is a chairperson of the faculty digitalisation group. His research interests focus on learning at the digitally intensive environments, education for sustainability, science education, student engagement, and professional learning.

Ida Viola Kalmark Andersen is a cand. scient. in Physics from the Niels Bohr Institute. She currently works as an upper secondary teacher in Roskilde, Denmark, and as a research assistant at the Department of Science Education at the University of Copenhagen. Her research focuses on network analysis of teaching situations.

Shulamit Kapon is an Associate Professor of Physics Education at the Faculty of Education in Science and Technology, at the Technion – Israel Institute of Technology. Professor Kapon is interested in the interrelations between teaching, instructional design and students' learning in physics education. Related areas of study include explanations and sensemaking, scientific and epistemic practices, and teacher education. One of the key goals of her work is to use her research to improve physics education in Israel and elsewhere.

Alla Keselman is Senior Social Science Analyst in the Division of Specialized Information Services, US National Library of Medicine, National Institutes of Health. Her research interests involve Science|Environment|Health, lay understanding of complex health concepts, scientific literacy as a component of health literacy, and consumer health informatics. She also coordinates activities of the National Library of Medicine K-12 Group that develops materials for students and teachers and conducts school outreach. Dr. Keselman holds a doctorate in human cognition and learning and a master's degree in biomedical informatics from Columbia University.

Won Jung Kim is a Ph.D. student in the Department of Teacher Education at Michigan State University, United States. She holds a master's degree in secondary education from Korea National University of Education. She taught secondary science for 12 years in South Korea. Her research focuses on justice-oriented pedagogies in support of students' sustainable and epistemically agentic engagement with science and socio-scientific issues.

Kerstin Kremer is a Professor of Biology Education at Leibniz University Hannover, Germany. She held professorships for science education at the Technical universities of Munich and Aachen and at the IPN in Kiel, Germany. She studied biology and chemistry to become a teacher. After attending teacher training and working as a school teacher, she completed her Ph.D. in biology education at the University of Kassel, Germany. Her main research interests are science communication in health and sustainability sciences, understanding of scientific inquiry and nature of science in school science.

Jari Lavonen is a Professor of Science Education at the University of Helsinki, Finland. He is currently a director of the National Teacher Education Forum and chair of the Finnish Matriculation Examination Board. He is a Distinguished Visiting Professor at the University of Johannesburg and a Honorary Chair Professor at the National Tsing Hua University, Taiwan. He has been researching both science and technology and teacher education for the last 31 years. His publications include

150 refereed scientific papers in journals and books, 140 other articles, and 160 books on education for science teachers and science education.

Florence Le Hebel has a Ph.D. in Geosciences and is Associate Professor at the University of Lyon (France). She teaches geosciences and science education. Her research projects focus on students' understanding of scientific literacy, in particular according to their socio-cultural and economic background. She conducts a project of secondary analysis of PISA data. She is involved in several international projects in science education and was notably involved in European projects in science education (STEAM, ASSIST-ME). She is currently preparing an habilitation thesis (« habilitation à diriger les recherches ») in science education at the Ecole Normale Supérieure de Lyon.

Birgitte Lund Nielsen is Senior Associate Professor and Research Leader at VIA University College. Her research in the field of technology-enhanced learning focuses in particular on how teachers can scaffold student exploratory dialogues in science classes mediated by various artifacts and learning technologies. Her research interests furthermore include teachers' professional learning, professional learning communities, teacher inquiry projects and professional agency. She has, for example, been involved in research following science teachers' meaning-making in collaborative professional development.

Andri Snær Magnason is an Icelandic writer born in Reykjavik. His book, *On Time and Water* is coming out in 27 languages in 2020 and 2021, exploring the language of climate change and why data and science has not reached our souls or policy. Andri is a writer of fiction, non-fiction, poetry, plays and documentary films. His book *LoveStar* won the Philip K. Dick special citation in 2014 and Le Grand prix de l'Imaginaire in France 2016, and his children's book, *The Story of the Blue Planet,* was the first children's book to win the Icelandic literary award and has been published in 32 languages. His book *Dreamland, a Self Help Manual for a Frightened Nation* has contributed to a new energy policy in Iceland and the vision of the Highland National Park in the Central Highlands of Iceland. Andri Snær Magnason ran for president in Iceland in 2016 and came third in the election.

George Malandrakis is an Assistant Professor in Environmental Education/ Education for Sustainability at the Department of Primary Education, Aristotle University of Thessaloniki, Greece. His research interests include, but are not limited to, students' understanding about various environmental issues, including footprints (ecological, energy, water), climate change, sustainable cities, and communities. He is also interested in pre-service and in-service teachers' training as well as the development of teaching and learning methods aiming at the conceptual development and the active engagement of learners towards more sustainable ways of living.

Silvija Markic is a Professor of Science Teaching and Learning at the Ludwigsburg University of Education, Germany. She finished her teacher training program for grammar school teacher of chemistry and mathematics and worked as a senior researcher at the Department for Chemistry Education at the University of Bremen. She works in the field of chemistry teacher education and researches in the didactics of chemistry. Her research interests include linguistic heterogeneity and cultural diversity in science classes, science teachers` beliefs and pedagogical content knowledge, cooperative learning and alternative teaching methods.

Sonya N. Martin is Professor of Science Education at Seoul National University in Korea where she leads the Sociocultural Approaches to Science Education Equity (SASEE) Lab. Since 2005, she has lived and worked as an academic and researcher in the United States, Korea, and Taiwan. She uses critical research methods and sociocultural theories to explore the intersections of culture and language in the context of school science and she seeks to expand students' science learning opportunities by supporting the professionalization of science teachers so they are positioned to effectively identify and understand barriers to students' access to and success in science. She is Editor-in-Chief of the journal Asia-Pacific Science Education

Olga Mayoral is teacher of the Department of Didactics of Experimental and Social Sciences in the Faculty of Teacher Training and deputy director of the Botanical Garden of the University of Valencia. She has a Ph.D. in biology and focuses on one hand on botany and biological conservation and on the other on Environmental Health Education, and outdoor education and pseudoscientific issues and critical thinking in education. She is involved in a research project of the Ministry of Science and Innovation, related to the inclusion of sustainability in teacher training and is the coordinator of the European project Young Innovators at the University of Valencia, funded by Climate-KIC.

Eilish McLoughlin is an Associate Professor in the School of Physical Sciences and Director of the Research Centre for the Advancement of STEM Teaching and Learning (CASTeL) at Dublin City University. Her research interests focus on the development of curriculum, instruction and assessment strategies to support student learning and inclusion in physics and STEM education, at all levels from primary school to Ph.D. level. In particular, she has led many research collaborations at local, national and international levels that focus on supporting teacher professional learning and collaboration in STEM education.

Susanne Menzel is a Full Professor for Didactics of Biology at Osnabrück University (Germany). In her research she focuses on explaining environmental perception and protection behaviours among young people with particular interest in the role of values and intercultural perspectives. In her empirical studies she uses qualitative and quantitative methods, often in combination.

Marisa Michelini is Full Professor in Physics Education in the Department of Mathematics, Informatics and Physics at the University of Udine in Italy. She teaches Physics Education for prospective primary teachers and for Master Students in Mathematics. She is president of GIREP and board member of EPS-PED and MPTL. Her physics education research interest are in content related research on modern physics, ICT for overcoming conceptual knots, IBL, informal learning, teacher education and innovation in university teaching/learning physics producing 660 peer review publications and supervising different Ph.D. thesis. In 2018 she received the award for physics education research by ICPE-IUPAP.

Athanasios Mogias is an Assistant Professor in Environmental Education at the Department of Primary Education, Democritus University of Thrace, Greece. His research interests and scientific publications are in the fields of Education for Sustainability and Ocean Literacy. The focus of his current research lies primarily on (a) elementary and secondary education students, as well as pre- and in-service school teachers, regarding their level of knowledge on ocean sciences issues, and their attitudes and behaviours towards the marine environment, (b) the design of relevant educational material, and (c) the implementation and evaluation of teaching interventions regarding ocean literacy issues.

Pascale Montpied has a Ph.D. in Neurobiology and was a researcher for several years at the National Institute of Health in Washington DC (USA). Pascale Montpied is currently a researcher at the CNRS (National Center of Scientific Research) in the ICAR laboratory at the Ecole Normale Supérieure in Lyon (France). Her research focuses on the acquisition of scientific culture by students and more specifically on the motivational dimension. She has been involved in several European projects in science education (mind the gap, STEAM, ASSIST-ME).

Audrey Msimanga was Professor of Science Education, Head of Education at Sol Plaatje University as well as a Visiting researcher at the University of the Witwatersrand. Her research spanned Biodiversity, Science Education and Teacher Education. In Science Education her research sought to understand the role of social interaction and classroom talk in learner meaning making and emerging from this the role of language in science education in multilingual classrooms. Her work was published in various local and international journals and she served as an Associate Editor for the Journal for Research in Science Teaching (JRST).

Joanne Mulligan is honorary professor of Mathematics Education at Macquarie University, Australia. She has contributed significantly to the field of mathematics education research focused on mathematical development in early childhood, primary and middle school years. Her research on number concepts and processes and the role of patterning and visualisation is internationally renowned. Her research involves large-scale empirically-based longitudinal designs and a range of quantitative and qualitative methodologies. From 2002 she built upon previous seminal work on mathematical imagery and representations and conceptualised a new theoretical approach to researching mathematical development, by proposing a generalised construct, Awareness of Mathematical Pattern and Structure (AMPS). She is widely published and an active member of the Australian mathematics education community.

Knut Neumann is Director of the Department of Physics Education at the Leibniz-Institute for Science and Mathematics Education (IPN) in Kiel, Germany. Before becoming a professor for physics education at the IPN, Knut was a post doc in the research group and graduate school "Teaching and Learning Science" at the University Duisburg-Essen. He obtained his Ph.D. from the Teacher University of Heidelberg. His research interests lie in the development of student competence in science and the role that teacher professional competence plays in it.

Jonathan Osborne is Kamalachari Professor in Science Education, Emeritus, in the Graduate School of Education, Stanford University. He was President of the US National Association for Research in Science Teaching (2006–7) and has won the Association's award for the best research publication in the Journal of Research in Science Teaching twice (2003 and 2004) and the Distinguished Contribution to Science Education Award in 2018. He was a member of the US National Academies Panel that produced the Framework for K-12 Science Education. He also chaired of the expert group for the science assessments conducted by the OECD PISA since 2012.

Penelope Papadopoulou is an Associate Professor in Biological Education at the Department of Early Childhood Education, University of Western Macedonia, Greece. Her research interests and scientific publications are in the fields of Biology teaching and learning, Environmental/Sustainability Education as well. The focus of her current research lies primarily in (a) Teaching and Learning Basic Concepts of Biology in Preschool and in Comprehensive Education, (b) Environmental/Sustainability Education in Theory and in Practice, (c) Anthropocentric Values in Biological and Environmental Education, and (d) Identifying factors of the Conceptual Ecology of Evolution Theory.

Jerneja Pavlin Ph.D., is Assistant Professor of Physics Education, focuses her research on the investigation of different approaches to physics and science education (e.g. inquiry-based learning, outdoor teaching, use of sub-microscopic 3D representations, didactic games, peer instruction etc.), aspects of scientific literacy and the introduction of contemporary science into physics education, from the development and optimization of experiments in modern materials to the evaluation of the developed modules. She has been involved in several national and international research projects.

Maurício Pietrocola is a science educator and professor at the University of São Paulo in Brazil. He received his doctoral degree from The University of Paris 7 (Denis Diderot) in 1992 and has since been the author of numerous publications.

His areas of work include curriculum development, pedagogical knowledge and innovative strategies of teaching and learning. His current focus is on connections between innovative education and risk taking, which contributes to an understanding of the failure of new educational innovations. In 2013–2014, Mauricio was a Fellow at the Advanced Research Collaborative at the Graduate Center of CUNY.

Gorazd Planinšič is a Professor of Physics at the University of Ljubljana, Slovenia. Since 2000 he has led the Physics Education program, which prepares almost all high school physics teachers in the country of Slovenia. During the last 10 years, his work has mostly focused on the research of new experiments and how to use them more productively in teaching and learning physics. Professor Planinšič is co-author of more than 80 peer-refereed research articles, and a co-author of College Physics: Explore and Apply textbook and the Active Learning Guide. He is a recipient of the GIREP medal for his contributions in Physics Education.

Sonja Posega Devetak MD, MS is a paediatrician with special interest in allergology at General and teaching hospital in Izola, Slovenia. As an European Academy of Allergy & Clinical Immunology (EAACI) member she participates in permanent postgraduate education for clinical allergists and she is an active member of Slovenian Association of Allergology and Clinical Immunology. She holds a master degree in biomedicine from the University of Ljubljana, Faculty of Medicine. She is a mentor to paediatric trainees and she is also engaged in teacher education at University of Ljubljana, Faculty of Education about the management of children with allergies.

Alberto J. Rodríguez is the Mary Endres Chair in Teacher Education and Professor of Cross-Cultural Science Education in the Department of Curriculum and Instruction at Purdue University in the United States. His research interests include: Critical culturally and socially relevant pedagogy and curriculum, emergent bi/multilingual education, transdisciplinarity in STEM education, and dismantling systemic inequalities through critical systemic thinking. Dr. Rodriguez recently received a Fulbright Grant to continue his work with pre-service science teachers in Costa Rica. He has also received various research awards for his work on addressing equity, diversity, and social justice issues in science teacher education.

Lilith Rüschenpöhler is a researcher in Chemistry Education at the Ludwigsburg University of Education in Germany. She studied chemistry, French, philosophy and political sciences in Bremen and Paris. In her dissertation, she analysed secondary school students' chemistry self-concepts with a special focus on the impact of gender and culture, and developed the concept of chemistry capital. Her research interests include science identities, chemistry capital, vocational orientation, chemistry education with refugee students and empowerment in chemistry education.

Katariina Salmela-Aro is Professor of Educational Sciences, University of Helsinki. Visiting Professor Io, UCL and School of Education, Michigan State

University. Past-President of the European Association for Developmental Psychology, and previous Secretary General ISSBD and expert OECDEducation2030. Director of several ongoing longitudinal studies among young people. Key research themes school engagement, burnout, optimal learning moments, experience sampling, motivation and related interventions. Associate Editor the European Psychologist journal. Published over 250 papers (h-index 73) and received several national/international grants Academy of Finland, Marie Curie ITN, Horizon2020 and EU Coordinator Marie Curie post-doc grant.

Paulo Sarriugarte cooperates with Jenaro Guisasola and Kristina Zuza in Physics Education in the University of the Basque Country.

Kellys Saucedo is Ph.D. in Education, in the Area Scientific, Mathematical and Technological Education at the University of São Paulo. She is Professor in Pedagogy Course. His current research focuses on emotions, teacher education and teaching situations. She is member of Group Training of mathematics and science teachers (FOPECIM) and Center for Research in Curricular Innovation (NUPIC).

Jaqueline Scheibstock studied biology and French to become a teacher for lower and upper secondary schools at the University of Vienna. The topic of her master thesis was students' conceptions about evolution. After teaching in secondary schools for a few years, she started to work in the biology education research at the Austrian Educational Competence Centre for Biology at the University of Vienna. She continued her work of the master thesis and made a longitudinal study on the development of evolution concepts of students who learn evolution in a carefully planned teaching and learning sequence.

Martin Scheuch is a trained vegetation ecologist with working experience in biodiversity and nature conservation. He is Professor for Biology Education at the University College for Agricultural and Environmental Education in Austria and is associated lecturer at the University of Vienna. He teaches biology teachers and environmental educators. His research interests are students' conceptions, he leads a longitudinal project about learning evolution. He is interested in out-of-school learning in biology and learning in the context of citizen science. In vocational education, he works with competency-based learning tasks for agricultural schools in Austria linked with the education for sustainable development.

Anja Schiepe-Tiska is a Senior Research Fellow at the Centre for International Student Assessment (ZIB) e.V. at the Technical University of Munich in Germany. She is part of the German PISA National Project Management Team and is responsible for the science domain and the background questionnaires. Her research focuses on multidimensional educational goals in STEM education and the individual, instructional, and school conditions of achieving different educational goals.

Bruce Sherin is a Professor, School of Education and Social Policy at the Northwestern University in the United States. His work focuses primarily on conceptual change in science – the process through which our everyday understanding of the natural world changes over time and with instruction. In early work, he engaged in the design and study of novel interventions for physics instruction. As part of that work, he explicated his theory of Symbolic Forms, which captures the conceptual structures that successful physics experts learn to see in equations. More recently, his work has focused on some methodological issues in the study of conceptual change. As part of this newer work, he applies techniques from natural language processing to interview protocols.

Dagmara Sokołowska Ph.D., is an adjunct at the Faculty of Physics, Astronomy and Applied Computer Science, Jagiellonian University. She is involved in physics/science education research in Inquiry-based Learning and Practitioner Inquiry at all levels of schooling – from primary to higher education. She participated in the EU projects on education: Fibonacci, SECURE, SAILS (7th FP), 3DIPhE (ERASMUS+), Akademickie Centrum Kreatywności, Wiking, Feniks (EU Structural Funds). Since 2014 she has been a member of GIREP vzw (International Research Group on Physics Teaching) Board. She is the author of the National Contest in Science for K1-K8 (Swietlik, eng. *Firefly*) in Poland.

Alberto Stefanel is a researcher in Physics Education Research Unit of the University of Udine, Italy. From 2015 he is Director of the Inter-Departmental Center for Educational Research of the University of Udine. His research field is Physics Education. His research activity is documented in more than 300 works regarding: teaching and learning modern physics in high school; cognitive studies on the role of informal learning environments and hands-on/minds-on labs in activating learning process of primary school pupils on thermal phenomena, electromagnetism, mechanical phenomena, sound, energy; role of ICT in Physics education; studies on teacher preparation and formation; teaching physics to university students of the bio area.

Tina Vesel Tajnšek MD is working as a paediatrician since 2008 in University Children's hospital in Ljubljana, Slovenia in the Department of allergology, clinical immunology and rheumatology. She gained actively also additional knowledge of allergology during annual congresses of European Academy of Allergy & Clinical Immunology (EAACI) including passing EAACI exam in 2013. She teaches children with allergy and their caregivers about allergic diseases and she cooperates also with University of Ljubljana, Faculty of Education regarding educating pre-service teachers at master level on allergic diseases.

Xin Tang is a Post-doc researcher at the Faculty of Educational Sciences, University of Helsinki, Finland. He received his doctorate in psychology in University of Jyväskylä, Finland. His research topics include motivation, engagement, grit, curiosity, and classroom practices. He is skilled in advanced statistical modeling (e.g., mixture modeling, growth modeling, and network analyses) and has published in high-profile journals (e.g., Teaching and Teacher Education).

Èlia Tena is a Ph.D. student at the research center CRECIM at Universitat Autònoma de Barcelona, Spain. Her research work is based on scientific practices in primary school education, especially focus on learning of inquiry based on models and modelling in air and pollution. She also has participated in national and European projects regarding STEM in early childhood and primary school education level and equity in STEM education.

Andrée Tiberghien is an Emeritus "directrice de recherche" au CNRS (National Center of Scientific Research). Her research activity focuses on classroom practices and the design of teaching resources. This activity is largely carried out within collaborative groups of teachers and researchers. She has initiated and is currently contributing to an educational research instrumentation project using video as data (ViSA). She is also involved in a project of secondary analysis of PISA data. She was a member of the Science Expert Group of PISA 2006, 2012, 2015.

Russell Tytler is Alfred Deakin Professor and Chair of Science Education at Deakin University, Australia. He has researched and written extensively on student learning and the role of representation as a multimodal language for reasoning in science, on teacher learning, socio scientific reasoning and interdisciplinarity, school-community partnerships, and STEM curriculum policy and practice. He is a Fellow if the Academy of Social Sciences Australia. He has written a number of influential government reports, been active in curriculum development, and has been a member of the Science Expert Group for PISA 2015 and 2024, and a member of the advisory group for the European PARRISE project focused on socio-scientific issues.

Linda Udby is associate professor at the Niels Bohr Institute at the University of Copenhagen, Denmark. Her research spans neutron scattering, e-learning at university level, and network analysis. Recently, she has taken up physics education research at the upper secondary level.

Loris Vergolini is a Research fellow at the Research Institute for the Evaluation of Public Policies of the Bruno Kessler Foundation (Italy). His research interests lie at the intersection between policy evaluation and social inequalities. His specific areas of expertise include the evaluation of educational policies, the consequences of anti-poverty measures and the analysis of the inequalities of educational outcomes in particular with respect to the choice of the field of study at the university and at the upper secondary school.

Wanda Viegas is Ph.D. in Genetics from Wageningen University, The Netherlands. She is Jubilate Full Professor with a long teaching activity in the areas of Genetics and Cell Biology at the Instituto Superior de Agronomia, University of Lisbon and a research activity oriented to genomes dynamic behaviour in terms of genetic and epigenetic changes mediated by genomes interactions and environmental factors, both in crops, model species and human cell lines. In this context, she coordinated / participated in several research projects, and also oriented/co-oriented dozens of Ph.D. and Masters students. She is the co-author of more than a hundred articles published in international journals.

Elisa Vilhunen is a Ph.D. student at the Faculty of Educational Sciences, University of Helsinki, Finland. Prior to this, she was a secondary school science teacher in Finland. She holds a MSc in biology from University of Oulu, Finland and a MA in education from University of Jyväskylä, Finland. Her research interest include affective and emotional aspects of teaching, studying and learning science.

Peta White is a Science and Environmental Education senior lecturer at Deakin University, Australia. Peta has worked in classrooms, as a curriculum consultant and manager, and as a teacher educator in several jurisdictions across Canada and Australia. Peta gained her Ph.D. in Saskatchewan, Canada where she focused on learning to live sustainably which became a platform from which to educate future teachers. Her current research interests follow three directions including science and biology education; sustainability, climate change, and environmental education; and collaborative/activist research.

Albert Zeyer Dr. med. et dipl. math., is a mathematical physicist and a physician. After working as a science teacher and a children's doctor, he was a research scientist at the University of Zurich and a professor of medicine at Bern University of Applied Sciences. He is now professor and head of science teacher education at the University of Teacher Education Lucerne, Switzerland. His research interests are Science|Environment|Health and public understanding of science. At ESERA, he is a co-coordinator of the Special Interest Group Science|Environment|Health and a member of the editorial board of the ESERA Science Education Research Series.

Kristina Zuza is lecturer in the Applied Physics Department of the University of the Basque Country (UPV/EHU). She is graduated in Physics (Astrophysics specialty) by the University of La Laguna and she got her Ph.D. in Physics Education by the UPV/EHU. Her research has been developed within DoPER (DOnostia Physics Education Research) and has as interest points studies students' difficulties understanding and the design and implement and evaluation of Teaching Learning Sequences at university level. She is involved in some national and European projects and she has several publications in high impact journals like PR-PER, AJP, EJP, IJSEd and Enseñanza de las Ciencias.

Chapter 1
Beauty and Pleasure of Understanding – Words of Introduction

Igal Galili

It is a special pleasure for me to introduce this conference organized around the theme "The Beauty and Pleasure of Understanding". The "beauty and pleasure of understanding" explains better than anything else my own engagement with science education and my perception of the essential meaning of scientific knowledge. This is the meaning, I argue, that should be shared with newcomers to the human endeavour of learning science. I will begin by pointing out the connection between the conference theme and the conference location of Bologna, Italy. I will describe three types of science content that are frequently argued to induce interest in science and desire of understanding. To elaborate, I describe how scientific knowledge can be regarded as a special culture. I will close by arguing that drawing upon the cultural structure of scientific theories can suggest new curricular directions and approaches to teaching that engage the beauty and provide the pleasure of understanding among students in science classes.

These words, more or less, were what I wrote in the abstract for the opening talk I gave at the ESERA 2019 conference. The theme seemed to me to be a striking departure from typical themes, indeed never observed before among the meetings of ESERA since it was founded in 1995. However, looking through the list of contributions to the conference, I did not see that many colleagues had addressed this theme directly. Through a long history of collaboration with the Bologna science education group, I was deeply familiar with our shared orientation towards the importance of "the beauty and pleasure of understanding". Thus, I felt the need to highlight here this aspect of science education that we consider to be of central importance, yet is often secondary in contemporary teaching. Indeed, the "beauty and pleasure of

I. Galili (✉)
The Faculty of Mathematics and Natural Science, The Hebrew University of Jerusalem, Jerusalem, Israel
e-mail: igal.galili@mail.huji.ac.il

O. Levrini et al. (eds.), *Engaging with Contemporary Challenges through Science Education Research*, Contributions from Science Education Research 9,
https://doi.org/10.1007/978-3-030-74490-8_1

understanding science", left unelaborated, could appear to be an empty platitude, such as "Life is good". So I was eager to unpack this theme in the short talk in the first day plenary session, a special honour. I cannot objectively evaluate if I was successful. In any case, I was asked to share my words from the conference and so I present an introductory essay that I have composed for the proceedings based upon my opening remarks.

Warmest greetings, colleagues and friends! I am greatly obliged to the organizers of this conference in Bologna for their invitation, especially in relation to the conference theme "The beauty and pleasure of understanding". I agree with the organizers that this theme is an expression of an essential aspect of science and thus also important for science education. I also agree that there are plenty of other important aspects of science and science education captured in the second part of the conference theme, "engaging with contemporary challenges through science education", an all-inclusive and popular claim about the significance of science education. Both parts are, of course, complementary.

The broad public recognizes the existential need for science for our very survival, helping us live effective and comfortable lives. Let us begin by noting these "pragmatic" benefits, listed in the lefthand column of Table 1.1. They are often sufficiently attractive to many students to interest them in science. Yet, the Bible hints at the fact that this list of benefits is incomplete; humans need more:

Table 1.1 Pragmatic (left) versus Spiritual (right) benefits of science education

Science education…	Science education…
– is required to develop, understand and use technology that promises to support personal and social well-being.	– provides an understanding of how Nature is organized as a whole, its law-like design, explaining "how it all works." This understanding is pleasing for its cosmic universal perspective (and is often poorly expressed as "fun").
– enables reliable solving of problems across the great variety of human activities	
– stimulates development of individual skills and abilities, creativity, the art of logic	– reveals to students a special beauty of causal design, the architecture of Nature as theory based, an intellectual edifice of perfect harmony (aesthetic value). These aspects remain unknown to those who do not learn science.
– familiarizes students with the rules of effective activity and knowledge that are objective and rational. It stimulates students' critical thinking, being responsible, cooperative, modest, open to criticism.	– introduces students to the beautiful idea of unified, inductive and deductive logic by which an infinite variety of phenomena can be reduced to a few governing principles. That is to say science expresses the amazing unity across variety and variety in unity.
– faithful to science's aspiration to objectivity, offers a universal picture of the world	– provides its students with intellectual depth, appreciation of sophisticated symmetry, the fundamental complementarity and parsimony in our grasp of reality
– provides a solid basis for individual prosperity, social respect, and successful careers	
– introduces ethical norms of social behavior and rules of productive cooperation	– through historical examples, promotes the non-pragmatic values of altruism, romanticism, devotion to the needs of society, learning from others.

Deuteronomy 8:3: … to teach you that *man does not live on bread alone* but on every word that comes from the mouth of the Lord.

In the righthand column of Table 1.1, I tried to enumerate benefits of another kind that could be considered "spiritual". While they may appear more esoteric, I argue they actually encourage us to construe science as its own distinctive cultural. This distinguishes science from a craft that has specific rules must be followed to be successful.

The pragmatic claims in the left hand column are taken for granted and commonplace. Their objective validity draws on the great achievements of our society. The "spiritual" features in the righthand column, however, are often considered to be emotional and illusive and therefore, optional, subjective, secondary, causing merely affective impact that may enhance the effect of teaching but dilutes, misleads and detracts attention from the "true" content of science learning, and results in the loss of valuable time.

Let me begin my argument by observing that science itself did not start for the purpose of providing practical benefits. Science started with searching for objective causes of natural phenomena, law-like regularities, and then introducing abstract concepts and models of the natural order. Rational objective knowledge – *episteme* – was invented in Classical Greece. There was no obvious need to do that since technological knowledge – *techne* – was not immediately related to natural philosophy. Why know about the arrangements of stars and planets in multiple spheres with complex structure? Why know about "effective" causes for seasons and the mechanism behind their cycle? Why know about elements and structure of matter? Why know how vision works? Yet, these and other questions about reality emerged very early and continue to provoke people through the ages without immediate and obvious practical benefits. Skipping a comprehensive analysis, we may listen to the scientists themselves[1] who, from the dawn of science, continuously addressed their drive, motivation and intention. For example, in modern times, this is how James Peebles, who won the Nobel Prize in physics in 2019, reacted to the announcement:[2]

> The prizes and awards, they are charming, much appreciated, but that's not part of your plans. You should enter science because you are fascinated by it.

"Prizes" can be understood in a broader sense as practical benefits. Scientists continuously repeat the idea which Leonardo expressed as:

> The noblest pleasure is the joy of understanding

The renowned scientist of the recent past, Henry Poincaré, refined it while reflecting on the history of science:[3]

---

[1] "To learn from the horse mouth" (Wong & Hodson, 2009).

[2] https://finance.yahoo.com/news/nobel-prize-awarded-physicists-changed-152100367.html

[3] Poincare (1908)

> The scientist does not study nature because it is useful to do so. He studies it because he takes pleasure in it, and he takes pleasure in it because it is beautiful. If nature were not beautiful it would not be worth knowing, and life would not be worth living. (p. 22)

> Is it merely that things which seem to us beautiful are those which are best adapted to our intelligence, and that consequently they are at the same time the tools that intelligence knows best how to handle? (p. 23)

> ...the Greeks loved the intellectual beauty hidden behind sensible beauty, and that it is this beauty which gives certainty and strength to the intelligence. (p. 24)

The history of science abounds with such confessions emphasizing the pleasure of understanding as a special type of emotional excitement caused by revealing the specific type of beauty the world possesses. Thus, pleasure and beauty are components of science as practiced. But are they essential? Let us proceed.

For centuries, the concept of beauty has been considered emblematic of Italian culture. Beauty attracts people universally; including beauty as revealed through science. But this reaches a level of refinement in Italy, in particular, that is noteworthy. Consider Florence, the place where people especially venerated beauty. In the sixteenth century, they placed the statue of David by Michelangelo in the central square of the city as a symbol of beauty (Fig. 1.1a). It stays there now – a lovely young fighter of perfect proportions. Yet, the youth was neither Apollo nor Alexander the Great, but the Biblical hero, King David. What is remarkable about this choice is the fusion of a warrior, demonstrating the power of force, courage and devotion to his people with something very different. David was a poet whose poetry, the book of Psalms, talking to and about God, has been in continuous use for three

(a) (b)

Fig. 1.1 (**a**) A fragment of the statue of David by Michelangelo (c. 1504) in Florence. (**b**) A fragment from the bas-relief on the sarcophagi of a professor in Bologna University by Masegne (c. 1383)

thousand years by people around the world in their everyday prayers. David apparently symbolized the symbiosis of internal and external beauty in ultimate harmony which seemingly left no place to add anything else. Was it so?

In fact, the people of Bologna did not agree with Florence and pointed to another dimension of beauty missing in the Florencian set – the beauty and pleasure of understanding, not less and possibly more divine in its nature. In the eleventh century, the people of Bologna established a new type of temple, the temple of knowledge, the university – Alma Mater Studiorum. Their heroes were people of knowledge and understanding: students and professors. Within the national tradition of artistic visualization, they produced the image of a student (Fig. 1.1b). In parallel to David, a young warrior-poet and emblem of the beautiful inside and out, the figure of the young student is delighted by the knowledge revealed to him; this became emblematic of Bologna. It is this image that we may consider as a visualization of the title of our conference – *The beauty and pleasure of understanding.*

Over the course of one thousand years, this university was decorated by a long gallery of renowned scholars of which I mention a few whose names I encountered: Giovanni di Casali, Giovanni Battista Riccioli, Francesco Maria Grimaldi, Giovanni Domenico Cassini, Luigi Galvani, Guglielmo Giovanni Maria Marconi and of course, Umberto Eco. They all illuminated the minds of numerous students who were introduced here to the unique beauty and pleasure of understanding. In line with this tradition, in 2002, the journal *Physics World* announced the choice made by its readers to consider the physics experiment performed in 1974 by three Bologna professors (Pier Giorgio Merli, Gian Franco Missiroli, and Giulio Pozzi) as the most… beautiful of all time. From all the possible characteristics that could be used to describe scientific products, a rather unusual description, "beautiful", was chosen for the experiment providing evidence of the amazing interference of an electron with… itself.

As we turn to science education, it is my special pleasure to pay tribute to University of Bologna professor Nella Grimelini Tomasini (Lella) who has raised the flag of the *Pleasure of Understanding* in physics education. Many people certainly join me now in sending her our deep appreciation, sincere gratitude and wishes of health and prosperity that she so much deserves.[4]

[4] In great sadness I mention now that Lella has recenly passed away. May her memory be blessed and stay with us.

It seems to me that the location of this meeting and the idea of highlighting the "spiritual" aspects of science education, which was made explicit in the title, is as surprising as blossoms in springtime. It is therefore paramount to capture this moment and draw attention to the importance of encouraging this perspective in science education. This spiritual commitment of the physics education group in Bologna (established by Lella) encourage us to explore this intellectual direction in science education.[5] They have raised the flag, however, profound questions emerged regarding its implications. What should we actually do to encourage students to experience the pleasure of understanding science? What content should we specifically address? Is there something teachers can do in addressing this content beyond the general claims of Table 1.1?

We need to show that even if pragmatic values may prevail for their existential benefits, their spiritual extensions are vital for science *understanding*. Such recognition cannot be spontaneous or intuitive. It requires clarification and specific restructuring of numerous curricular components and their underpinnings be disciplinary, cognitive, philosophical, historical and sociological, which all contribute to our understanding of understanding. What should be included and how? To provide my answer, let me briefly present results of a comprehensive study[6] dealing with this topic. This line of work suggested a special organization of the science curriculum, which I call *discipline-culture*. Within this perspective, I intend to answer those questions raised above.

Scientific knowledge is comprised of big clusters of knowledge elements which are internally coherent. These clusters can be structured and hierarchically ordered. The elements of each cluster share a certain historical thread, methodological tools of production of new elements, adopting some and rejecting others. They create a colony or a *culture*. These groups comprise the fundamental theories known to us, each providing a specific picture of the world (mechanics, electromagnetism,

[5] It is a pleasure to mention the students and colleagues of Lella whom I had the honour to cooperate with in research: Olivia Levrini, Barbara Pecori, Marta Gagliardi, Eugenio Bertozzi, Paola Fantini, Giulia Tasquier.

[6] Tseitlin and Galili (2005, 2006), Levrini et al. (2014), Galili (2017)

Fig. 1.2 Singing angels from the Ghent Altarpiece (c. 1432) by van Eyck brothers

quantum, etc.). One may imagine science as a polyphony of different perspectives on Nature, together comprising the Book of Nature as Galileo put it. In a sense, the situation can be well animated by the artistic image of angels singing in divine, but different, voices from the same book (Fig. 1.2). Scientific theories create a family of cognate knowledge systems describing nature. They share certain concepts and differ in others.[7]

We may identify this dialogue of theories as a special culture – the culture of science.[8]

Furthermore, the traditional disciplinary perspective considers a scientific theory to be structured by its nucleus (fundamental principles, concepts, paradigmatic model) and its body (derivations, implications, working models, conceptions, experiments) coherent with the nucleus. The discipline-culture perspective upgrades the disciplinary persective with the third type of knowledge elements, the periphery. The latter includes elements sometimes at odds with the nucleus, representing open problems, competitive principles and accounts by other theories. In this way, a

[7]The idea of family resemblance is due to L. Wittgenstein, while the many-faceted somewhat contradictive accounts of nature in the discourse regarding the world may remind the idea of carnival by M. Bakhtin in his literature critique.

[8]The concept of culture ascribed to a family of fundamental theories may remind the *culture* used in biology to designate a colony of micro-organisms of the same kind.

culture includes the potential to change itself. In a way, the presented approach bridges the opposition between *discipline* and *culture* as defined by Kant in 1781.

My colleagues and I have argued that the obtained triadic structure, nucleus-body-periphery, is more faithful to the reality of knowledge exploration than a disciplinary portrait of a theory is. It also happens to be effective in representing scientific knowledge in the context of education.

This discipline-culture perspective implies a pertinent restructuring of introductory curricula. It will emphasize principles, connect them to phenomena and guide the construction of explanations beyond technical manipulation. It will encourage us to make explicit the limits of the validity of each theory by pointing to alternative accounts, either correct (from the more advanced theories) or wrong (from the rejected theories). For instance, the ideas of mechanics of Aristotle, Einstein and Bohr all appear in the periphery of the theory in which the nucleus incorporates the Newtonian laws of motion. The triadic identification of curricular elements can match variation in the interests, skills and preferences students naturally display, expanding their willingness to learn and explore. Being exposed to the subject matter of all three aspects, each individual combines his/her interests in different proportions of efforts and desire.

Thus the first group of students shows a special interest to the *nucleus*, the theoretical paradigm. These students are interested in the big ideas, and they take a holistic perspective on what that theory tells us about the world, a kind of philosophical standpoint. They are not much interested in solving standard problems. They easily decide to leave that to others and rely on scientists to justify this knowledge. It looks as if such students ask us "Show me God", the overall design of the world. Einstein, Newton, Kepler, Aristotle shared the same focus in their interest with respect to understanding. In a sense, we could call such students "philosophers" or "enlightened".

In contrast, the second group shows interest and readiness to deal primarily with concrete problems, mastering the power of control over nature, and seeking immediate benefits. They show much less interest in the justification of the principles used as long as they help them to reach the goal. They are interested in the *body* knowledge as a tool box for experimenting, solving new problems, realization of knowledge potential. Using simplified examples: understanding and addressing climate change – yes; the analysis of the nature of inertia and principle of equivalence (as interested Mach and Einstein) – no. In a sense, we may call these students "practitioners" or "consumers".

Finally, the students of the third group take a sort of critical stance. They question the claims of the nucleus: "Why these principles not the others? Where did you get them from? Was there any choice?" This was Einstein's and Leibnitz' interest. "Are these principles universal, unique, applied outside of science? Where do these laws fail?" If we, as their teachers, ignore this kind of questions, those students are disappointed and lose their interest: "I do not like science, I prefer something more human…" They may dream easily, miss their teacher's instructions, and fail on exams. They "do not care" and could be "trouble-makers" in class. They could be referred to as – "revolutionaries". This all may change if the teacher addresses the *periphery*. We may, then, observe a radical change in their attitude and witness them joining the rest of the class equipped with a different motivation.

As to the attitude of science teachers, they often ironically patronize the "philosophers" trying to encourage them to be more serious, invest more effort… We are often satisfied with the "practitioners" and usually provide them with all available support. As for the "revolutionaries" (who loves trouble makers?!), they disturb the smoothness of lecturing. Their questions may puzzle us and they might expose us unprepared. We call them to order, trying to explain that such questions lead us astray, impede understanding, break the thread of explanation and take time away from what they need most training in. We may promise answers in some unclear future, just not now.

These are contrasting preferences and each individual combines all three of them in some proportions. Looking back over my own experience in teaching introductory physics courses for many years, I may mention that the students who identify strongly as "philosophers" usually became scientifically literate, enlightened citizens able to consider problems of the society saturated with scientific content. The students whom we identified as "practitioners" normally became involved in technology and applied sciences and medicine. They became proficient consumers of science, "normal" explorers (Kuhn called them "puzzle resolvers"). As to the "trouble-makers", some of them indeed switched to humanities and activities outside of science, but there were a few among them, who proceeded to higher degrees and joined the researchers at the frontiers of science and high-tech. These were the students who produced new knowledge (Kuhnian "revolutionaries"). In any case, I do not intend to create a fully deterministic picture, but just to share my experience. Reading the memoirs of scientists, especially those known for their contributions to modern science (e.g. Galileo, Heisenberg, Einstein, Poincare, Weinberg), may provide additional support to this three-fold perspective on a scientific theory as a *discipline-culture*.

I return now to the question how science teaching can stimulate, encourage and instigate the perception of pleasure in understanding and a sense of the beauty of science and scientific knowledge. For that, we need to recognize the structure of science as a culture and recognize the preferences in the non-homogeneous population of students at schools and universities. Accordingly, our curricula and teachers should talk in three voices addressing the nucleus, body and periphery of the theories considered. This equally addresses ontological (content) and epistemological (methodology) aspects. The new approach discharges the claim of "two cultures" (science versus humanities)[9] which implied a simple dichotomy of students, good and bad at science. The reality is much richer and more interesting, allowing a wider population to enjoy learning science. Cognitive resonance between the emphasis of instruction with their intellectual preferences will allow students to enjoy science class, appreciate the beauty of scientific knowledge and identify their own areas of interest. This approach involves various aspects of the humanities (epistemology,

[9] Snow (1959)

logic, history, aesthetics, world view) intertwined with the science content; an approach that will help more of our students have deep experiences of pleasure in understanding. This approach suggests a framework for addressing the dual nature of science mentioned in Table 1.1. The enjoyment of learning cannot be of the same kind for all, and it is not unique for each. To illustrate, let us look again at the same piece of art mentioned earlier to be emblematic of Bologna (Fig. 1.1b), this time taking a wider view. In light of the introduced perspective, we may speculate about the factors that cause different students to delight (or not) in what we teach them (Fig. 1.3).

A final comment is on the colloquial claim "Science is fun". Its rather uncertain meaning may easily miss the intellectual depth of *the pleasure of understanding* science. "Fun" has the connotation of being light, amusing, superficial, and fleeting, which poorly matches being analytic, appreciating aesthetics, and delighting in the beauty of science and experiencing pleasure with the understanding of complexity. Though there is no need to engage in a crusade against casual "fun", it would be good if a teacher who proclaims that "Science is fun!" is aware of what is deeply enjoyable about science that is not captured by this phrase (similar to us when enjoying cola not forgetting about good wine).

I conclude with the belief that the paradigm of *discipline-culture*, by revealing the structure of scientific knowledge, the nature of knowledge elements and scientific dialogue, can transform the *pressure for understanding* to the *pleasure of understanding* science. It creates a bridge between the realms of science (related to objective pragmatic benefits) and the humanities (related to subjective and spiritual values) often perceived to be in opposition. Understanding the formal disciplinary content does not exclude, but is enormously enriched by the relevant philosophical background. Together, they result in the pleasure of understanding science. Indeed, science can bring fun, but mainly, it can enrich us with much more – *the pleasure of understanding* which is a serious business, because it reveals us the *genus* of science, and this is truly, exciting.

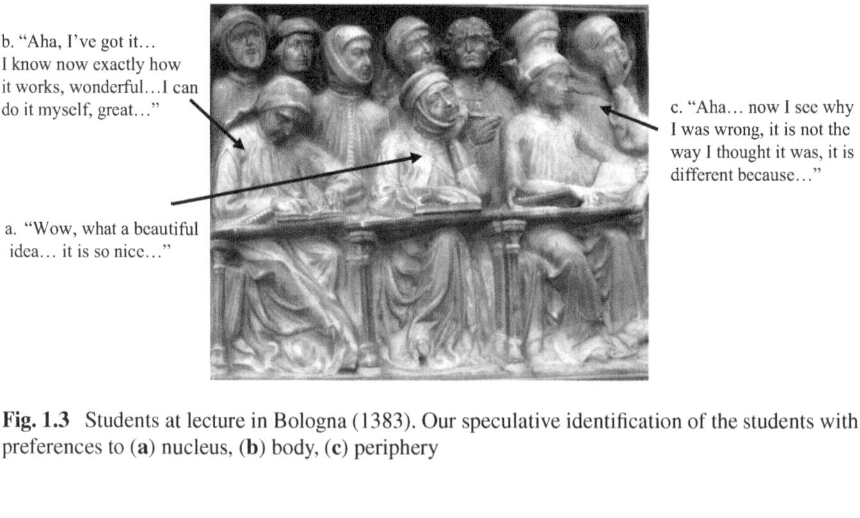

Fig. 1.3 Students at lecture in Bologna (1383). Our speculative identification of the students with preferences to (**a**) nucleus, (**b**) body, (**c**) periphery

Good luck and enjoy your endeavours hopefully charged with the spirit of Bologna.

References

Galili, I. (2017). Scientific knowledge as a culture – A paradigm of knowledge representation for meaningful teaching and learning science. In M. R. Matthews (Ed.), *History, philosophy and science teaching research. New perspectives. Ch. 8* (pp. 203–233). Springer.

Kant, E. (1781). *The critique of pure reason.* In *Britannica great books.* The University of Chicago Press.

Levrini, O., Bertozzi, E., Gagliardi, M., Grimellini-Tomasini, N., Pecori, B., Tasquier, G., & Galili, I. (2014). Meeting the discipline-culture framework of physics knowledge: An experiment in Italian secondary school. *Science & Education, 23*(9), 1701–1731.

Poincare, H. (1908). *Science and method.* Dover.

Snow, C. P. (1959). *The two cultures and scientific revolution.* Cambridge University Press.

Tseitlin, M., & Galili, I. (2005). Teaching physics in looking for its self: From a physics-discipline to a physics-culture. *Science & Education, 14*(3–5), 235–261.

Tseitlin, M., & Galili, I. (2006). Science teaching: What does it mean? – A simple semiotic perspective. *Science & Education, 15*(5), 393–417.

Wong, S. L., & Hodson, D. (2009). From the horse's mouth: What scientists say about scientific investigation and scientific knowledge. *Science Education, 93*(1), 109–130.

Part I
Meeting Societal Challenges

Chapter 2
The White Noise of Climate Change (the Language of Climate Change)

Andri Snær Magnason

How do you talk about something that is bigger than language? What words do you choose when a scientist shows you how everything will change in the next 100 years. How the glaciers will vanish, and become ocean, how the ocean will rise and swallow coastal areas, while the pH of the oceans, the acidity will change more in than it has in the previous 50 million years? How do you understand 50 million years? How do you understand that 0.3 in the logarithmic scale of the pH levels? 50 million years are too big to register, 0.3 in pH too small and abstract to understand. A person born today can measure in his lifetime greater changes in the oceans than, not only the whole evolution of man, but ten times that. Such change is not just historical, you could say its mythological.

We are shaking the foundations of the planet but we have normalised the situation. We yawn when a climate conference is mentioned, mock the private jets sarcastically, like there was something normal for humans to discuss if the Earth shall become 1.5C, 2C or 5C warmer. Gengis Khan, Ramses II, Cæsar, Napoleon and Stalin, never thought they could melt glaciers and make the oceans rise. The times we are living is not just a new chapter in history books, but in geology books. Changes that previously took a hundred thousand years happen in one hundred. Such speed affects all life on Earth, the roots of everything we think, choose and produce. It affects everyone we know and love. The changes that are more complex than most of what our minds typically deal with, they surpass most of the language and metaphors we use to navigate our reality. Even when you walk on a glacier and lie down with hundreds of meters of ice below your feet, it's impossible to imagine that this mass might be gone in the lifetime of your own child.

We see headlines and think we understand the words: "glacial melt," "record heat," "increasing emissions." But they pass us by while we respond to smaller

A. S. Magnason (✉)
Writer, Reykjavík, Iceland

O. Levrini et al. (eds.), *Engaging with Contemporary Challenges through Science Education Research*, Contributions from Science Education Research 9,
https://doi.org/10.1007/978-3-030-74490-8_2

words. We have a stronger reaction when a single animal is shot, rather than to data about the loss of 60% of all the planets wildlife since 1970. Like a scientist can not see a black hole, because the enormous gravity absorbs all light, we could say that the scale of global warming makes all meaning collapse. We can not scale up words, say the issue is big to the 12th degree. If you write IN CAPITAL LETTERS WITH THREE EXCLAMATION MARKS!!! you will be discredited. They should directly alter our actions and choices. But it seems that 99% of the words' meanings disappear into white noise.

History shows us how long it can take new words to settle in. *Soul*, *sin*, *conscience*, were dominant words in our culture for centuries and pure power in the hands of the church and the ruling class. But these words had not always existed. *Skaldic poetry* of the Viking age used language and metaphors based on Norse mythology: They referred to Earth as "Oðin's bride" and to heaven as "dwarf's helmet." When Christianity came the poet was in trouble. How do you explain God, the creator of heaven and earth when the language forces you to say, creator of "Dwarfs helmet" and "Odins bride"? The new world could not be explained without the old world.

When I was a child, I learned that Iceland had yearned for freedom and independence for 600 years as a colony of Denmark, until finally 1918 the struggle succeeded. But in 1809 a peculiar event happened. British soap merchants and the local Danish authorities in Iceland had a dispute, resulting with the merchants taking the Danish authorities hostage. Jørgen Jørgensen a Danish translator on the British ship became the temporary ruler of Iceland. He was deeply inspired by the French Revolution and declared: "Iceland is a free country, independent from Danish rule." Jörgen declared he would be in control until we had a constitution and a parliament the following year: "Where the poor and common people shall have an equal share in the government with the rich and powerful." This sounds like an obvious wish of a colonised nation. But there was a problem. His words had little or no meaning as the literature of liberty, democracy and brotherhood had hardly been translated to Icelandic. No Icelander had yet asked for freedom or independence. We were given freedom but had no wish to take it. The independence heroes in Iceland were born around 1809 they did not get these ideas until they sat drinking in the bars of Copenhagen around 1830. It took 100 years of poems, songs and essays until the words of independence were fully charged.

Now we have many new words that are as new as democracy was in 1809. The term "ocean acidification" was created in 2003. It explains how the pH level of the oceans are dropping because they absorb about 30% of the CO_2 entering the atmosphere. The pH is expected to drop from 8.1 to 7.7 during this century and this change could disrupt the entire ecosystem. Ocean acidification is therefore one of the biggest words in the world. I feel like I understand the word, but probably not. Like the meaning of the word nuclear bomb before or after Hiroshima. Like the word Holocaust in the 1930s vs the 1970s when it had been loaded with a different meaning and millions personal experiences.

Big. Large. Enormous. I use these words every day about things in my daily life. But how can I scale them up to capture a whole planet, everything that lives and breathes. Maybe only our oldest stories can capture what is happening. History takes place when humans are doing the human things. Fighting over land, power, and ideas. Mythology takes place when the fundaments of nature are changing. When the moon was placed in the sky. When fire was stolen from the gods. Mythology tells us the times of the great flood, beginnings and ends of worlds.

Robert Oppenheimer is probably the true mythological figure of the twentieth century. In Greek mythology, Prometheus brought fire to mankind by stealing it from the highest peak of Mount Olympus; Oppenheimer dived into the smallest unit of matter and brought the world's leaders the nuclear bomb, the godly force to blow up the whole the planet. Oppenheimer himself realized the mythical context. Seeing his explosion he remembered a line from the Bhagavad Gita: *Now I am become Death, the destroyer of worlds*. Oppenheimer caused nightmares for a whole generation but our current threat comes from good old Promethius. He angered the gods by stealing fire and they probably knew that we could not control it. We see words like 415 ppm, "emissions" and 35GT of CO2. But CO2 is invisible and 35 Gigatons sounds like very much of nothing. Then older words are helpful, like fire. Our emissions come from invisible fires, hidden in cars, factories and energy plants. We see traffic but not the fire.

When the Eyjafjallajökull volcano erupted in Iceland in 2010 and closed down the European airspace it emitted about 150,000 tons of CO2 every day. Human emissions are about 100 million tons per day. If we translate human activities into volcanic eruptions we can see that human emissions are like 666 of these volcanoes. (Sorry for the number loaded with meaning, I was just doing math: 100,000,000/150,000), If you ask a geologist, when so many volcanoes erupted for decades he will only find eras of major disruption in all of earths systems. We have kindled the greatest flames the earth has ever seen. We might ask like Oppenheimer: *Have we become the destroyers of our world?*

Time is running out. Nature has left geological speed of change. Our reactions are still on a geological scale. We seem disconnected with years like 2050, 2100. When I talk in universities, I ask the students, born around 2000 to do a simple calculation: When is someone still alive that you will love? They ask what I mean and I explain. "If you become 90, you might have a favorite 20 year old in your life, perhaps a grandchild, born 2070. If that person becomes 90, the person you might love the most in your life, is still talking about you in the year 2160."

Democracy holds the belief that we can best understand complex matters, and to vote for them. Scientists have pointed out that, we have up to now voted against the oceans, the glaciers, our climate, against future generations. Can we choose to steer the world in the right direction?

Education is not something floating in mid air, disconnected from the world we live in. Slowly in the last decades, education has merged into serving a system and language where the end of education is to produce capable people to serve the market forces of expanding economies and extracting resources from the earth to make products. Education has become "investment" in the international competition of nations and corporations. Natural sciences have been subjected to serve and sometime greenwash these activities. Humanities are like nature, hard to measure and therefor worthless. Someone that studies poetry or history but becomes an engineer, has lost valuable time according to that mentality.

But a new reality emerged. The world has to cut all CO2 emissions in the next decades. If we do not succeed, we might lose everything. Everything. At the same time we have to remove about 1000Gt of CO2 in the atmosphere, that will continue to cause warming during the century, even if we manage to slow down emissions. This is the main goal of all politics and all education until we have succeeded. When a student asks, why are we studying algebra? The answer is simple: We have 1000GT of CO2 to remove and nobody knows how to do it. But why are we studying ethics? Because someone that has done too much algebra might stumble upon a solution that is not ethical. But why are we studying poetry? Because we are human and poetry is something humans have always needed, and will always need.

Chapter 3
Prediction and Adaption in Science|Environment|Health Contexts

Albert Zeyer, Nuria Álvaro, Julia Arnold, Deidre Bauer, Iztok Devetak, Sonja Posega Devetak, Valentín Gavidia, Kerstin Kremer, Olga Mayoral, Tina Vesel Tajnšek, and Alla Keselman

3.1 Introduction

The term Science|Environment|Health (S|E|H) stands for a pedagogy of mutual benefit between environmental, health and science education (Zeyer & Dillon, 2014). Complexity is an important aspect of most S|E|H issues (Zeyer et al., 2019). However, non-complex contexts are still very predominant in science education (Fensham, 2012). One reason is that in the natural sciences, prediction plays a crucial epistemological role. Every causal explanation has the form of a prediction, and

A. Zeyer (✉)
Science Teacher Education, University of Teacher Education, Lucerne, Switzerland
e-mail: albert.zeyer@phlu.ch

N. Álvaro · V. Gavidia
Department of Experimental and Social Sciences Education, University of Valencia, Valencia, Spain
e-mail: valentin.gavidia@uv.es

J. Arnold
School of Education, FHNW University of Applied Sciences and Arts Northwestern Switzerland, Muttenz, Switzerland
e-mail: julia.arnold@fhnw.ch

D. Bauer · K. Kremer
Institute for Science Education, Leibniz University, Hanover, Germany
e-mail: bauer@idn.uni-hannover.de; kremer@idn.uni-hannover.de

I. Devetak
Faculty of Education, University of Ljubljana, Ljubljana, Slovenia
e-mail: iztok.devetak@pef.uni-lj.si

O. Levrini et al. (eds.), *Engaging with Contemporary Challenges through Science Education Research*, Contributions from Science Education Research 9,
https://doi.org/10.1007/978-3-030-74490-8_3

every prediction is also an explanation (Hempel, 1965). Yet, for systems theoretical reasons, complex systems do not allow for full prediction. "Don't predict, adapt!" is therefore a famous slogan in complexity talk, coined by Per Bak, an influential theoretical physicist in the 1990s (Bak, 1996).

3.2 The Symposium Contributions

What may adaption look like in complex systems and what role can science play in it? This paper features an invited symposium of the ESERA conference 2019 where S|E|H examples were presented in which the relationship between prediction and adaption is important.

The first contribution was theoretical and discussed the concept of dual-process theories as a potential framework for discussion and conceptualisation. The second contribution presented findings concerning the quality of life of students with food allergies. The third contribution investigated how students conceive present and future impacts of a vegetarian diet on sustainability issues. The fourth contribution analysed the Spanish mandatory curriculum in view of environmental health.

Notice that in order to give enough room for basic considerations and tentative reflections that emerged from the symposium, we refrain here from providing methodological information or detailed results of the presented research.

3.2.1 Contribution 1: A Dual-Process Approach to Prediction and Adaption

Albert Zeyer.

S. P. Devetak
Department of Paediatrics, General and Teaching Hospital Izola, Izola, Slovenia

O. Mayoral
Department of Experimental and Social Sciences Education, University of Valencia, Valencia, Spain

Botanical Garden of the University of Valencia, University of Valencia, Valencia, Spain
e-mail: olga.mayoral@uv.es

T. V. Tajnšek
Department of Allergology, Rheumatology and Clinical Immunology, University Children's Hospital, University Medical Centre, Ljubljana, Slovenia
e-mail: kc29260@ukclj.si

A. Keselman
National Library of Medicine, National Institutes of Health, Bethesda, MD, USA
e-mail: keselmana@mail.nlm.nih.gov

Table 3.1 Characteristics of type 1 and 2 processing, adapted from Stanovich and West (Stanovich & West, 2000)

	Type 1 processing	Type 2 processing
Properties	Associative	Rule-based
	Holistic	Analytic
	Automatic	Controlled
	Relatively undemanding of cognitive capacity	Demanding of cognitive capacity
	Relatively fast	Relatively slow
	Acquisition by biology, exposure, and personal experience	Acquisition by cultural and formal tuition
Task construal	Highly contextualised, personalised conversational, and socialised	Decontextualised, depersonalised, asocial

3.2.1.1 Dual-Process Theories

Dual-process theories assume that decision making is an interplay between analytical processes and adaptive processes (Schlosser, 2015). Table 3.1 provides characteristics of dual-process theories as they have been clustered from an impressive body of data that emerged from at least a dozen dual-process theories. Many labels have been used to characterise the duality captured in Table 3.1 Finally, the most neutral terms have become widely accepted, which are type 1 and type 2 processing (Stanovich & West, 2000).

From Table 3.1, it is important to observe for the following that the task construal of type 1 processes is characterised as "highly contextualised, personalised, conversational, and socialised", while type 2 processes, in contrast, are described as "decontextualised, depersonalised, and asocial".

3.2.1.2 The Interplay of Type 1 and Type 2 Processing

So far, there is no commonly accepted view on how the interaction between the two processes works (Schlosser, 2015). Many dual-process psychologists favour a default-interventionist theory (Evans, 2008). Here, type 1 processes provide intuitive judgements by default. These must then be endorsed, or corrected, by type 2 processing. Other researchers support a parallel-competitive approach, which assumes that both types of processing are equally important for good decision-making (e.g., Stanovich & West, 2000).

Systems theory adds an interesting perspective to this, which may be particularly relevant to S|E|H issues. It suggests that this debate cannot be resolved out of context. In ordered systems, default-interventionist approaches may be state of the art, because type 2 processing may eventually solve any reasonable S|E|H problem by a scientific (predictive) approach.

In complex S|E|H systems, however, the predictive power of type 2 processing is limited. "Don't predict, adapt!" can be interpreted as the suggestion not to overrate the role of type 2 processing for decision-making, but to rely more upon type 1 processing. The slogan is certainly too extreme, but it is inspiring. It points out that type 1 thinking may be an important and underestimated resource in complex situations. A parallel-competitive approach, or even a parallel-complementary approach, may then be most adequate.

3.2.2 Contribution 2: Quality of Life of Students with Food Allergies

Iztok Devetak, Sonja Posega Devetak, and Tina Vesel Tajnšek.

3.2.2.1 Introduction

Food allergies are a serious concern in schools. Approximately 5% of children have food allergies (Sicherer, 2011). About 18% of food-related allergic reactions in children occur in school situations (Mahl et al., 2005). Scientific knowledge predicts that every contact with the allergen may cause an allergic reaction that might eventually result in an anaphylactic shock.

In principle, this requires the child to completely avoid the allergen and the administration of appropriate therapy including adrenaline if accidental ingestion and severe allergic reaction occur. However, it is known that such a regime may substantially lower the quality of life for children with allergies and also for their parents. Indeed, quite a number of domains of health-related quality of life appear to be affected, including social relationships, emotions, school experience, and finances. Adolescents with food allergies also reported social isolation and depression (Morou et al., 2014).

It is in this context that the authors of contribution 2 interpreted the term of adaption. They suggest that children with allergies, their parents, and their teachers adapt successfully if they have the same health-related quality of life as their healthy peers. The authors presented a model of key medical competencies that would help teachers to adapt successfully in this sense (Fig. 3.1).

3.2.2.2 Pre-service Teachers with and Without Food Allergies

If the predictive power of basic medical knowledge (see top circle in Fig. 3.1) was very high, i.e. if every potentially allergic situation could be definitively assessed in terms of "unproblematic" or "anaphylactic shock", then a predictive approach would always definitively solve the problem. Decision-making would then be

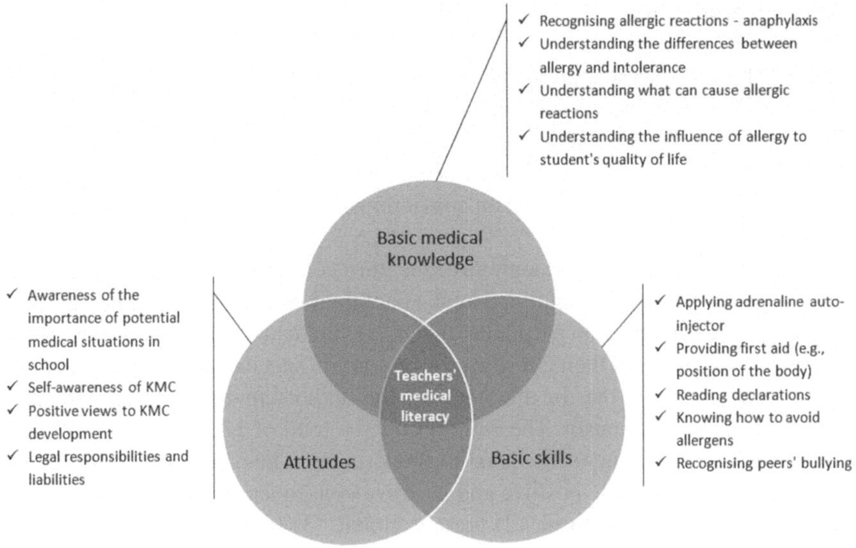

Fig. 3.1 The model of teachers' Key Medical Competencies (KMC) (Devetak et al., 2019)

decontextualised, depersonalised, and asocial, and the teachers would just have to meticulously follow the rules.

However, potentially allergic situations are always complex, and only in trivial cases is there a binary decision. Therefore, involved teachers need what the authors call attitudinal and basic skills (down left and right circle in Fig. 3.1), which are contextualised, personal, and social.

The authors explained that these skills are acquired by exposure and experience, and they hypothesised that teachers who had a food allergy themselves would turn out to be more skilled than others without this experience. They reported data from a study where they had investigated 239 pre-service primary and lower secondary school teachers by a "key medical competencies score" based on Fig. 3.1. The results suggested that those student teachers who had a food allergy themselves had better key medical competencies in managing children with allergies than their colleagues without an allergy. Indeed, those pre-service teachers with allergies achieved significantly higher scores than those without allergies.

3.2.3 Contribution 3: Pupils' Perceptions About Sustainability-Related Impacts of Their Consumer Behaviour

Deidre Bauer, Julia Arnold and Kerstin Kremer.

3.2.3.1 Introduction

Empirical findings show that consumers make decisions based on the assumption that a specific action has certain impacts on the well-being of others (e.g., Roberts, 1996; Vermeir & Verbeke, 2006). These findings are also reflected in various behavioural models. In the theory of planned behaviour, for example, it is argued that any reasoned decision-making is based on convictions about behavioural outcomes (e.g. Ajzen, 1991).

Based on these findings, the authors of the third contribution assumed that, when it comes to reasoned action-taking, people adapt their decisions to their perceptions about potential outcomes. In particular, they were interested in adolescents' perceptions of consequences of their consumer behaviour on specific sustainability dimensions (Bauer et al., 2018). In this context, the authors investigated adolescents' perceptions of vegetarianism. The authors coded a total of 1224 statements of 125 pupils from two secondary schools in north Germany. The participants were asked to list at least two possible positive and negative consequences of living on a vegetarian diet for (1) present local, (2) present global, (3) future local, and (4) future global generations. Figure 3.2 presents the results of a qualitative content analysis in percentage share of subcategories, and in three sustainability dimensions (social, economic, ecological).

3.2.3.2 Present and Future Consequences

For each sustainability dimension, Fig. 3.2 shows that adolescents were able to intuitively name a considerable number of possible impacts of vegetarianism, both under close and distant conditions. However, when it came to distant effects, they had trouble relating their behaviour to the concrete well-being of others. Rather they saw the greatest consequences on the abstract level, particularly concerning the impact for future generations in the social or economic dimension.

The authors concluded that in order to help adolescent consumers reasonably adapt their choices to the requirements of sustainability, education needs to provide ways to make concrete behaviour-well-being relations visible for both close and

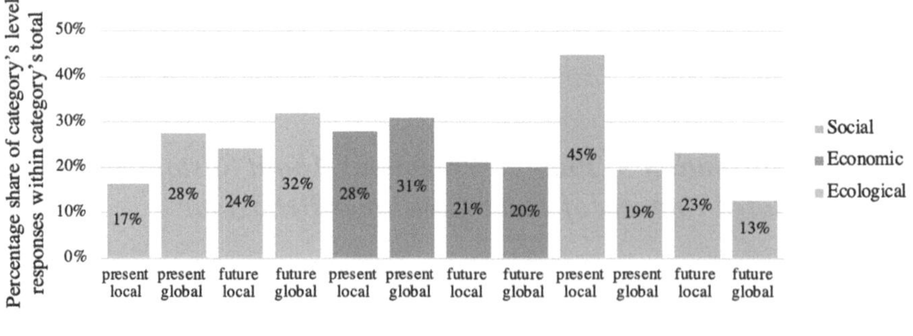

Fig. 3.2 Percentage of category responses about impacts of vegetarianism in each distance level compared to the category's total responses

distant temporal and spatial impact levels. Students would then not only surrender to abstract predictive rules. They rather would make sense of concrete situations and find a way to adapt personally, socially, and contextually. Scientific sustainability knowledge would help with the sense making by interpreting the sustainable character of vegetarian strategies.

3.2.4 Contribution 4: Development of Adaptive Didactic Resources for Decision-Making on Environmental Health Problems. First Step: Curricular Analysis

Valentín Gavidia, Nuria Álvaro and Olga Mayoral.

3.2.4.1 Introduction

The World Health Organisation (WHO) states that health should be promoted and protected through an environment that enables sustainable actions at individual, community, national and global levels (Gil et al., 2015). The authors of the fourth contribution have been analysing the Spanish curriculum in terms of environmental health in an encompassing, still ongoing project. In previous publications, they documented Spanish students' competences on environmental health at the end of compulsory education (e.g. Zeyer et al., 2019).

In this symposium, they presented aspects of the Spanish mandatory curriculum (Real Decreto 126/2014 and 1105/2014). The authors were interested in assessing whether the curriculum provides students with the necessary competences to face environmental health problems, and particularly if the curriculum addresses the complexity of these problems and promotes students' participation in decision-making.

3.2.4.2 Missing Aspects in the Spanish Mandatory Curriculum

The analysis was based on a Delphi study. Thirteen experts with diverse backgrounds were asked about the environmental health topics desirable in the Spanish curriculum. The results of this Delphi study are summarised in Table 3.2.

Starting with this list, the authors identified a number of topics that were missing in the mandatory Spanish curriculum, like "the importance of the indoor (home, work…) and outdoor environment for health", or an "assessment of the importance of soil in people's life". Generally, they stated a considerable lack of content about the quality of water and its ecosystems, the advantages and disadvantages of consumption, issues of consumerism, and the consequences of catastrophes on people's health.

Table 3.2 Summary of the list of environmental health topics expected for the curriculum, grouped by their problematic situations

Derived from water, air and soil pollution	Derived from overconsumption	Derived from catastrophes	Interrelation between all environmental problems
On water: Main pollutants. Health problems associated with water pollution. Water as a finite resource. Water treatment and purification.	Production and distribution of goods and services. Fashion and advertising. Relationship between consumption and environment. Fair Trade. Sustainable development.	Natural: earthquakes, tsunamis, etc. Anthropogenic: wars, hunger, migrations, etc. Risk factors.	Sustainable Development Goals (SDG)
On air: Radiation and most common substances that can cause health problems. Useful strategies to avoid the problems associated with air pollution.			
On soil: Health risks derived from soil pollution. Desertification. Waste production and its environmental impact.			

3.3 Discussion

How should science education account for the inherent unpredictability of complex (living) systems and what are the consequences for decision-making in Science|Environment|Health contexts? Though this issue has already been discussed in systems theory literature for decades, to our knowledge, it has not yet been introduced in science education and science education research so far (e.g., Yoon et al., 2017).

The results of this symposium show that there is still a long way to go in answering these questions satisfactorily. In particular, the symposium's discussion exposed two obviously confusing questions.

1. What does it really mean to explicitly take unpredictability into account in science education?

Actually, it is true that the three symposium examples were referred to as being complex. But, at the same time, predictability assumptions were always present, at least implicitly or in the background.

2. What is adaption actually, and how can science knowledge be used to truly adapt?

Because of the implicit presence of predictability assumptions in all three examples, adaption was often used as *surrender to prediction*. Is adaption only this, or is it more?

During the discussion at the symposium, *Albert Zeyer* pointed out that one important aspect had been neglected in all of the presentations. It is the temporal relation of the two terms prediction and adaption. A prediction is always made *ex ante*, while adaption always takes place *ex post*. One is always predicting a future state, but one always adapts to a state in the present or in the past. If scientific prediction *ex ante* is limited in complex systems, can scientific knowledge have another role that may help with adaption *ex post*? In a recent paper in *Science*, Hofmann et al. (2017) asked this question, too. They pointed out that scientific understanding can be seen in two ways. *Understanding as prediction*, a standard concept in natural sciences, accounts for observed empirical regularities. *Understanding as interpretation*, more often used in social sciences, is the subjective feeling of having made sense of something. "Although these two notions of understanding are frequently conflated," they write, "neither one necessarily implies the other: It is both possible to make sense of something *ex post* that cannot be predicted *ex ante* and to make successful predictions that are not interpretable" (Hofmann et al., 2017, p. 3).

Zeyer suggested, that this distinction could help to structure the discussion. Indeed, three of the symposium contributions provide examples for complex S|E|H issues and discuss decision-making from different points of view. In all three, scientific *understanding as prediction* is obviously present. However, careful analysis, in every case, provides indications for an important role of *understanding as interpretation*, too.

For example, as *Iztok Devetak, Sonja Posega Devetak, and Tina Vesel Tajnšek* pointed out, in complex contexts such as school life, prediction can never be absolute. Therefore, students with an allergy, along with their teachers, may have to cope at any time with situations *ex post* that could not have been predicted *ex ante*. Though scientific knowledge was not able to predict the outcome, it may still help students and teachers to make sense of the present situation. Figure 3.1 can thus be seen in terms of *understanding as interpretation*. As described above, the authors had found that pre-service teachers with an allergy were better at adapting than those without relevant expertise. This may be an indication that while *understanding as prediction* is a type 2 process, i.e. systematic, decontextualised, asocial, and non-personal, *understanding as interpretation* may rather be a type 1 process, i.e. intuitive, contextual, social, and personal.

Deidre Bauer, Julia Arnold, and Kerstin Kremer stated that they had included both an *ex ante* and an *ex post* perspective in their research design. In fact, when students reflected on a vegetarian lifestyle, they *interpreted* present consequences, but they *predicted* future outcomes. Again, the researchers found that students processed present consequences in a type 1 manner, i.e. intuitively, personally, and socially. For the future consequences, however, the researchers found type 2 like argumentations, i.e. they were analytical, impersonal, non-social, and free of context.

Valentín Gavidia, Nuria Álvaro and Olga Mayoral, eventually, pointed out a double temporal perspective in their curriculum analysis. While the Delphi list of desirable content can be used both predictively and interpretatively, the Spanish teachers' request for teaching materials that help students to adaptively respond, indicates an *ex post* point of view.

Take, for example, the issue of "catastrophes and their impact on peoples' health", which, as the authors pointed out, was largely missing from the Spanish mandatory curriculum. *Understanding as prediction* predicts *ex ante* the possibility of a sudden catastrophe and provides preventive strategies for avoiding risks. These strategies are type 2. Take a pandemic as an example. To prevent a pandemic, everybody has to wash their hands four times a day – whether or not they personally prefer to do so, and independently of context and of social approval.

Once the pandemic is here, scientific knowledge still helps to predict the outcome. However, because the situation is highly complex, these predictions are uncertain. However, science also helps in interpreting the situation *ex post*. These considerations are type 1 like – i.e. personal, social, and contextual. They are about sense making: Must I be afraid to go to school (personal)? Are my grandparents in danger (social)? Should we all wear a face mask on the bus, but not in the streets (contextual)?

The pandemic example particularly underlines the core message of complementary dual-process theory. One process alone never results in a good decision. Decision-making is always an interplay between two processes. This implies, for the role of science in decision-making and in the light of the two ways of understanding scientific knowledge, that it may be involved always in two complementary ways, in prediction *ex ante*, and in interpretation *ex post*.

In ordered systems, the first way may be more important and more successful. In complex S|E|H situations, however, a constant back and forth between understanding scientific knowledge *as prediction* and understanding scientific knowledge *as interpretation* may be more helpful. Perhaps, as the symposium participants conjectured, this is the dual role of science in a complementary concept of dual-process decision-making.

3.4 Conclusion

Consider, for a moment, tango dancing as a metaphor for complex living systems. A tango school that only taught a person how to lead and not how to follow would be considered weird. For a good dance, the rules of tango have to be used *ex ante* for leading (prediction), and *ex post* for following (interpretation). Each of the dancers has to know both ways of understanding, although they preferentially practice one.

In science education, however, because natural sciences are so deeply grounded in a culture of prediction, understanding scientific knowledge *as interpretation* is usually still a neglected perspective. In an S|E|H pedagogy, both understandings are essential, because in complex living contexts a sole leader attitude may be not only overrated, but even harmful in the long run.

We believe that the approach of *understanding science as prediction* and *understanding science as interpretation* is fairly new in science education. Yet, there are some science education concepts that pave the way for such a dual-process approach although they do not do so explicitly.

Here, we think of the distinction between vision I and vision II of scientific literacy proposed by Roberts (e.g., Dillon, 2007), or the framework of informal reasoning (e.g., Sadler, 2004). Also, a hermeneutic approach to science education (e.g. Eger, 1992) emphasises starkly the process of sense making in science education and thus *understanding science as interpretation.*

To our knowledge, none of these widely discussed concepts includes the temporal direction of *prediction ex ante* and *interpretation ex post* and links it to systems theory and complexity considerations. An S|E|H pedagogy of complex living systems brings these points to the fore.

Acknowledgement The authors of the third contribution, Deidre Bauer, Julia Arnold, and Kerstin Kremer wish to thank Kerstin Schütte from IPN-Leibniz Institute for Science and Mathematics Education, the University of Kiel, Germany, for her valuable content-related and structural input.

References

Ajzen, I. (1991). The theory of planned behavior. *Organizational Behavior and Human Decision Processes, 50*, 179–211.

Bak, P. (1996). *How nature works. The science of self-organized criticality*. Springer.

Bauer, D., Arnold, J., & Kremer, K. (2018). Consumer-intention formation in education for sustainable development: An adapted model based on the theory of planned behavior. *Sustainability, 10*, 34–55.

Devetak, I., Posega Devetak, S., & Vesel Tajnšek, T. (2019). Developing pre-service teachers' competences of the management of students' allergies in the school environment. In E. McLoughlin, O. Finlayson, S. Erduran, & P. Childs (Eds.), *Contributions from science education research, selected papers from the ESERA 2017 conference*. Springer Verlag.

Dillon, J. (2007). An organic intellectual? On science, education, and the environment. In K. Tobin & W.-M. Roth (Eds.), *The culture of science education: Its history in person* (pp. 311–322). Sense Publishers.

Eger, M. (1992). Hermeneutics and science education. *Science & Education, 1*, 337–348.

Evans, J. S. B. T. (2008). Dual-processing accounts of reasoning, judgment, and social cognition. *Annual Review of Psychology, 59*, 255–278. https://doi.org/10.1146/annurev.psych.59.103006.093629

Fensham, P. J. (2012). Preparing citizens for a complex world: The grand challenge of teaching socio-scientific issues in science education. In A. Zeyer & R. Kyburz-Graber (Eds.), *Science|environment|health. Towards a renewed pedagogy for science education* (pp. 7–30). Springer.

Gil, A., Ruiz-López, M. D., Fernández-González, M., & Martínez de Victoria, E. (2015). Guía FINUT de estilos de vida saludable: más allá de la Pirámide de los Alimentos. *Nutrición Hospitalaria, 31*(5), 2313–2323.

Hempel, C. G. (1965). *Aspects of scientific explanation and other essays in the philosophy of science*. Free Press. ISBN 0-02-914340-3.

Hofman, J. M., Sharma, A., & Watts, D. J. (2017). Prediction and explanation in social systems. *Science, 488*(February), 486–488.

Mahl, A., Wahn, U., & Niggemann, B. (2005). Anaphylactic reactions in children – A questionnaire-based survey in Germany. *Allergy, 60*, 1440–1445.

Morou, Z., Tatsioni, I. D. K., & Papadopoulos, N. (2014). Health related quality of life in children with food allergy and their parents: A systemic review of the literature. *Journal of Investigational Allergology and Clinical Immunology, 24*, 288–297.

Real Decreto 1105/2014, de 26 de diciembre, por el que se establece el currículo básico de la Educación Secundaria Obligatoria y del Bachillerato. BOE, 3 de enero de 2015.

Real Decreto 126/2014, de 28 de febrero, por el que se establece el currículo básico de la Educación Primaria. BOE, 1 de marzo de 2014.

Roberts, J. A. (1996). Green consumers in the 1990s: Profile and implications for advertising. *Journal of Business Research, 36*, 217–231.

Sadler, T. D. (2004). Informal reasoning regarding socioscientific issues: A critical review of research. *Journal of Research in Science Teaching, 41*(5), 513–536. https://doi.org/10.1002/tea.20009

Schlosser, M. (2015). Agency. In *The Standford Encyclopedia of Philosophy* (Fall 2015). Standford University. https://plato.stanford.edu/archives/fall2015/entries/agency/

Sicherer, S. H. (2011). Epidemiology of food allergy. *The Journal of Allergy and Clinical Immunology, 127*, 594–602.

Stanovich, K. E., & West, R. F. (2000). Individual differences in reasoning: Implications for the rationality debate? *The Behavioral and Brain Sciences, 23*(5), 645–665.

Vermeir, I., & Verbeke, W. (2006). Sustainable food consumption: Exploring the consumer "attitude – Behavioral intention" gap. *Journal of Agricultural and Environmental Ethics, 19*, 169–194.

Yoon, S. A., Goh, S.-E., & Park, M. (2017). Teaching and learning about complex systems in K–12 science education: A review of empirical studies 1995–2015. *Review of Educational Research, 88*(2), 285–325.

Zeyer, A., & Dillon, J. (2014). Science|environment|health—Towards a reconceptualization of three critical and inter-linked areas of education. *International Journal of Science Education, 36*(9), 1409–1411.

Zeyer, A., Álvaro, N., Arnold, J., Benninghaus, J. C., Hasslöf, H., Kremer, K., Lundström, M., Mayoral, O., Sjöström, J., Sprenger, S., Gavidia, V., & Keselman, A. (2019). Addressing complexity in science | environment | health pedagogy. In E. McLoughlin, O. Finlayson, S. Erduran, & P. Childs (Eds.), *Contributions from science education research, selected papers from the ESERA 2017 conference*. Springer Verlag.

Chapter 4
Inquiry Based Learning and Responsible Research and Innovation: Examples of Interdisciplinary Approaches at Different Schooling Levels

Claudio Fazio, Amélia Branco, Mojca Čepič, Cláudia Faria, Odilla E. Finlayson, Cecília Galvão, Luís F. Goulão, Eilish McLoughlin, Jerneja Pavlin, Dagmara Sokołowska, Wanda Viegas, and Marisa Michelini

4.1 Introduction

The Inquiry-Based approach to Science Education (IBSE) is presented in scientific literature (Hake, 1998; Sharma et al., 2010) and in research projects reports (e.g. ESTABLISH, 2010; SAILS, 2014) as a credible solution to the reported lack of efficacy of more 'traditional' educative approaches, especially when focused on

C. Fazio (✉)
Dipartimento di Fisica e Chimica – Emilio Segrè, Università di Palermo, Sicily, Italy
e-mail: claudio.fazio@unipa.it

A. Branco
Instituto Superior de Economia e Gestão, Universidade de Lisboa, Lisbon, Portugal
e-mail: ameliab@iseg.ulisboa.pt

M. Čepič · J. Pavlin
Faculty of Education, University of Ljubljana, Ljubljana, Slovenia
e-mail: mojca.cepic@pef.uni-lj.si; jerneja.pavlin@pef.uni-lj.si

C. Faria · C. Galvão
Instituto de Educação, Universidade de Lisboa, Lisbon, Portugal
e-mail: cbfaria@ie.ulisboa.pt; cgalvao@ie.ulisboa.pt

O. E. Finlayson
School of Chemical Sciences and CASTeL, Dublin City University, Dublin, Ireland
e-mail: odilla.finlayson@dcu.ie

L. F. Goulão · W. Viegas
Instituto Superior de Agronomia, Universidade de Lisboa, Lisbon, Portugal
e-mail: goulao@isa.ulisboa.pt; wandaviegas@isa.ulisboa.pt

O. Levrini et al. (eds.), *Engaging with Contemporary Challenges through Science Education Research*, Contributions from Science Education Research 9,
https://doi.org/10.1007/978-3-030-74490-8_4

Research and Innovation (RRI) (Sutcliffe, 2006; IRRESISTIBLE, 2013; Ark of Inquiry, 2013).

IBSE is an inductive approach to science teaching that considers direct experience at the center of learning. IBSE activities and strategies actively involve students in the identification of relevant evidence, in critical and logical reasoning on the evidence collected and in reflection on their interpretation. Students learn to conduct investigations, but are also led by the teachers to understand the typical processes scientists use to build knowledge. IBSE strategies are credited to improve student understanding in many conceptual fields, due to their strongly contextualized nature, that focuses on the interdependence of situation and cognition, and on the ways they facilitate learning processes (Fazio, 2020).

Responsible Research and Innovation is an approach that tries to discuss and assess potential implications and societal expectations with regard to the products of research and innovation and their impacts on the environment, with the aim to foster the design of inclusive and sustainable innovation. A pedagogical focus on RRI themes can improve the student interest in science, as they can be actively involved in discussing cutting edge topics and research results and possibly in working together with researchers to better align both the research processes and their learning outcomes with the values, needs and expectations of today's society.

Current science education is mainly discipline-based, and even IBSE/RRI approaches are very often developed from the perspective of a single subject such as physics, chemistry, biology, and earth sciences. However, it is important to recognize that disciplinary distinctions are artificial: every natural phenomena can be discussed and associated with the core ideas of a variety of disciplines. Teaching and learning practices should carefully take this into account, as each big idea in a specific discipline might be a prerequisite for a big idea in another area of science, and this should be a key idea conveyed to the students during the learning activities. Moreover, professional development programmes for both in-service and pre-service teachers should take this issue into account, so as to prepare teachers to reshape their teaching strategies and orient them towards an approach that appreciates the value of a interdisciplinary approach to science education.

In this paper, we integrate four studies that discuss the effects of Inquiry-Based (IB) and/or RRI based approaches to science education from an interdisciplinary

E. McLoughlin
CASTeL and School of Physical Sciences, Dublin City University, Dublin, Ireland
e-mail: eilish.mcloughlin@dcu.ie

D. Sokołowska
Smoluchowski Institute of Physics, Jagiellonian University, Krakow, Poland
e-mail: dagmara.sokolowska@uj.edu.pl

M. Michelini
Dipartimento di Scienze Matematiche, Informatiche e Fisiche, Università di Udine, Udine, Italy
e-mail: marisa.michelini@uniud.it

point of view and following a vertical approach that explores the impact of IB and RRI based activities on students across a range of different educational and cultural contexts. Firstly, the impact of a series of inquiry-based learning activities on the acquisition of scientific knowledge and in the development of pupils' positive attitudes toward science in pupils aged 10–13 is discussed. Special attention is paid in this first study to medium-term retention of learning outcomes through inquiry. Three basic skills typical of IBSE are specifically targeted in this study: putting forward hypotheses, planning investigations and drawing conclusions. Then, a discussion on important pre-conditions and necessary steps needed for an introduction of cutting edge topics, typical of RRI, and new research findings to the classroom, and the role of the inquiry that has to be offered to students to effectively construct knowledge is done, together with some considerations on opportunities and challenges new findings in science bring to the science education at the pre-university level. Following this, an innovative Doctoral Degree in Sustainability Science, which seeks transformative education that challenges the attitudes of both professors and students, is presented. Main drivers include building a new body of interdisciplinary knowledge leading to the application of science to address real problems, and the integration of knowledge and innovation with the participation of society and citizens. Issues related to the involvement of students in a project work based on an inquiry perspective, and dealing with different dimensions are discussed, also in the aim to understand if this methodological approach is perceived by students as important to their learning as professionals and citizens. Finally, an approach adopted by three large scale European projects (ESTABLISH, SAILS and OSOS) that focus on enhancing science education curricula, pedagogy and assessment practices and supporting science educators in embedding IBSE and RRI principles in science education is presented. The opportunities and challenges for integrating IBSE and RRI principles in science education are discussed using examples of classroom practices designed and implemented in these three projects, and the ways IBSE and RRI aspects are conceptualized and explored.

4.2 Study 1

Intensive research on the effectiveness of teaching and learning by means of an inquiry approach has been carried out since 1980s, but there is still no consensus between researchers about the impact of teaching through inquiry on the increase of scientifically relevant skills and content knowledge. An inference quite common across different studies is that IB learning can be more effective than other instructional approaches provided that learners are supported and guided adequately (Lazonder & Harmsen, 2016). However, some studies seem to confirm that the effectiveness of learning through inquiry decreases with age (Hattie, 2008). It is quite astonishing that the rich spectrum of research papers exploring the impact of IB approaches on learning outcomes encompasses only a few studies in a medium and long term (e.g. Metz 2008), given the interest and significance these studies can have to teachers considering inclusion of IB teaching-learning methods in their teaching practice. Thus, to address this issue we will briefly discuss here a research on retention of learning outcomes in a medium term (Sokolowska 2018), focused on a 10-hour IB learning approach implemented in ten classes of pupils aged 10–13, during their science lessons.

4.2.1 Research Method

The aim of the study was to investigate the effects of implementation of a series of guided-inquiry activities on students' learning outcomes. Two rural and two urban schools in Poland were selected based on official school ranking. The implementation of the inquiry activities took place in all classes at the same time, over a 5–7 weeks duration. The IB intervention involved 170 students from ten classes, interacting with IB approaches for the first time in their school career. Each class was provided with ten lessons of guided inquiry: in each the teacher posed a problem to be investigated and provided resources, and students planned investigations, conducted experiments, collected data and drew research-based conclusions.

Three different tests were implemented to measure learners' acquisition and retention of knowledge through guided-inquiry. The tests were designed according to students' grade, and each encompassed multiple choice and open-ended questions, tasks involving reasoning and scientific inference, tasks requiring explanation of phenomena or observations, as well as one task of designing the experiment (consisting of formulating hypothesis, choosing adequate materials and tools, planning the investigation and drawing conclusions). In each class a first test, named here T1, was administered at the end of April 2015, just after completing the guided-inquiry lessons. The same test was administered again, as test T2, 6 months later (see Fig. 4.1 for more research design details).

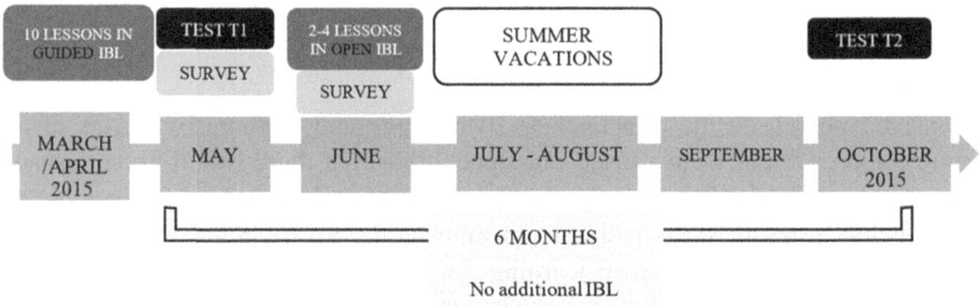

Fig. 4.1 Research design schema for an investigation of the medium-term retention of learning outcomes acquired through guided IB learning (IBL)

4.2.2 Findings

In order to describe the change of individual learning gains over a medium-term (6 months), we calculated the normalized change factor, c (Marx & Cummings, 2007). Mann-Whitney test (double-tailed) indicated no statistical significant difference in normalized change between males ($c_{Mdn} = -0.079$) and females ($c_{Mdn} = -0.086$) at the level of $p < 0.05$ (Sokolowska, 2018). ANOVA Kruskal-Wallis one-way analysis of variance by ranks revealed no statistically significant difference of a normalized change across three ability groups, defined as Level 1 (L1) – students with average scores, below 70%, Level 2 (L2) – students with average scores, about 70–80% and Level 3 (L3) – students achieving average scores, above 80% in their regular science classes (Sokolowska, 2018).

While studying solely the retention of three basic research skills (formulating hypotheses, planning investigations and drawing conclusions), we found the two latter similarly difficult for learners, and the decrease of planning skills as the only statistically significant change in these three skills over the period of 6 months. Class observations and examination of students' worksheets, as well as answers given by students in Survey 1 and Survey 2 also confirmed poor development of planning skills.

4.3 Study 2

Some examples of the introduction of contemporary science to the pre-university classroom are reported in literature (e.g. Garcia-Carmona & Criado, 2009; Pavlin et al., 2010, 2013; Mandrikas et al. 2019). Front-end science can be, and has to be, introduced to students at the pre-university level, given its relevance for technology we meet every day. As new findings in science often consider phenomena not included in regular curriculum, a pedagogical approach that provides experience through exploration and inquiry-based learning is crucial.

It is well known that students acquire information on subjects they are curious about through informal channels. But how extensive, and well grounded, is this information?

We investigated these issues among approximately 400 students enrolled in pre-service teacher education programmes at the University of Ljubljana. They include pre- and primary school and art and STEM subjects teachers, and social and special education programmes. Participants have different social status, abilities, personal motivations and interests – therefore forming a good representative sample for the population of students involved in high school in general.

We chose five interdisciplinary science topics: liquid crystals, hydrogels, bio-diesel, gels and osmosis, and microwaves in anisotropic materials. The first-year students filled-in questionnaires which included a set of short questions on those topics, allowing identification of the level of familiarity and knowledge related to them. Here we present some results of an analysis on two issues: familiarity with the

Table 4.1 Self-assessed knowledge about contemporary topics. The questionnaire was filled in by N = 257 first year students, age between 19 and 20. *The data for liquid crystals comes from a preliminary study with different participants (Pavlin 2010)

Topic	I am familiar with the name [%]	I know little/some about the topic [%]
Liquid crystals*	33.0	5.2
Hydrogels	18.7	2.3
Microwaves	75.1	6.6
Polarization of light	55.6	6.6
Biofuels	88.4	11.6
Osmosis	90.9	24.9
Diffusion	90.9	23.2

name of the topic, and a self-assessed level of knowledge about it. The familiarity with the name was investigated by the yes/no question: *Have you already heard about liquid crystals/hydrogels/…?* Self-assessed knowledge offered more options for answering the question: *How much do you know about liquid crystals/hydrogels/…?* nothing/very little/little/some.

Results are given in Table 4.1 Not surprisingly, the findings showed that familiarity with the topics ended with the recognition of names and the familiarity with the name provided the context.

4.3.1 Experiments in Teaching Units

To effectively introduce new findings in science to pre-university education, a close collaboration among researchers and teachers is required. The specialist researchers are the main source of knowledge about new topics, the educators find appropriate ways to reconstruct these topics to align them to school curriculum, and adapt them to the cognitive level of students, in cooperation with the researcher. Both of them develop experiments to provide learning experience for students. The teachers act as "critical friends", who introduce the reality of the classroom to the team, and provide the "in-vivo" testing of newly developed teaching-learning units.

The experimental support for the introduction of new findings is crucial. Very often, researchers participate in events to increase the popularization of science and prepare lectures with lots of nice photos and stories about discoveries. However, often the load of new information and the pace of the lecture does not allow a non-expert to develop an effective understanding. So, students have to experiment themselves, possibly through an IB approach to actively construct new knowledge during the process. As an introduction of a new phenomenon usually meets an absence of preliminary knowledge, the IB approach can provide a learning experience for the unfamiliar topic. Although it is widely believed that the laboratory equipment needed to teach new findings is not accessible in schools, simplified qualitative experiments can be developed. Figure 4.2 presents examples of experiments that students meet during the two successfully introduced topics: hydrogels and liquid crystals.

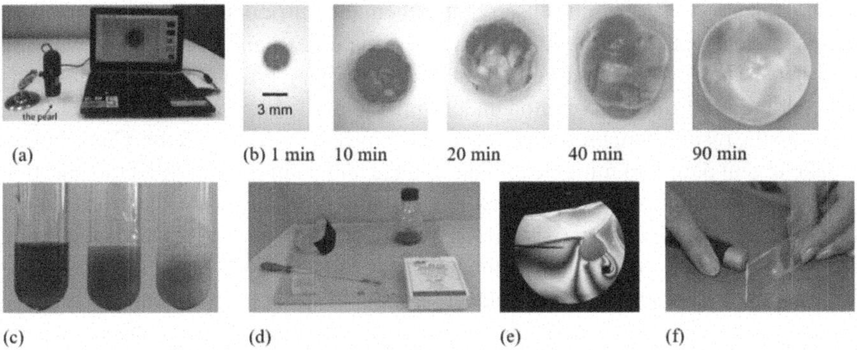

Fig. 4.2 The growth of a hydrogel pearl observed under an USB microscope. (**a**) The simple setup which allows for observation of hydrogel growth. (**b**) The shape of the surface changes drastically in time stimulating discussion and testing on possible reasons for the behaviour. (**c**) Three phases of a liquid crystal. (**d**) Equipment for making a cell for observation of liquid crystals (**e**) and propagation of light through them (**f**)

4.4 Study 3

The challenges which today society faces are complex and multidimensional, and lead to new paradigms associated with sustainability. To be promoters of change, the twenty-first century professionals must be endowed with solid scientific knowledge and, most importantly, must hold the capacity to incorporate it in order to understand the interactions between global, natural, social and human systems, and how such interactions affect the sustainability contexts (UNESCO, 2015).

Building a new scientific area of Sustainability Science requires assimilation of knowledge and mastery of tools that are seldom addressed by individual disciplines and scientific areas with an integrated approach. Today, disciplinary science deeply contributes to understand the function of the various pieces that make up our world but has gaps in understanding how these parts relate to each other (Pellegrino & Hilton, 2012).

The required change to reach sustainability needs a responsive pedagogical model that is also attentive to the transition to new forms of skill acquisition and knowledge-building. A Doctoral Degree in Sustainability Science can address this challenge by offering an innovative program which seeks transformative education that challenges the attitudes of both professors and students, combining a multidisciplinary composition of its Faculty members, gathering a wide range of disciplinary knowledge committed with a shared responsibility between natural sciences and social sciences in the coordination, organization and teaching of each curricular unit and in thesis supervision. This is a collaborative experience that relies on sharing all the class materials and establishing a bidirectional permanent work channel between students and professors through an ITC e-learning platform. Taking as an example the organization of four multidisciplinary thematic curricular units, we created an inquiry scenario, using a local multidimensional context, and the students, working in groups, contributed to solve real and contemporary problems. Curricular units were paired according to common underlying themes to cope with the same starting point, so the results here presented relate to two project work problems. Contributing to take urgent action to mitigate climate change and its impacts framed the question addressed under the scope of the first pair of curricular units. Organizations should be aware of the implications of reducing their greenhouse gas emissions and in a scenario of need to reduce emissions by 2/3 of today's levels, students were challenged to work on a multilevel explanation model for the problem and to present a critical proposal for an adaptation program for the primary sector in the specific ecosystem of the Tagus "Lezírias".

4.4.1 Methods

This research intends to understand if the PhD students involved in a project work, based on an inquiry perspective, dealing with different dimensions, can offer contribution to solve a problem on sustainability. Understanding if this methodological approach is perceived as important to their learning as professionals and citizens is also a research aim. We worked with 14 doctoral students with different academic backgrounds that were simultaneously working as consultants, experts in environmental institutions, or on their own businesses. These students were organized in five groups, each one responsible for researching two specific dimensions per work, which are part of the real problem to be solved. The dimensions involved were: a) Technologies and innovation; b) Economics, management and marketing; c) Social practices; d) Policies, institutions and governance; e) Human and environmental health. Ethics and values dimension was transversally addressed.

The research aims were: To understand 1) the potentialities and advantages of project work, based on an inquiry perspective, in trying to solve a real and multidisciplinary problem; 2) the difficulties experienced by students with IB methodology; 3) the students' opinion concerning the potentialities and advantages or the drawbacks of project work methodology to their professional and personal development; and 4) the perceived trade-offs of working a limited number of dimensions and therefore relying on the colleagues' complementary work to reach the wider perspective that contributes to a result.

Data were collected by direct observation of the working sessions, a questionnaire applied to the students at the end of the units, and individual reflections. Additionally, the evaluation of the final work of the groups was based on several rubrics created for the assessment of the oral presentation and written work. These rubrics were discussed with students and teachers from the beginning of the course.

4.4.2 *Findings*

The Sustainability Science Doctoral Course was designed to progress stepwise in two main phases. In the first phase, curricular unit syllabus and teaching were multidisciplinary, although an environment that prompts interdisciplinary thinking among students was promoted. During the first edition all professors were fully engaged with the pedagogical model assumed, and were active on preparing extended summaries and selected core materials to frame their topics, and producing dedicated e-learning materials, including professional video recording of short lessons. All materials were made available in the ICT e-learning platform to which students were granted access one week before the corresponding session. Field visits related to project work problems were organized and were successful in joining a multidisciplinary group of professors. Most importantly, "in class" sessions always took place with the simultaneous participation of a minimum of three professors with distinct knowledge backgrounds and academic competencies, resulting in debates that successfully challenged concept and methodological confrontations. Evaluation of project works through a methodology rooted in individual evaluation by professors from distinct disciplinary areas, followed by discussions to reach consensus grades, showed great potential to pave the way for their interdisciplinary thinking.

4.5 Study 4

UNESCO's recent report "Rethinking Education: Towards a common global goal?" (UNESCO, 2015) reminds us that the changes we face in the world today are characterized by new levels of complexity and contradiction. Citizens need a deeper understanding of global societal challenges and their implications for themselves, their families and their communities. This requires a broader vision of an active, engaged and responsible citizenship for the twenty-first century (Hazelkorn et al., 2015) and recommends that *"Science educators, at all levels, have a responsibility to embed social, economic and ethical principles into their teaching and learning in order to prepare students for active citizenship"* (Hazelkorn et al., 2015 p. 35). In particular, this report advocates that "*education policies and systems should support schools, teachers, teacher educators and students of all ages to adopt an inquiry approach to science education as part of the core framework of science education for all*" (Hazelkorn et al., 2015 p. 29). These objectives are further highlighted in the OECD Education 2030 framework, which aims to build a common understanding of the knowledge, skills, attitudes, and values necessary to shape the future towards 2030 (OECD, 2018). These reports highlight that, in order to equip today's learners with agency and a sense of purpose, and the competencies they need, to shape their own lives and contribute to the lives of others requires that changes are made in science education curricula, pedagogy and assessment practices in the classroom. Over the past decade, several large-scale projects have focused on addressing these challenges. They have supported teachers in adopting the principles of Inquiry-Based Science Education (IBSE) (Bevins & Price, 2016) and RRI (Sutcliffe, 2006) in their classroom practices.

4.5.1 Methods

This research examines how IBSE and RRI are conceptualized and implemented across a diverse range of educational and cultural contexts under three large-scale European projects namely, ESTABLISH (2010), SAILS (2014) and OSOS (2017). The ESTABLISH (2010) and SAILS (2014) projects adopted an understanding of inquiry as the intentional process of "diagnosing problems, critiquing experiments, and distinguishing alternatives, planning investigations, researching conjectures, searching for information, constructing models, debating with peers, and forming coherent arguments" (Linn et al., 2004). ESTABLISH (2010) designed a pedagogical framework for the development of IBSE units that were used to support teacher's use of inquiry based approaches in the second level science classroom. Each IBSE unit was presented over six parts: (1) Science topic; (2) IBSE character; (3) Pedagogical Content Knowledge; (4) Industrial Content Knowledge; (5) Learning Path(s) and (6) Student Learning Activities. A total of 18 IBSE units were developed across Physics, Chemistry, Biology and Integrated Science topics. SAILS (2014) presented a Framework for Inquiry and Assessment that addressed two key questions in science education: what to assess and how to assess? SAILS (2014) developed 19 Inquiry and Assessment Units that exemplified a range of strategies and tools to assess science inquiry skills. OSOS (2017) designed a framework that supported schools in embedding RRI principles and adopting an Open School Model to embed strategies that link education content to wider societal goals and engage learners to become responsible citizens.

The research questions were: (1) How are IBSE and RRI principles conceptualized in science education and (2) What is the impact on teachers of using IBSE and RRI principles in the classroom? Data was collected from participating teachers in all three projects using questionnaires that had a combination of Likert scale and open response questions that were used at the beginning and end of the professional development programmes for science teachers.

4.5.2 Findings

Analysis of science teacher's responses to questionnaires prior to participation in professional development programmes, revealed that teachers use of IB practices was low and some of the obstacles to their use of inquiry was "*uncertainty of how to ask higher order questions that promotes thinking*", "*managing a classroom where each student group is doing different activities is difficult*" and "*feeling uncomfortable with teaching areas of science that I have limited knowledge of and of asking questions that I do not know the answer to*" (N = 458) (ESTABLISH, 2010). In general, the teachers that participated in the teacher education programmes of the ESTABLISH (N = 2090) and SAILS (N = 2500) projects, indicated increased understanding, attitudes and confidence of utilizing inquiry approaches. In addition, the SAILS approach strengthened teacher's assessment practices through developing their understanding of the role of assessment (SAILS, 2014). The SAILS approach exemplified how assessment practices can be embedded into inquiry lessons and illustrated a wide variety of assessment opportunities and/or assessment processes that are available to science teachers. Data from teachers that participated in OSOS project demonstrated how adopting RRI principles in science education promoted the development of strategies that link student learning (including knowledge, skills, attitudes and values) to wider societal goals and engaged learners in becoming responsible citizens (OSOS, 2017).

4.6 Discussion

This paper highlights the importance of bridging the gap between science education research, the use of educational practices and the varied perceptions and conceptualizations of teachers, students, parents and other stakeholders for enhancing science education. Particularly, the studies here discussed are aimed at analysing the impact of Inquiry-Based (IB) and/or RRI-based approaches to science education on lower secondary school, pre-university and doctoral student learning, and the ways new IBSE and RRI aspects are conceptualized, owned and implemented by science teachers involved in professional development programmes. The issue of conceptualization and appropriation (Levrini et al., 2015) of contents and skills is particularly important in science education and can be easily found also in other similar contexts, such as mathematics education, where teacher beliefs, the relationships between these beliefs and practice, and belief change after an experience have been widely studied (e.g. Ernest, 1989; Speer, 2005; Liljedahl, 2010). The four studies discussed in this paper highlight the benefits that IB- and RRI-based approaches can have on students/teachers across a range of science education levels.

IB approaches can be used to engage young students in science and develop their investigation skills, as described in study one. Students aged 10–13 years, often face difficulties in designing and planning investigations as observed during IB implementation and examination of their worksheets. Instead of writing a straight plan, the students often apply a trial-and-error approach when conducting the experiments. They seem to find it unnatural to stop in the middle of the inquiry process and write a rigorous plan before conducting an experiment. In order to improve development of planning skills and, at the same time, preserve learners' active involvement in IB learning, it may be more effective to let them do investigation first, and ask them to design a coherent investigation plan afterwards. In addition, an IB approach was found to be inclusive and did not seem to favor any gender and a higher retention of learning over an extended period after instruction was also evident.

IB approaches can be used to introduce cutting-edge and innovative topics with students at the pre-university level. The introduction of such topics often faces challenges, such as a lack of teachers' content knowledge or availability of appropriate laboratory equipment, and the need to rethink the pedagogical approaches used by teachers to deal with more traditional topics. However, as discussed in study two, students learned the basic concepts (Pavlin, 2013), and were enthusiastic about dealing with new contemporary topics. Thus an IB approach can be effective in allowing students to access knowledge about new findings in science and open up new perspectives of science as a relevant subject with a vivid research. This new knowledge may also stimulate awareness of the impact that science can have on the society as a whole.

Project work based on an IB approach can offer mature (doctoral level) student's opportunities to solve problems on sustainability and foster a unifying view of science, as discussed in study three on the PhD program in Sustainability Science. Such an approach can offer transformative experience that challenges the attitudes of both professors and students, combining a multidisciplinary composition of its Faculty members, gathering a wide range of disciplinary knowledge committed with a shared responsibility between natural sciences and social sciences in the coordination, organization and teaching of each curricular unit and in thesis supervision. Students' proposals were realistic and viable, and were complementary enough to contribute to collectively respond to the global problems. The use of approaches from different areas of knowledge was clearly planned and the project methodology was well understood by the students. Some difficulties concerning team work and information management were observed, particularly during the first activities, but significant improvements were attained when the groups addressed the second problem.

The widespread use of IB- and RRI-based approaches across diverse science teacher education contexts, as outlined in study four highlighted that: (1) Learning science through inquiry can result in better understanding and more broadly applicable scientific knowledge along with the development of transferable skills and competencies; (2) Many models of IBSE exist, so it is important to adopt an approach that achieves learning outcomes in terms of knowledge, skills, attitudes &

value; (3) Teaching and assessment need to be considered as a dynamic and iterative process, so as effectively support IBSE; (4) Sustained collaboration is crucial in science education – between teachers and educators and across borders, both classrooms and countries and (5) Schools need to be facilitated to act as shared sites of science learning for which leaders, teachers and the local community share responsibility for embedding social, economic and ethical principles into science education in order to prepare students for active citizenship.

The collective findings of these four studies offer insights to the educational research community on how to conduct and improve IB- and RRI-based activities to maximize their impact on modifying student attitudes toward science learning.

References

Ark of Inquiry: Inquiry Awards for Youth over Europe (2013). http://www.arkofinquiry.eu/homepage. Accessed 26 October 2020.

Bevins, S., & Price, G. (2016). Reconceptualising inquiry in science education. *International Journal of Science Education, 38*(1), 17–29. https://doi.org/10.1080/09500693.2015.1124300

Ernest, P. (1989). The knowledge, beliefs and attitudes of the mathematics teacher: A model. *Journal of Education for Teaching, 15*(1), 13–33.

ESTABLISH (2010). European science and technology in action: Building links with industry, schools and home www.establish-fp7.eu. Accessed 26 October 2020.

Fazio, C. (2020). Active learning methods and strategies to improve student conceptual understanding: Some considerations from physics education research. In J. Guisasola & K. Zuza (Eds.), *Research and innovation in physics education: Two sides of the same coin* (pp. 15–35). Springer. https://doi.org/10.1007/978-3-030-51182-1_2

García-Carmona, A., & Criado, A. M. (2009). Introduction to semiconductor physics in secondary education: Evaluation of a teaching sequence. *International Journal of Science Education, 31*, 2205–2245.

Hake, R. R. (1998). Interactive-engagement versus traditional methods: A six-thousand-student survey of mechanics test data for introductory physics courses. *American Journal of Physics, 66*, 64–74.

Hattie, J. A. C. (2008). *Visible learning. A synthesis of over 800 meta-analyses relating to achievement*. Routledge.

Hazelkorn, E., Ryan, C., Beernaert, Y., Constantinou, C.P., Deca, L., Grangeat, M., Karikorpi, M., Lazoudis, A., Casulleras, R.P., & Welzel-Breuer, M. (2015). Report to the European Commission of the expert group on science education. Science education for responsible citizenship. European Commission. http://ec.europa.eu/research/swafs/pdf/pub_science_education/KI-NA-26-893-EN-N.pdf. Accessed 26 October 2020.

IRRESISTIBLE (2013). Including Responsible Research and innovation in cutting Edge Science and Inquiry-based Science education to improve Teacher's Ability of Bridging Learning Environments. http://www.irresistible-project.eu/index.php/en/. Accessed 26 October 2020.

Lazonder, A. W., & Harmsen, R. (2016). Meta-analysis of inquiry-based learning: Effects of guidance. *Review of Educational Research, 86*, 681–718.

Levrini, O., Fantini, P., Tasquier, G., Pecori, B., & Levin, M. (2015). Defining and operationalizing appropriation for science learning. *Journal of the Learning Sciences, 24*(1), 93–136.

Liljedahl, P. (2010). Noticing rapid and profound mathematics teacher change. *Journal of Mathematics Teacher Education, 13*(5), 411–423.

Linn, M. C., Davis, E. A., & Bell, P. (2004). *Internet environments for science education*. Lawrence Erlbaum Associates.

Mandrikas, A., Michailidi, E., & Stavrou, D. (2019). Teaching nanotechnology in primary education. *Research in Science & Technological Education*. https://doi.org/10.1080/02635143.2019.1631783

Marx, J. D., & Cummings, K. (2007). Normalized change. *American Journal of Physics, 75*, 87–91.

Metz, A. M. (2008). Teaching statistics in biology: Using inquiry-based learning to strengthen understanding of statistical analysis in biology laboratory courses. *CBE Life Sciences Education, 7*, 317–326.

OECD Learning Framework 2030, (2018). http://www.oecd.org/education/2030/learning-framework-2030.htm. Accessed 26 October 2020.

OSOS (2017). Open Schools for Open Societies, www.openschools.eu. Accessed 26 October 2020.

Pavlin, J., Vaupotič, N., Glažar, S. A., Čepič, M., & Devetak, I. (2010). Slovenian pre-service teachers' conceptions about liquid crystals. *Eurasia Journal of Mathematics, Science and Technology Education, 7*(3), 173–180.

Pavlin, J., Vaupotič, N., & Čepič, M. (2013). Liquid crystals: A new topic in physics for undergraduates. *European Journal of Physics, 34*, 745–761.

Pellegrino, J. W., & Hilton, M. (Eds.). (2012). *Education for life and work: Developing transferable knowledge and skills in the 21st century*. National Academic Press.

SAILS (2014). Strategies for assessment of inquiry learning in science, www.sails-project.eu. Accessed 26 October 2020.

Sharma, M. D., Johnston, I. D., Johnston, H. M., Varvell, K. E., Robertson, G., Hopkins, A. M., & Thornton, R. (2010). Use of interactive lecture demonstrations: A ten year study. *Physical Review Special Topics - Physics Education Research, 6*, 020119.

Sokołowska, D. (2018). Effectiveness of learning through guided inquiry. In D. Sokołowska & M. Michelini (Eds.), *The role of laboratory work in improving physics teaching and learning* (pp. 243–255). Springer.

Speer, N. (2005). Issues of methods and theory in the study of mathematics teachers' professed and attributed beliefs. *Educational Studies in Mathematics, 58*, 361–381.

Sutcliffe, H. (2006). *A report on responsible research & innovation*. www.diss.unimi.it/extfiles/unimidire/243201/attachment/a-report-on-responsible-research-innovation.pdf. Accessed on 26 October 2020.

UNESCO. (2015). Rethinking education. In *Towards a common global good?* Paris, United Nations Educational, Scientific and Cultural Organization.

Chapter 5
International Perspectives on Science Education Research in Multicultural and Multilingual Contexts

Mariona Espinet, Sonya N. Martin, Alberto J. Rodríguez, Saouma BouJaoude, and Audrey Msimanga

We dedicate this chapter to the memory of Audrey Msimanga who passed suddenly away in June 2021. Prof Msimanga was a dedicated scholar, devoted colleague and mentor, and caring friend.

5.1 Introduction

The authors of this chapter came together as a panel in the ESERA2019 conference to examine and discuss the issue of cultural and linguistic diversity in science education research across the globe. We come from different countries where there are long-standing and open conflicts in education related to culture and language preservation. We all have experience in doing science education research in either multilingual or multicultural science classrooms. Building from our collective

---

M. Espinet (✉)
Universitat Autonoma de Barcelona, Cerdanyola del Vallès, Catalonia, Spain
e-mail: mariona.espinet@uab.cat

S. N. Martin
Seoul National University, Seoul, Republic of Korea
e-mail: sonya.n.martin@gmail.com

A. J. Rodríguez
Distinguished Professor in Education, Department of Curriculum & Instruction, College of Education, University of Houston, Houston, Texas, USA
e-mail: ajrodr31@central.uh.edu

S. BouJaoude
American University of Beirut, Beirut, Lebanon
e-mail: boujaoud@aub.edu.lb

A. Msimanga (deceased)
The University of the Witwatersrand, Johannesburg, South Africa

O. Levrini et al. (eds.), *Engaging with Contemporary Challenges through Science Education Research*, Contributions from Science Education Research 9,
https://doi.org/10.1007/978-3-030-74490-8_5

dialogue, in this chapter we first provide some data on cultural and linguistic diversity in science classrooms and science education research journals globally. We later discuss ways to develop a more culturally and linguistically relevant science education research that addresses the complexities of our diverse science classrooms. Although there are many ways to do so, we have chosen to focus on science education research methodologies, international networks and collaborations, and policy development.

5.2 Cultural and Linguistic Diversity in Science Classrooms

Students' cultural and linguistic diversity continues to increase globally. Certainties about clear distinctions in ethnic group, culture, religion, and language are untenable since individual students can no longer be bound to particular forms of diversity. This shift has been termed *superdiversity*, and it is changing the way educational research is conducted in relation to multicultural and multilingual classrooms (Barwell, 2016). The diversity issues described in this chapter stem from changes in schools and societies resulting from new migration patterns and outcomes. We use the original concept of superdiversity rather than intersectionality since the former refers specifically to migration-related categories whereas the latter focuses on the complex relationship between race, gender, class and sexuality (Vertovec 2017). Table 5.1 presents some data on linguistic diversity in science classrooms of six different countries in relation to official languages, students' nationalities, language education policy, language use in science classrooms, and science teacher preparation in multicultural and multilingual science teaching. This table shows examples of how science classroom diversity worldwide calls for science education researchers to undertake research on superdiverse science classrooms. Theoretical frameworks and research methodologies that are more culturally and linguistically relevant are needed.

5.3 Cultural and Linguistic Diversity in Science Education Research Publications

Expanding the representation of cultural and linguistic diversity in our research publications that also reflects the diversity of students in schools, scholars in academia, and research topics in science education is an issue of critical importance that has often been overlooked. Examining current trends in publications in science education journals can enable the science education research community to better appreciate to what extent research describing multicultural and multilingual contexts is being shared (Martin & Siry, 2011; Martin & Chu, 2015; Martin, 2020). To address this need, 2177 papers published between 2011 and 2018 in four top-ranked science education research journals were analysed to identify how many of these papers included one or more words from a list of more than 100 indicators that

Table 5.1 Linguistic diversity in science classrooms across the globe

	South Africa	United States of America	Lebanon	Catalonia, Spain	Republic of Korea	Chile
Official languages in the country	Eleven official languages: English, Afrikaans, and nine African languages	No official language at the federal level. Thirty-one states have made English the official language. Hawaii, South Dakota, and Alaska have other official languages	One official language: Arabic	Three official languages: Catalan, Spanish, and Aranese	Korean	No official language, though Spanish is the most common. Six indigenous languages
Students' national origins*	99.5% south African (76.4% black, 9.1% white, 8.9% Coloured, 2.5% Asian); 0.5% other	86.3% USA, 2.4% Mexico, 2.2% Middle East, 2.1% Central America, 1.7% Africa, 5.2% other (2018)	85% Lebanese, 9% Syrian, 4.6% Palestinian, 1.4% other	88% Spanish, 4% Morocco, 2.6% Latin America (19 countries), 5.4% other (2017)	~98% Korean, ~2% bi-ethnic (Korean + other ethnicity and non-Koreans)	97.8% Chileans, 2.2% Latin American countries (7 countries)
Language education policy	Since 1997 schools can choose any of the 11 languages as a language of instruction.	No national plurilingual or even bilingual education policy. Some school districts offer bilingual education, depending on state policies, parent advocacy, and critical mass of students	According to the 1989 constitutional amendments, schools are free to choose the language of science and math instruction.	Immersion schooling in Catalan was established in 1983. Since 2018 there has been a plurilingual education policy overcoming bilingualism.	The 2012, Bilingual Education Act established training for 125 teachers to provide limited bilingual language support.	Language education policy as part of a national intercultural education policy to revitalize the cultures of indigenous communities

(continued)

Table 5.1 (continued)

	South Africa	United States of America	Lebanon	Catalonia, Spain	Republic of Korea	Chile
Language use in science classrooms	In spite of policy provisions for use of any and all 11 languages in the classroom, English is the preferred language of instruction even for non-English speakers.	English is the predominant language of instruction. Some schools may offer bilingual support in other languages (mainly Spanish), but it varies by state/school district.	Arabic is first language. English or French is the language of science and math. Armenian and Aramaic are taught in some schools.	Catalan is the language of instruction. Science is taught either in Catalan or in a foreign language such as English in CLIL programs.	Only Korean language is used in classrooms. Some schools and classrooms provide sheltered Korean language instruction.	Spanish is the language of instruction. Few schools teach in Mapudungun or Rapanui.
Science teacher preparation on MC & ML Ed.	No specific training but some institutions have introduced courses/modules on language in science	Most programs only offer one multicultural education class to meet teacher certification standards.	Most programs do not explicitly address multilingual and multicultural issues.	Science teachers do not have specific training other than one general multicultural education class.	Science teachers do not have any specific training.	No national programs for science teachers but small projects in Auracania region

[a]Due to space limitations, national ethnic diversity is not reflected in this table although it is important for the description of students' diversity in science classrooms

Table 5.2 Journal publications related to multicultural or multilingual research

Journal	Total number of publications (2011–2018)	Total number of publications related to equity issues	Percentage related to linguistic diversity
JRST	416	71 (17%)	15 (3.6%)
Sci Edu	351	41 (11.7%)	11 (3.1%)
RISE	393	23 (5.9%)	10 (2.5%)
IJSE	957	77 (8.1%)	18 (1.9%)
TOTAL	2117	212 (10.0%)	36 (2.8%)

could appear as part of the title, abstract, or keywords (Martin, 2019). These indicators represented a wide range of topics focused on race, ethnicity, socioeconomics, language, gender, and religion and indicators related to concepts related to equity including ability, diversity, inclusion, and more. Two hundred and twelve papers were identified, representing a small total percentage of all papers published (10%). However, by focusing more narrowly on identifying what percentage of all papers related to equity and diversity issues addressed the topic of linguistic diversity in science education, the analysis revealed that less than 3% of all publications dealt with this topic (See Table 5.2).

To understand why so few science education papers that deal with issues related to linguistic diversity specifically have been published in these journals over nearly a decade, it is important to consider what kinds of structures may limit the scope of research and opportunities for research that deals with policy and collaboration, including lack of government policies requiring teacher education programs, professional development, curriculum, and material resource development to focus attention on equity issues (as reflected in Table 5.1). Without policy to support and inform decision making, it may be difficult to get the funding needed to engage in research on these topics. Additionally, most science education researchers may lack expertise in multicultural and multilingual research as the theories and methods underpinning these types of inquiry are not generally supported by traditional science education research trajectories. It could also be difficult to find outlets for publishing research focused on equity. Similarly, it can be difficult to find the right balance between the "science" and equity issues being addressed. In the past, some journals were less receptive to studies focusing on non-content-related topics; however, this analysis of journal publications has shown an increase in the percentage of equity-related publications since 2015, with at least one special issue in a journal dedicated to linguistic diversity in 2019 (see *Research in Science Education*, 49(4) and another in 2020 (see *International Journal of Science Education*, 42(14)) highlighting the complexities of multilingual contexts in science education (Salloum et al., 2020). While these developments are a positive step in the right direction, few structures currently exist to support collaborative efforts and networks for researchers to more directly address equity and language issues in their research. There is only one organization, which was established 3 years ago, focused on equity and social justice in science education research: *Science Educators for Equity, Diversity and Social Justice* (SEEDS; seedsweb.org). Finally, researchers in science education lack professional organizations that can support the development of partnerships with equity and diversity scholars in other fields (language, sociology, anthropology, etc.).

5.4 Towards More Culturally and Linguistically Relevant Science Education Research

We advocate that science education research should be more culturally and linguistically relevant to be able to deal with the compexities of our superdiverse science classrooms. By *relevant* science education research we mean research that focuses on science curriculum and pedagogy connected to students' everyday lives and real local/global issues. It also integrates students' cultural and linguistic diversity, ways of knowing, and needs into the curriculum and pedagogy. This approach aligns with Aronson and Laughter's (2016) synthesis by embracing under the term "relevant" a group of approaches to multicultural education. What follows is a discussion of three aspects of relevant science education research that need change: (a) research methodologies and methods, (b) research networks and collaborations, and (c) education and research policies. We briefly present each aspect and provide some questions to activate reflection.

5.5 Rethinking our Science Education Research Methodologies and Methods

What research methodologies and methods could then enable us to best "see" and understand the lived experiences of the Other—the people whose culture, language(s), and socioeconomic status may be so different from our own? Based on our experiences doing research on equity, diversity, and social justice with marginalized populations we share some insights herein. These insights are articulated in more detail in recent publications (Rodriguez & Morrison, 2019; Tolbert et al., 2018; Rodriguez, 2016) and can be useful to frame the reflection on how to develop more culturally and linguistically relevant science education research. To address the question posed above, we suggest that all research methodologies and methods are appropriate for conducting research in culturally diverse educational contexts. Research methods (tools) and methodologies (research frameworks) are only necessary to organize schema for individuals to conduct research. Just like a carpenter chooses a hammer to hit a nail instead of a screwdriver, so it is up to the researcher to employ the right tools to investigate the desired research questions.

What we need to rethink then is how our *worldviews* and *positionalities* influence all aspects of our research work. That is, what research questions we choose to pursue, with whom and in which context we choose to collaborate, with which populations we choose to conduct our work, and why we choose certain topics and not others. To start this important reflexive process, we can ask ourselves two basic questions: For whom do we conduct research? and Whose interests are being represented in our work? As mentioned earlier, we provide more comprehensive discussions of these questions elsewhere. Due to space limitations, we briefly provide below some practical suggestions for alternative ways of thinking about and conducting research in collaboration with the Other. The three aspects of culturally relevant research—caring, relevance and rigor, and relational responsibility—are meant to be enacted throughout the entire research enterprise and they are closely linked to one another.

5.5.1 *Caring*

Our knowledge of how sociocultural, historical, and institutional factors influence the participation and success of traditionally underrepresented students in science has increased significantly over the decades. We now know a great deal about how some groups are consistently marginalized due to their socioeconomic status, abilities in dominant language(s), family structure, ethnicity, skin color, gender expression and/or sexual orientation, and physical ability. We also know that good intentions and well-intended neo-liberal policies focused on "tolerance," "acceptance" and "diversity" have failed due to the recalcitrant disinclination to address the root causes of oppression. No one can dismantle these complex webs of oppression on their own, but we can more effectively contribute to this goal by embracing a *research ethics of caring*. This approach involves re-conceptualizing our research methodology and methods so that they provide multiple spaces for mutually beneficial collaboration and social transformation. In other words, deficit perceptions of the Other as "lacking" and "in need of saving" are substituted by respectful understanding of the Other as partners with unique voices and agency in the research enterprise. In this way, the focus shifts from seeking to investigate the Other so that they can be moulded into existing oppressive and dominant practices to working with the Other to expose and to transform those practices.

To help us reorient our thinking using a research ethics of caring, we can ask ourselves these questions: Who/what do we care about when conducting science education research? Why do we conduct research on a chosen topic? Is it to advance research and practice, benefit educators and their students, increase scientific literacy, advance knowledge in our field, or secure our own academic advancement at our institutions. Should any one of these answers matter the most, and if so, who or what is most negatively compromised by that answer?

5.5.2 *Relevance and Rigor*

To begin the shift to a research ethics of caring, we need to recognize that traditional, masculine, Western notions of "objectivity," "rigor," and dichotomous framings are forms of colonized thinking that only serve existing power structures.

Instead, if we focus on *relevance*—how our research is relevant to the needs of the individuals collaborating with us—then *rigor* takes on a different meaning. That is, rigor becomes a construct in service of the people involved in our research and not a construct in service of a presumed detached, objective community of researchers: a community that too often pretends to exist outside of the very same world in which it conducts its research.

These questions could help us begin to make relevance our primary focus: In what ways is our science education research contextualized and responsive to the participants' needs? In what ways are participants involved as collaborators (voice and agency)? In what ways are our claims (impact/research findings) measured by the benefits in the lives of the people involved in our research from their point of view?

5.5.3 Relational Responsibility

None of the suggestions shared thus far are possible without first recognizing our positionalities as privileged intellectuals. As middle-class science educators and researchers, we hold unique positions of power that by default give us the *responsibility* to establish a more *relational* and mutually beneficial collaboration with the Other. Thus, instead of seeking to suppress our humanity through colonial and detached notions of "objectivity" and "rigor," we should seek to embrace the humanity represented in a research ethics of caring. In this way, we position ourselves not as the only purveyors of knowledge, but as members of multiple communities that influence, and are influenced by, the people and the sociocultural context in which we work.

When we consider these questions, we can appreciate the importance of relational responsibility: When we begin a research project, in what ways are we seeking to establish meaningful, respectful, and mutually beneficial collaborations with the participants? What are the benefits from the participants' points of view? In what ways are we reflecting and acting on our privileged positionalities so that we do not just hear but *truly listen* to the participants' voices and understand their experiences? In what ways do we recognize and act upon how we may be implicated and benefit from the very webs of oppression that we write about in our research?

5.6 Reorienting our International Research Networks and Collaborations

A more culturally and linguistically relevant science education research demands that researchers reorient their ways to collaborate at both the national and international levels. Current trends in globalisation make collaboration in science education research imperative to understanding both local and global issues in the teaching and learning of science and identifying relevant, common, and specific solutions. Thinking about other researchers' contexts may also help us to think differently about our own contexts and blind spots. During this current time of dwindling research funding, collaborative research could also provide a means to mobilise multiple expertise to improve funding success. The potential areas for collaboration arise from the very variability that a global rather than local gaze brings.

5.6.1 The Research Focuses of Collaboration for Language Diversity

Science teaching and learning in multilingual classrooms happens within a universal political context in which access to the dominant language provides access to the culture of power. We need research that confronts predominant discourses that do not acknowledge the *value of multilingualism* and that frame multilingualism not as a deficiency but as something positive, and even preferable to monolingualism.

We need to expand our understanding of the *dynamics of language* as a resource and not a barrier in science teaching and learning. Some ideological and epistemic considerations include challenging discourses of deficiency of non-dominant languages that undervalue the currency of non-dominant languages in science learning. We need to challenge the use of the dominant language as a measure of ability in science learning. Collaborative research can help by augmenting empirical evidence from a diversity of contexts to demonstrate how science can be learned in any language.

Our collaborations must address ways in which *pedagogical use of language* in the science classroom can include rather than exclude learners, invariably those from low socioeconomic backgrounds. Collaborations have potential to critique diverse and discordant language policies and curricula across the global social, economic, political, and linguistic contexts elaborated in this chapter. Developing collaborative relationships that can lead to transformation in policy and practice will require structures to ensure equitable participation.

5.6.2 The Challenges of Collaboration

The need for careful management of intra-national collaboration in geographical, economic, and policy contexts that are nearly homogenous cannot be taken for granted. This need becomes greater in inter-national or cross-national collaborations, even within the same socio-political and economic zones where there are many variable contexts. Policy contexts, however, are lagging behind. Partner countries' policy frameworks are an important affordance for dealing with language issues in teaching and learning. Critiques and comparisons cannot be made without careful consideration of the diversity of legal and educational contexts that inform and are shaped by the diversity of policy frameworks. This calls for research collaborations that are fair and ethical so that their findings have the integrity to be relevant locally and globally. In addition to the variation in education and language policy across nations, there is the matter of variable funding policies and requirements by funders.

5.6.3 The Tensions in Collaboration

The tensions in the quest for inclusion in collaborative research manifest in diverse ways. For instance, the inclusion—or largely the exclusion—of scholars in certain racial, ethnic, or socio-economic categories in collaborations and/or publications is particularly interesting in the way it plays out both in similar and different ways in different contexts across the globe. In Africa, for instance, policy developments in individual African countries and the external policies governing the funding often masquerade as inclusive of academics in African countries. Over the past few years, there has been an increase in interest by both the global North, the global West and the global East in collaborating with African scholars, particularly black academics and researchers (Department of Science and Technology, 2020). With a shift from research *on* Africans to research *with* Africans, inclusion of African scholars has become a more visible requirement in policy frameworks and nation-to-nation funding agreements. However, questions still abound on the ethics and relevance of the intended research and the nature of the collaborations so fostered: whether and how policies and funding calls, for instance, are structured in fair and ethical ways to meet real needs and goals of the "overseas" partner and/or their country; the power relations embedded in the fine print of funder's policies; whether and how overseas academics can exercise agency and negotiate themselves into the partnership for truly mutual benefit; how the conduct of the research on the ground respects, recognizes, and protects all participants while meeting the requirements of an international funder (usually faceless) and/or the collaborating partner. Meaningful collaborations must demonstrate genuine shifts from "extraction"' from research sites to equal, fair, and ethical sharing of the research in terms of agency and ownership of the research, decisions on data management and dissemination together with clear agreements on envisaged intellectual rights (Suresh, 2012).

Research funding practices and legal frameworks in global collaborations must include an explicit requirement for proper representation of marginalized groups in studies with a focus on marginalized populations. Token inclusion of equity and diversity discourse in proposals just to secure funding without any accountability clause by the funding agency can only lead to continued large expenditures without any of the anticipated impact on teacher preparation and/or student learning, particularly in poor communities. Research on the benefits of such funding continues to show persistent lack of meaningful change in spite of substantial investment in education research (Heinze, 2008).

Meaningful cross- and inter-national collaborative research calls for respectful and mutually beneficial negotiations. To maximize the benefits of collaboration, we should consider the following questions: How do the benefits differ for collaborating partners? How can benefits be maximized for all partners? What terms and conditions need to be agreed to upfront in view of diversity of policy and economic contexts? What are other possible issues of concern that must be included in the various agreement documents?

5.7 Influencing Language Education Policies

A more culturally and linguistically relevant science education research demands that researchers get involved in the construction and implementation of educational policies as part of their research work. Language education policies need to be seen as dynamic processes in which all community members engage in active negotiations in order to replace more top-down approaches to policy research and practice (Menken & Garcia, 2010). From this perspective, influencing policy implies participation not only in the construction of the policy document but also in the interpretation, negotiation, and ultimately re-construction of the policy implementation process.

One of the ways to influence policies related to language learning and use and equitable access and success in education is by developing and actively pursuing strategies to make citizens aware of the changing nature of the demographic composition of societies. According to Pujolar (2007), globalization and increased mobility have altered the linguistic make-up of almost all contemporary societies. The changing demographics of many countries as a result of legal and illegal immigration or displacement of people because of war and natural disasters has resulted in many countries that were initially considered monolingual becoming multilingual countries. Bermingham and O'Rourke (2018) propose that "multilingualism has become the norm rather than the exception and more and more, individuals find themselves engaging in a language or languages other than their 'native' or 'national' one" (p. 143). However, the governments of many countries whose populations include relatively high percentages of immigrants from various ethnicities still maintain the primacy and purity of the dominant language (Golden, 2001), as reflected in Table 5.1. Consequently, parents of immigrant children and human rights groups support providing immigrants and refugees voice and agency to decide their own fates, including their rights to maintain their mother tongue because it is the carrier of the culture and a means to stay connected with their families who still reside in their homeland (Kwon, 2017).

Another way to influence language education policies is by reconceptualizing the role of language in the teaching and learning of different curriculum areas. In the context of learning science, language is viewed as a mediating artifact and cultural tool in science and science classrooms. Karlsson et al. (2019) detailed the role of language in science education. It can either be a bridge that provides students with access to science or an obstacle that alienates students from the context within which scientists work and science develops. The role of language is even more prominent and multilayered in multilingual science education. This has resulted in a persistent concern in multilingual science education because of the pervasive achievement gap in science between language learners, who are typically immigrant children, and their counterparts (Ünsal et al., 2016).

In addition, several other endeavors can be used to have an impact on policy through science education research. Examples of these endeavors include conducting research that has an impact on practice, such as action research and design-based research; being involved with policymakers in setting research agendas; involving teacher education institutions in setting research agendas; conducting research on effective instructional methods geared towards teaching and learning in multilingual contexts, and communicating research findings to those who are interested in their practical implications, such as policymakers and teachers.

To empower science education research communities to be able to change language education policies to become more oriented towards equity and social justice we should consider the following questions for reflection: What implicit and explicit language education policies are in place? How are these policies preventing marginalized community students from access to educational resources and success in science education? How do science and science education researchers develop their agency in the implementation of language education policies in particular superdiverse contexts?

5.8 Conclusions

What steps could researchers take to undertake science education research that is more culturally and linguistically relevant? Starting with the most basic question of who benefits from our research can encourage us to begin a process of honest and transformative introspection that could lead to new research collaborations and opportunities for addressing equity, diversity, and social justice issues locally and globally. With these goals in mind, we have suggested that when conducting science education research in culturally and linguistically diverse contexts, we need to focus more on how our research benefits the participants. This involves listening more closely to their needs and exploring ways in which our research processes and findings can help address those needs. In this chapter we have identified three ways to conduct more culturally and linguistically relevant science education research, but some other equally important areas were not addressed. One of them is the development of design based-research to identify science teaching methods, materials, and additional learning opportunities that would benefit students' learning in multicultural and multilingual science education contexts. Another area deals with the provision of culturally and linguistically relevant science teacher professional development, which at the moment is lacking in many countries. International associations such as ESERA[1] (Europe), EASE[2] (Asia), REDLAD[3] (Latin America), ASERA[4] (Australia), SAARMSTE[5] (Africa), and NARST[6] (USA) can play an important role in providing the support for the development of a most needed agenda on culturally and linguistically relevant science education research.

References

Aronson, B., & Laughter, J. (2016). The theory and practice of culturally relevant education: A synthesis of research across content areas. *Review of Educational Research, 86*(1), 163–206.

Barwell, R. (2016). Mathematics education, language and superdiversity. In A. Halai & P. Clarkson (Eds.), *Teaching and learning mathematics in multilingual classrooms* (pp. 25–43). Sense.

Bermingham, N., & O'Rourke, B. (2018). Language awareness amongst "new speakers" in a multilingual classroom. In C. Hélot, C. Frijns, & K. Van Gorp (Eds.), *Language awareness multilingual classrooms in Europe* (pp. 143–168). de Gruyter, Mouton.

Department of Science and Technology. (2020). *National research foundation strategy*. Republic of South Africa.

Golden, D. (2001). "Now, like real Israelis, let's stand up and sing": Teaching the national language to Russian newcomers in Israel. *Anthropology and Education Quarterly, 32*(1), 52–79.

Heinze, T. (2008). How to sponsor ground-breaking research: A comparison of funding schemes. *Science and Public Policy, 35*(5), 302–318.

Kwon, J. (2017). Immigrant mothers' beliefs and transnational strategies for their children's heritage language maintenance. *Language and Education, 31*(6), 495–508.

Karlsson, A., Nygård Larsson, P., & Jakobsson, A. (2019). Multilingual students' use of translanguaging in science classrooms. *International Journal of Science Education, 41*(15), 2049–2069.

Martin, S.N. (2019). *Science education in multicultural and multilingual contexts panel presentation to the 2019 European science education research association (ESERA) conference*. Bologna, Italy, August 26–30, 2019.

Martin, S. N. (2020). Asia-Pacific Science Education (APSE): Challenged to lead in uncertain times. *Asia-Pacific Science Education, 6*(1), 3–13.

Martin, S. N., & Chu, H. E. (2015). Asia-Pacific Science Education (APSE): Expanding opportunities for publishing science education research. *Asia-Pacific Science Education, 1*(3), 1–18.

Martin, S. N., & Siry, C. (2011). Networks of practice in science education research: A global context. *Journal of Research in Science Teaching, 48*(6), 592–623.

Menken, K., & Garcia, O. (Eds.). (2010). *Negotiating language policies in schools*. Routledge.

Pujolar, J. (2007). Bilingualism and the nation-state in the post-national era. In M. Heller (Ed.), *Bilingualism: A social approach* (pp. 71–95). Palgrave.

Rodriguez, A. J., & Morrison, D. (2019). Expanding and enacting transformative meanings of equity, diversity and social justice in science education. *Cultural Studies in Science Education, 14*, 265–281.

---

[1] European Science Education Research Association (esera.org)

[2] East-Asian Association for Science Education (theease.org)

[3] Red Latinoamericana de Investigación en Didáctica de las Ciencias Experimentales

[4] Australasian Science Education Research Association (asera.org.au)

[5] Southern African Association for Research in Mathematics, Science and Technology Education (saarmste.org)

[6] National Association of Research in Science Teaching (narst.org)

Rodriguez, A. J. (2016). For whom do we do equity and social justice work? Recasting the discourse about the other to effect transformative change. In N. M. Joseph, C. Haynes, & F. Cobb (Eds.), *Interrogating whiteness and relinquishing power: White faculty's commitment to racial consciousness in STEM education* (pp. 241–251). Peter Lang.

Salloum, S., Siry, C., & Espinet, M. (2020). Examining the complexities of science education in multilingual contexts: Highlighting international perspectives. *International Journal of Science Education, 42*(14), 2285–2289.

Suresh, S. (2012). Research funding: Global challenges need global solutions. *Nature, 490*, 337–338.

Tolbert, S., Schindel, A., & Rodriguez, A. J. (2018). Relevance and relational responsibility in justice-oriented science education research. *Science Education, 102*(4), 796–819.

Ünsal, Z., Jakobson, B., Molander, B.-O., & Wickman, P.-O. (2016). Science education in a bilingual class: Problematising a translational practice. *Cultural Studies of Science Education, 13*, 317–340.

Vertovec, S. (2017). Talking around super-diversity. *Ethnic and Racial Studies, 42*(1), 125–139.

Chapter 6
Policy and Pedagogy: International Reform and Design Challenges for Science and STEM Education

Richard A. Duschl, Doris Jorde, Eilish McLoughlin, and Jonathan Osborne

6.1 Beyond Knowledge – 21st Century Competencies: Skills, Character and Meta-Learning

The models and frameworks for education are changing, and rapidly. Globalization, rapid technological changes, and emerging markets along with the national standardization of education systems are raising important questions and issues about educational goals and outcomes. Policy, standards, and research syntheses documents, while addressing important epistemic, equitable and ethical complexities for the design of STEM learning environments and ecosystems, are nonetheless serving as disruptive agents posing significant policy and pedagogy challenges. Moreover,

---

Panel Participants:
Richard Duschl, (Chair)Southern Methodist University, USA; Costas Constantinou, University of Cyprus, Cyprus; Doris Jorde, University of Oslo, Norway; Jonathan Osborne, Stanford University, USA/UK; Eilish McLoughlin, Dublin City University, Ireland; Audrey Msimanga, Sol Plaatje University, South Africa; Fang-Ying Yang, National Taiwan Normal University, Taiwan.

R. A. Duschl (✉)
Caruth Institute for Engineering Education, Southern Methodist University, Dallas, TX, USA
e-mail: rduschl@smu.edu

D. Jorde
Department of Teacher Education and School Research, University of Oslo, Oslo, Norway
e-mail: dorisj@ils.uio.no

E. McLoughlin
CASTeL and School of Physical Sciences, Dublin City University, Dublin, Ireland
e-mail: eilish.mcloughlin@dcu.ie

J. Osborne
Graduate School of Education, Stanford University, Stanford, CA, USA
e-mail: osbornej@stanford.edu

O. Levrini et al. (eds.), *Engaging with Contemporary Challenges through Science Education Research*, Contributions from Science Education Research 9,
https://doi.org/10.1007/978-3-030-74490-8_6

an emerging concern is that the introduction of policy agendas such as those found across Asia/Pacific nations, in the European Union Science Framework and the US Next Generation Science Standards (NGSS) as well as those among state and local educational systems are being challenged by (1) innovation and improvement efforts borne out of workforce needs and (2) research on learning, teaching, and designing curriculum, assessments, and learning environments.

Learning progressions, incorporation of engineering into science standards, characterizations of scientific evidence, styles of reasoning, scientific practices, student engagement in knowledge construction, and teacher professional learning communities are the topics and themes taken up in a *Journal of Research in Science Teaching* Special Issue on the NGSS (NGSS Lead States, 2013). The *JRST* editors sought to engender scholarly reflections on educational policy and reform efforts. The lead editors state:

> Given the significance of the NGSS for the field of science education … .it is incumbent upon the science education research community to engage in critical examination of the NGSS, its underlying framework (NRC 2012), and its cascading effects… Our goal in putting together the issue was to encourage thoughtful, critical, and constructive examination of the NGSS … [that] can and should inform international policy around science standards, state and district-level decision making, design of curricula and assessments, and classroom implementations. (Sadler & Brown, 2018)

Richard Duschl's motivation for proposing the NARST sponsored ESERA session, reported here, was his personal reaction to the articles in the *JRST* NGSS Special Issue. While he agreed with many of the comments and positions taken by the authors and editors, he was struck by implications for policy and policy processes. As a member of the National Assessment of Education Progress (NAEP) 2009 SCIENCE (NAGB, 2008) redesign planning committee, Chair of the committee that produced the NRC *Taking Science to School* (2007) synthesis research report and then served as a member of the NGSS Leadership Writing Team, co-chair of the Earth/Space Sciences writing group he became informed about guidelines and polices, as well as the audiences of policymakers (ministries, departments of education, schools, etc.) that needed to be considered and adhered to when preparing documents and protocols for national standards and tests.

As such, while the *JRST* NGSS Special Issue authors' and editors' comments and criticisms are well founded, many of them could not be considered when preparing the NAEP and NGSS documents. The cardinal rule was to avoid any language regarding how to teach, how to sequence instruction, or otherwise attempt to guide instructional implementation decisions. That was to be left up to local decision making of the States and Districts. Thus, of the many criticisms leveled in the *JRST* NGSS Special Issue while cogent for subsequent implementation and design recommendations for States and Districts, it raises questions and issues about the framing and writing of policy documents as well as the adopted development procedures and objective-setting goals therein. Thus, there are questions about how reform documents attending to standards and assessment are constituted. Others have weighed in on this, too. Ault's *Challenging science standards* (2015) and Rudolph's *How we teach science: What's changed and why it matters* (2019) both examine 20th century deliberations regarding how accountability-driven standards determine what science should be taught and how fluctuating curriculum designs favoring either content/knowledge or process/inquiry over the last century has influenced the policies, practices, and images of how science is done.

Standards documents are inherently political documents, inasmuch as they are forged out of numerous compromises and tradeoffs to accommodate differences of opinion regarding 'what counts' as the right curriculum, instruction, and assessment models. In the case of the United States Next Generation Science Standards (NGSS, 2015), determining and negotiating the design and content of science standards were influenced right from the beginning by several factors. One factor was the Framework document (NRC, 2012) that set down the 'Three Dimensional' teaching and learning guidelines for K-12 science education: Science & Engineering Practices, Disciplinary Core Ideas, Crosscutting Concepts, as mentioned in the above quote.

A second factor was the influence of the first implementation of science education standards in the 1990s. By 2010, two-thirds of US States had developed State Science Standards guided by the National Academy of Science (NRC, 1996) *National Science Education Standards* (NSES). The other one-third of the States developed Science Standards guided by the American Association for the Advancement of Science (AAAS 1993) *Benchmarks for Science Education.* While surveys showed that the NAS and AAAS frameworks had 80% agreement with respect to conceptual content, one salient difference was the organization of Standards by grade levels in the NSES and by grade bands in the Benchmarks. The drafting of new NGSS needed to recognize this geographic distribution problem, and hence uptake of the new Three Dimensional guidelines by incorporating both the grade level NSES learning goals (K, 1, 2, 3, 4, 5... 12) and the grand band Benchmarks learning goals (K-2,3-5, 6-8,9-12). The NGSS leadership team and writers were instructed that the new NGSS framework, in order to accommodate State Department of Education adoption and transitioning, would need to embrace both grade level and grade band organizational formats. A third factor in shaping the NGSS documents was the politically charged atmosphere around States rights and the US tradition of local control of education. The dissemination and adoption of the Core Common Standards in mathematics and in English Language Arts was met with resistance by many States.

A moderated 'Symposium' was assembled to bring together seven panelists with expertise and experiences in international/national policy, standards, assessment, and/or leadership experiences in science learning and learning environment design. Panelists were asked to come prepared to examine and discuss the challenges/opportunities, tensions/agreements that arise when making policy and pedagogical decisions at school, district, state, and national levels.

Four reports bridging the domains of policy and pedagogy were examined and discussed (See Table 6.1). Two thorough and comprehensive reports focused on future 21st Century educational systems and on curriculum knowledge, literacy, and skill guidelines: OECD's *The Future of Education and Skills 2030*; and the Center for Curriculum Redesign's *Four-Dimensional Education: The Competencies*

Table 6.1 Four education policy reports

Report 1 – Rapid globalization and technological development pose social, economic and environmental challenges and opportunities for human development. Countries need help designing instructional systems that prepare students for an uncertain future world.	OECD 2030 Policy and Pedagogy: International Reform and Design Challenges of Science and STEM Education http://www.oecd.org/education/2030/
Report 2 – Students are not being prepared to "fit in with the world of the future, empowering them to actively work to improve it further." Education is not adapting quickly enough to a future consisting of greater volatility, uncertainty, complexity, and ambiguity	Center for Curriculum Design (2015) Four-Dimensional Education: The competencies learners need to succeed. https://curriculumredesign.org/our-work/four-dimensional-21st-century-education-learning-competencies-future-2030/
Report 3 – The NAE report examines international large-scale assessments (ILSA) and asks two questions: "What do the results of such assessments tell us about the strengths and the weaknesses of a nation's education system?" and recognizing that national education contexts and systems vary widely, "What do these assessments really tell us?"	National Academy of Education (2018) International Education Assessments: Cautions, conundrums, and common sense. https://naeducation.org/methods-and-policy-uses-of-international-large-scale-assessments/
Report 4 – Increasing the motivation and achievement of students studying STEM subjects poses challenges for European education systems. To augment the findings of the 2018 STEM Education Policies Report, Scientix used STEM Education Practices Survey, looking to assess how STEM teachers organize teaching practices.	Science, Technology, Engineering and Mathematics Education Practices in Europe. Scientix Observatory report. December 2018, European Schoolnet, Brussels http://www.scientix.eu/documents/10137/782005/STEM-Edu-Practices_DEF_WEB.pdf/b4847c2d-2fa8-438c-b080-3793fe26d0c8

Learners Need to Succeed. The OECD report poses two questions: What knowledge, skills, attitudes, and values will today's students need to thrive in and shape their world? How can instructional systems develop these knowledge, skills, attitudes, and values effectively? The CCR report presents a rethinking about the 'What' of education and does so with 'actionable' recommendations in mind regarding Four-Dimensions: Knowledge, Skills, Character, Meta-Learning. The two reports share commitments to the development of literacies and competencies: Health Literacy, Numeracy, Digital Literacy, Data Literacy, Global Literacy, Information Literacy, Environmental Literacy, among others skills.

The third report from the US National Academy of Education addressed International Large-Scale Assessments (ILSA) results which can be alarming and followed closely by academics, policy makers, business and industry leaders and members of the press. The report grew out of two workshops: (1) Methodological issues related to design, analysis and reporting of ILSAs; (2) Reporting, interpretation and policy uses for ILSAs. Of particular interest, is the international benchmarking and comparisons among nations. The fourth report from the European Union reports findings from two comprehensive surveys examining STEM teaching policies and practices and the transformation of education processes. Thirty-eight European nations participated and the research was conducted by European Schoolnet, a network of 34 Ministries of Education, and Scientix, the community for science education in Europe.

Panel members when reviewing their assigned policy reports were asked to summarize issues and recommendations. Additionally, they were asked to generate a set of questions and issues that would be shared at the NARST/ESERA- Bologna Invited Panel. Three panel members were assigned the role of commentators and asked to reflect on how the reports did or did not address issues from their regions of the world – South Africa, Asia/Pacific, European Union.

6.2 Policy Reports

Report 1 – *The Future of Education and Skills 2030 Project* **OECD (2018), Professors Jonathan Osborne & Audrey Msimanga** The report argues for a vision of education that will be needed for students in 2030 and the following decades. The report envisions a context where students will have to "abandon the notion that resources are limitless and are there to be exploited", rather "they will need to value common prosperity, sustainability and well-being." To achieve this goal, "they will need to be responsible and empowered, placing collaboration above division, and sustainability above short-term gain". Meeting such a goal, the report argues, will require curricula to evolve. The singular focus on curriculum is possibly rather narrow given that pedagogy may yet be transformed by technology, particularly the use of artificial intelligence, and Natural Language Processing to improve assessment which is the tail that drives much of what happens in classrooms.

The report sees three challenges that need to be met. The first challenge is environmental and the demands of living in a context of changing climate and depleted resources. The second is economic and the challenge of an ever-changing society arising from new emerging technologies and the sense of risk associated with lack of stability and changing contexts. The third is social – a product of increasing migration, urbanization and widening inequity. In this context, the report argues that education is about more than developing the capability of students for employment but "the need to equip students with the skills they need to become active, responsible and engaged citizens".

To navigate through a "complex and uncertain world", this report places an emphasis on the need to develop the capability of students' sense of agency. Two factors are prioritized for developing agency – the use of personalized learning environments and the building of a solid foundation in literacy and numeracy – in particular, digital and data literacy.

The basis of the learning framework they advance to achieve all of this is essentially a competency based model to which systems of education are increasingly moving (Koeppen et al., 2008; National Research Council, 2012). Competencies are seen as being an amalgam of knowledge, skills, and attitudes and values.

When it comes to the first of these elements, disciplinary knowledge is seen as important but epistemic– that is knowing how to think like a scientist, historian or mathematician – is also considered to be important. Likewise, some procedural knowledge will be required – knowing how something is done. For instance, in the case of science, knowing how to design and evaluate an appropriate investigation. The OECD thinks this is best developed by problem solving, and design and systems thinking. Pre-eminence is given to three competencies – the ability to create new value, to reconcile tension and dilemmas, and to take responsibility. To meet this challenge, the report advances a set of design principles which are giving students agency, ensuring rigor, providing focus through a relatively small number of topics in each grade, ensuring coherence such that any curriculum reflects the logic of the discipline, alignment between curricula, teaching and assessment, transferability of skills across disciplinary contexts and an element of choice.

The process of designing such curricula must empower teachers; ensure that the student experience has relevance which will require interdisciplinary learning; be based on constructing a curriculum which is "adaptable and dynamic"; and engage teachers, students and other relevant stakeholders to ensure ownership.

It is impossible to escape the feeling that this is an aspirational list. Taken seriously though, there are a number of challenges for those involved in science curriculum development. Current curricula, with the exception of the Next Generation Science Standards, are not competency based. Even the Next Generation Science Standards fail to specify the procedural and epistemic knowledge that should be attained. Too often curricula are overloaded with content, providing no opportunity for student agency, and placing little emphasis on competencies which are transferable such as the ability to read and interpret informational text, developing the facility to analyze and interpret data critically, or evaluate competing experimental designs. As for coherence, the school science curriculum has been searching for a narrative that might bind the sciences ever since its inception and current efforts are still wanting (Osborne et al., 2018). What would it mean to focus on fewer topics at each grade and how would these be selected?

When it comes to technology and developing data literacy, much science education still has not engaged fully with the affordances of what is offered by platforms such as Tuva Labs or the various tools emerging from the Concord Consortium. Whether science education is simply failing to prepare students for the needs of the coming decades and how it might change are clearly questions to be discussed at this symposium.

In the ensuing discussion, participants raised a number of issues. One is that the singular focus on curriculum may be rather narrow, given that pedagogy may be yet transformed by technology and the greater use of artificial intelligence. Another is the question of how these ideas could be transformed into a set of design principles that could be applied across different contexts. Inevitably with such calls, there is the issue of how decisions will be made to excise content and focus on fewer topics possibly of an interdisciplinary nature without damaging the coherence and the underlying logic of the discipline. And, given that competencies are knowledge dependent and acquired in a specific context how can they be transferred across disciplines? In short, while the report offers some challenges to contemporary curriculum and guidelines, it falls short of providing the structure necessary for immediate action.

Report 2 – *Four-Dimensional Education: The Competencies Learners Need to Succeed,* **Center for Curriculum Redesign (2015) Professors Richard Duschl & Fang-Ying Yang** The Keywords for the CCR report shed light on the CCR's ambitions: Curriculum, Standards, Competencies, Competency, Computer-Based Learning, Deeper Learning, Knowledge, Skills, Character, Metacognition, Meta-Learning, 21st Century Education, Education Technology, EdTech, Social-Emotional Skills, 21st Century Competencies, Education Redesign, 21st Century Curriculum, Pedagogy, Learning, Jobs, Employment, Employability, Eduployment, Education 2030, Mindset.

The CCR report proposes adopting a four-component sequence of reforms 1 – Educational Goals; 2 – Standards/Assessments; 3 – Curriculum; and 4 – Professional Development. The recommended Theory of Change for achieving goals is to begin with an initial focus on steps 1&2: Goals, Standards and Assessments and then Curriculum and Professional Development.

The three main drivers for the CCR Educational Goals and Standards Steps1&2 are (i) Personal development of individuals, (ii) Challenges of society, and (iii) Shifting needs of local and global workforces. The broader CCR agenda is to bring about reforms for how precollege and further education might address interdisciplinary Modern Knowledge agendas; "It is the job of standards and curricula to instill competencies to choose content that has depth, and to approach it intelligently. We must realign education goals, standards, and curricula to reflect our changing knowledge and the dynamic transformations happening in our world." (p. 26).

The CCR maintains "that our current, knowledge-focused curriculum does not adequately prepare students for today's workforces, much less tomorrow's and that students should practice applying their knowledge using skills." (p. 41). Thus, the 'Beyond Knowledge' competencies framework incorporates **Knowledge** "What we know and understand" but adds in **Skills** "How we use what we know", **Character** "How we behave and engage in the world", and **Meta-Learning** "How we reflect and adapt". The CCR recommendation is to focus on Modern (Interdisciplinary) Knowledge topics and themes such as Global Literacy, Information Literacy, Systems Thinking, Design Thinking, Environmental Literacy, Digital Literacy and actionable skills that focus on four Cs: Creativity, Critical Thinking, Communication, Collaboration.

The CCR report identifies two 'Tensions' regarding the realignment of education goals, standards, and curricula within the regimes of accreditation and standardized testing: that testing may create a focus on external goals of performance that sorts students and undermines attainment of personal learning goals; and that reforms may create an economic focus on education (e.g., students as customers and institutions as businesses) that shifts dynamics further away from personal mastery of learning competencies toward extrinsic goals and competition between students and among educational institutions.

One of the issues raised about the CCR proposed curriculum reforms concerns the frameworks and methodologies for designing education systems. Missing are considerations for the Macro, Meso, and Micro levels within educational systems as characterized by Improvement Science (Bryk et al., 2015) and the Research + Practice Partnerships that undergird Design-Based Implementation Research (Bevan et al., 2018; Fishman & Penuel, 2018). The decision to focus on Educational Goals and Standards/Assessment first and foremost, immediately raises questions and issues about the synergy between policy and pedagogy. Both the what (Standards and Assessment), and the how (Curriculum and Professional Development) need to change together over time. Not one and then the other. Leaving out the how as part of the initial conversations omits promising frameworks and methodologies for designing educational systems (e.g., R + P (Research + Practice) Partnerships, (Bevan et al., 2018) Design-Based Implementation Research (DBIR) (Fishman & Penuel, 2018) as well as important stakeholders' engagements with curriculum design/redesign efforts with Learning Progressions (Duschl, 2019) and Improvement Science/Network Improvement Communities (Bryk et al., 2015).

A related second issue is not co-developing standards and assessment along with curriculum materials and teacher professional development. Stakeholders' such as teachers and members Network Improvement Communities should be at the table. Many of the same OECD 2030 curriculum, instruction, and assessment issues and questions regarding knowledge, skills, and values and attitudes pertain here, too. Questions arose pertaining to leadership and teacher Professional Development; to coordinating and implementing the design of Curriculum, Instruction and Assessment when adopting Evidence Center Design and Learning Progression frameworks; and to creating curriculum contexts that adopt Twenty-First Century Information Literacy Tools; Systems Thinking; Design Thinking; Environmental Literacy; and the 4 C Skills: Creativity, Critical Thinking, Communication, Collaboration.

Report 3 – *International Education Assessments: Cautions, Conundrums and Common Sense* **National Academy of Education (2018) – Professors Doris Jorde & Costas Constantinou** The report summarizes two workshops to examine the future directions for International Large-Scale Assessments (ILSAs) from a variety of disciplinary aspects (including educational policy, journalism, research design and statistical analysis). Participants agreed that ILSAs provide valuable resources for countries. However, one needs to consider interpretations at all levels. The purposes of ILSAs (summary chapter, p. 69) include:

1. Describe and compare student achievement and examine relevant contextual factors across nations

2. Track changes over time in student achievement, contextual factors and their mutual relationships, within and across nations.
3. Disturb complacency about a nation's educational system and to spur educational reforms.
4. Create de facto international benchmarking by identifying top performing nations and jurisdictions, or those making unusually large gains, and suggesting ways to learn from this array of practices.
5. Evaluate the effectiveness of curricula, instructional strategies, and educational policies, while understanding that many of them are deeply contextualized.
6. Explore casual relationships between contextual factors (e.g., demographic, social, economic, and educational variables) and student achievement.

There is general Agreement on purposes 1, 2, & 3 but widespread Concerns and Disagreements on purposes 4, 5, & 6. Especially concerning to committee members was the pursuit of establishing casual relationships from ILSA data. Given the large number of factors affecting student achievement, as well as the fact that nations are so very different from one another with respect to size of population, history, culture, and politics; seeking casual relationships would never be a realistic goal for ILSA's as presently designed.

The report also took up issues with the ways media are reporting results, which can often be misleading. When sharing results with the public, there are only a few questions that are of importance: 1) Why did our country do so badly? 2) Why did another country do better? and 3) What is the other country doing that we can try in our country? Educational researchers assert that the tests are not able to provide such information. Nonetheless, this is what is communicated to the public.

In a web-seminar that launched the report in 2018, panel discussions brought up additional issues and concerns about the use of ILSA's. Again, the misuse of casual inference was discussed as a problem with this type of testing. However, results could be used to alert policy makers about promising topics that need more rigorous types of experimental designs (RCT or quasi) for the country.

The end comment is that "ILSA's are here to stay. Indeed, not only are they here to stay, they are likely to become even more salient to educational policy discussions as the world becomes increasingly globalized. For this to be a good outcome, technical issues must be addressed and policy makers, the press, and the public must be more aware of the data's limitations." (p. 77).

The general conclusion of the report is that ILSA's are here to stay. However, it is important that the "users" of the tests understand the nature of the data produced – possibilities and limitations. Used in the correct way, the data allows nations to follow their own trends in student achievement and to look critically at policy and areas of the curriculum demanding change. Benchmarking against other nations may be a valuable tool for learning about what works (including curriculum and policy), but only if used correctly, taking into consideration country context. There is consensus that longitudinal research using RCT or quasi experimentation is required if data is to provide information on casual relationships. Finally, helping the media understand the nature of the data is important for all countries.

Report 4 – *Science, Technology, Engineering and Mathematics Education Practices in Europe.* **Scientix Observatory Report. December 2018, European Schoolnet, Brussels Professor Eilish McLoughlin** The Scientix report STEM Education Practices in Europe draws on the analysis of 3780 responses (representing over 4500 classes) to the STEM Education Practices Survey, answered by teachers in 38 European countries (Scientix Observatory, 2018). The aim of this report was to provide a grassroots, European-wide perspective on how STEM teachers organise their teaching, in terms of resources and pedagogical approaches used, on the current state of teachers' professional development and support, and on their opinions and attitudes, particularly in relation to their school environment and their openness to cooperation with STEM industries. It must be noted that the findings presented in this report were based on teachers' self-reporting regarding their practices, needs and opinions on various aspects of STEM education. The report's findings were discussed under five areas addressing (i) pedagogical approaches used in STEM teaching, (ii) access to and use of resources and materials, (iii) professional development and support for STEM teachers, (iv) teachers experience and educational level in STEM teaching and (v) teachers' attitudes and influence of the environment.

Generally, the STEM teachers reported the use of a variety of pedagogical approaches, with very high use of formative and summative assessment methods, collaborative learning, differentiated instruction and project/problem-based approaches. The high reporting of formative assessment is encouraging, indicating that teachers are mindful of the need to monitor and evaluate learning outcomes and not exclusively focused on final evaluations. However, the report highlights a high use of traditional direct instruction compared with other, student-centred pedagogies, such as flipped classroom, Inquiry-based science education (IBSE) or peer teaching. The fact that STEM teachers report considerably more traditional instruction than IBSE is of particular concern – given that the use of IBSE has been widely promoted across Europe as a more effective pedagogy than traditional direct instruction. Mathematics classes, in particular, appear to be delivered through more teacher-focused, less innovative, and less contextualized pedagogies than the other STEM disciplines.

In terms of access to and use of resources, teachers reported, except when teaching ICT, an extensive use of paper-based materials in their teaching, followed by audio/video materials and slideshow presentations. In addition to reporting low use of ICT tools and specialised software/equipment in their STEM classes, teachers also indicated low use of resources for personalised learning and special needs learning. The majority of teachers surveyed do not subscribe to information channels – either of national and international educational projects – as a source of STEM resources or utilize resources published by companies operating in STEM fields.

According to the European Commission's Eurydice report on Teaching Careers in Europe (2018), in most European educational systems teachers' continuous professional development (CPD) is either compulsory or considered a professional duty (it is compulsory, but the number of hours is not defined). Additionally, in many educational systems, a certain number of hours or credits in CPD is required for career progression. The majority of teachers surveyed indicated they had not completed professional development of any kind during the previous two years. Teachers reported that they generally update their knowledge online and in their own time and rely on technological and pedagogical support from their peers that teach the same or other STEM subjects.

The extent of STEM teachers experience in the classroom and the educational level of the students were both reported to have an effect on teacher's use of innovative pedagogical strategies. With more experience, teachers were more willing to integrate more constructivist pedagogical approaches in their classes and limit the use of traditional direct instruction. A steady decrease in the use of student-centred pedagogies as the students approach the end of upper second level education, i.e. as national end-of-second-level evaluations approach, was also observed.

Overall, STEM teachers identified the pressure to prepare students for exams, inadequate school space organisation, lack of pedagogical strategies to teach STEM in an attractive way and insufficient technical support for teachers as the key factors that impact their teaching practices. Teachers generally indicated openness towards collaboration with STEM industries and towards bringing more innovation into their classrooms and expressed that this is best achieved when STEM teachers and their school administration share a common vision about innovative STEM teaching.

Issues and questions that arise from this report include:

- What are appropriate models of professional learning to provide continuous support to teachers to embed more student-centered pedagogies in all STEM classrooms, in particular how can mathematics teaching be reformed?
- What policy changes are needed for curriculum innovation to support a more integrative approach to STEM teaching?
- How can national policies be reformed to promote the use of diverse pedagogies and formative evaluation methods – particularly at end-of-secondary level education?
- How can the teachers that engaged in these innovative projects be supported to mentor peers in IBSE and other innovative pedagogies?

6.3 Summary

Individually, the four reports place emphases on different aspects of domains of teaching and learning. For example, the first report, addresses three challenges: environmental, economic and social. The second report is grounded within three drivers: personal development, societal challenges, and the needs of the local and global workforces. The third report focuses on student achievement however, contextual factors are prominent. The fourth report is based on a five-component model that includes pedagogy, curriculum, teacher professional development and cooperation with STEM industries. These differences in foci, however, are not surprising given the fact that the reports were developed in different geopolitical contexts which have unique educational agendas, visions, and goals.

Collectively, some similarities exist across the reports. One similarity is that all reports consider the social nature of learning and they touch upon contextual aspects of learning as well as societal challenges. Two reports (OECD, CCD) make references to the context in which students learn, while two others (NAE, Scientix) make references to the sociopolitical contexts in which schools are functioning. Moreover, curriculum materials have a prominent role in all reports even if some receive less attention than others. Interestingly, technology is not present in all reports as part of neither the curriculum nor any innovative pedagogies. Third, assessment and evaluation issues appear in all four reports, are but discussed in different and unique ways.

When thinking about future education policy and practices, one panel discussion was framed in terms of 3 key 'stages' of education:

- Stage 1, Primary & Secondary Schooling (Grades 1–12), formative learner and generalized learning;
- Stage 2, Higher or Further Education, Degrees & Certifications, Undergraduate and Postgraduate;
- Stage 3, Career Education, World of Work, Professional & Licensing Bodies, Lifelong Learner.

Each Stage has a different focus of education and serves a different purpose. How Stages 1 & 2 interface with Stage 3 though is significant for the future designs of educational models and systems. A concern that was raised is educational reform thinking in the four reports is tending towards workplace oriented skills and competences. If Stage 3 is where such skills and competences inform education systems, then how do we envisage Stages 1 and 2 to continue to work in ways that instill the 'habits of mind' for reasoning in/about disciplinary knowledge. A second concern pertains to the strong focus on the mind – where is the heart? How will future education frameworks that are attending to rapid technological, environmental, and workforce developments also deal with matters of ethics and values? How do we guard against the potential for these workforce developments to widen the equity gap? In particular, the need to consider variance in economic and political stability among developing nations that are experiencing persistent conflict and issue of migration.

Another rich discussion among panelists and the audience focused on teacher professional development issues. One challenging problem nations are facing is how to equip teachers with knowledge, values, skills, and attitudes that will help promote students' competences for solving personal, social, and global issues. There are some enthusiastic teachers who believe in the reforms and are organizing teacher learning groups to develop new instructional models that reflect the new education frameworks. But levels of understanding about the new frameworks is limited and many teachers are waiting to see what is going to happen when the new curriculum are put in place. Teachers view the new learning frameworks and standards as sound but with respect to implementation in classrooms there is a lot of confusion and questions.

Yet another issue regarding teacher education is how college and university STEM faculty will adapt. Within colleges of science and engineering, university teaching is still traditional, focusing mainly on disciplinary content knowledge and on problem solving skills. Only few of STEM faculty are aware of the new education frameworks. The focus of many faculty is on developing critical thinking, reflective thinking and problem solving. But there is tension with senior professors who reject the curriculum reform in pre-college levels because they feel students will not learn enough discipline knowledge from the new curriculum. If the universities do not explicitly support the educational reforms, then high school teachers might be discouraged to take up the reform agendas. After all, a major goal for high school teachers is to prepare their students to attend top ranking universities.

References

AAAS. (1993). *Benchmarks for science literacy.* Project 2016. New York: Oxford University Press.

Ault, C. R. (2015). *Challenging science standards: A skeptical critique of the quest for unity.* Rowman & Littlefield.

Bevan, B., Penuel, W. R., Bell, P., & Buffington, P. (2018). Learning, generalizing, and local sense-making in research-practice partnerships. In B. Bevan & W. R. Penuel (Eds.), *Connecting research and practice for educational improvement: Ethical and equitable approaches* (pp. 18–30). Routledge.

Bryk, A., Gomex, L., Grunow, A., & LeMahieu, P. (2015). *Learning to improve: How America's school can get better at getting better.* Harvard Education Press.

Duschl, R. (2019). Learning progressions: Framing and designing coherent sequences for STEM education. *Disciplinary and Interdisciplinary Science Education Research, 1*, 4. https://doi.org/10.1186/s43031-019-0005-x

European Schoolnet, Brussels. (2018). *Science, technology, engineering and mathematics education practices in Europe. Scientix Observatory Report 4.*

Fadel, C., Bialik, M., & Trilling, B. (2015). *Four-dimensional education: The competencies learners need to succeed.* Center for Curriculum Redesign.

Fishman, B. J., & Penuel, W. R. (2018). Design-based implementation research. In F. Fischer, C. E. Hmelo-Silver, S. R. Goldman, & P. Reimann (Eds.), *International handbook of the learning sciences.* Routledge.

Koeppen, K., Hartig, J., Klieme, E., & Leutner, D. (2008). Current issues in competence modeling and assessment. *Journal of Psychology, 216*(2), 61–73.

National Assessment Governing Board. (2008). *NAEP 2009 science framework development: Issues and recommendations.* US Department of Education.

National Research Council. (1996). *National Science Education Standards.* National Academies Press.

National Research Council. (2012). *A framework for K-12 science education: Practices, crosscutting concepts, and core ideas.* National Academies Press.

NGSS Lead States. (2013). *Next generation science standards: For states, by states.* National Academies Press.

OECD. (2018). *The future of education and skills 2030* Project. Paris: Organisation for Economic Co-operation and Development. http://www.oecd.org/education/2030/

Osborne, J. F., Rafanelli, S., & Kind, P. (2018). Toward a more coherent model for science education than the crosscutting concepts of the next generation science standards: The affordances of styles of reasoning. *Journal of Research in Science Teaching, 55*(7), 962–981.

Rudolph, J. L. (2019). *How we teach science: What's changed and why it matters*. Harvard University Press.

Sadler, T., & Brown, D. (2018). Editorial. Introduction to the special issue: A critical examination of the next generation science standards. *Journal of Research in Science Teaching, 55*(7), 903–906.

Scientix Observatory. (2018). *Science, technology, engineering and mathematics education practices in Europe*. European Schoolnet.

Singer, J., Braun, H., & Chudowsky, N. (Eds.). (2018). *International education assessments: Cautions, conundrums, and common sense*. National Academy of Education.

Chapter 7
PISA 2015: What Can Science Education Learn from the Data?

Jonathan Osborne, Cory Forbes, Knut Neuman, Anna Schiepe-Tiska, Mylène Duclos, Florence Le Hebel, Andrée Tiberghien, Pascale Montpied, Valérie Fontanieu, Sara Dozier, Davide Azzolini, Nicola Bazoli, and Loris Vergolini

7.1 Introduction

Jonathan Osborne
Stanford University, Stanford, CA, USA
e-mail: osbornej@stanford.edu

For many national governments, the outcome of the triennial OECD PISA tests matter. For instance, in a survey conducted of the impact of PISA for 17 countries, PISA was seen to be 'very influential', 11 others identified it as 'moderately influential', and only five countries saw PISA as 'not very influential' (Breakspear, 2012). The Director of PISA, Andreas Schleicher, sees PISA as a tool for identifying poor

J. Osborne (✉) · S. Dozier
Stanford University, Stanford, CA, USA
e-mail: osbornej@stanford.edu; osbornej@stanford.edu

C. Forbes
University of Nebraska-Lincoln, Lincoln, NE, USA
e-mail: cory.forbes@unl.edu

K. Neuman
IPN, Kiel, Germany
e-mail: neumann@leibniz-ipn.de

A. Schiepe-Tiska
Technical University of Munich, Munich, Germany
e-mail: schiepe-tiska@tum.de

M. Duclos · A. Tiberghien · P. Montpied
ICAR UMR5191, UMS LLE, ENS Lyon, Lyon, France
e-mail: mylene.duclos@ens-lyon.fr; andree.tiberghien@ens-lyon.fr;
pascale.montpied@ens-lyon.fr

O. Levrini et al. (eds.), *Engaging with Contemporary Challenges through Science Education Research*, Contributions from Science Education Research 9,
https://doi.org/10.1007/978-3-030-74490-8_7

performance in any countries' education system. Indeed performance on PISA has been shown to correlate with economic growth (Hanushek & Woessmann, 2012). However, PISA is not without its critics. In a series of articles, Meyer and colleagues argue that PISA has become part of "a pervasive normalizing discourse, legitimizing historic shifts from viewing education as a social and cultural to an economic project engendering usable skills and 'competences'."(Meyer, Tröhler, Labaree, & Hutt, 2014). Labaree has argued that PISA assesses what nobody teaches (Labaree, 2014).

Nevertheless, PISA provides an unparalleled set of data from over 70 countries which can be explored for the insights they provide in and between countries. The data consists of three sets of data from:

1. The cognitive test which is a measure of three core competencies drawing on content, procedural and epistemic knowledge in personal, local and global context;
2. The non-cognitive questionnaire which asks a range of questions including students' experience of teaching; and
3. A set of log files of keystrokes used by the students in answering the computer-based test.

Using these data, analyses are conducted by the OECD from which substantive claims are made about the strengths and weaknesses of certain forms of teaching – claims which policy makers attend to. The OECD's analysis of the data states "After accounting for students' and schools' socio-economic profile, greater exposure to enquiry-based instruction is negatively associated with science performance in 56 countries and economies. Perhaps surprisingly, in no education system do students who reported that they are frequently exposed to enquiry-based instruction score higher in science" (OECD, 2016b, p71) How valid are these claims is the question that Forbes, Neuman and Schiepe-Tiska in the first paper and Dozier in the third paper both ask? The focus of their question is on enquiry-based teaching – a topic of some controversy (Furtak et al., 2012). The focus of the second paper is to try and unpick how student performance is related to their economic, social, and cultural status (ESCS). This is an example of the kind of analyses that PISA data afford

F. Le Hebel
ICAR UMR5191, UMS LLE, ENS Lyon, Lyon, France

University of Lyon, Lyon, France
e-mail: florence.le-hebel@ens-lyon.fr

V. Fontanieu
IFE, ENS Lyon, Lyon, France
e-mail: valerie.fontanieu@ens-lyon.fr

D. Azzolini · N. Bazoli · L. Vergolini
FBK-IRVAPP, Trento, Italy
e-mail: azzolini@fbk.eu; bazoli@irvapp.it; vergolini@fbk.eu

showing as it does, that the features of the test that distinguish the low performers from the mean are not the same as those that distinguish the high performers.

The final paper exploits the new data set made possible by the use of computer-based testing in 2015. As well as students' actual responses, the computers stored their keystrokes as a set of log files. Using this big data set, Azolini, Bazoli and Vergolini explore what we can learn about student effort and persistence and its association with outcomes and how it varies by country.

7.2 Science Teaching and Learning: Analysis of Pisa 2015 Data from the United States and Germany

Cory Forbes
University of Nebraska-Lincoln, Lincoln, NE, USA
e-mail: cory.forbes@unl.edu

Knut Neuman
IPN, Kiel, Germany
e-mail: neumann@leibniz-ipn.de

Anna Schiepe-Tiska
Technical University of Munich, Munich, Germany
e-mail: schiepe-tiska@tum.de

7.2.1 Introduction to the Study

Consistent with global science education reform (GDSU, 2013; NGSS Lead States, 2013), today's students must not only develop an understanding of scientific knowledge, but also learn to negotiate its intersection with social, cultural, and economic values to concretely identify, analyze, and problem-solve real-world problems. This skillset and associated knowledge base – *scientific literacy* – is a core focus of the triennial international survey designed to evaluate education worldwide by testing the skills and knowledge of 15-year-old students. PISA measures students' scientific literacy as the use of scientific knowledge to identify questions, acquire new knowledge, explain scientific phenomena, and draw evidence-based conclusions about science-related issues (OECD, 2016a).

But how can scientific literacy be best cultivated in science classrooms? The relationship between science teaching and learning remains a fundamental concern of the field of science education. Historically, science education scholars have delineated a continuum of science teaching and learning from more teacher-directed to more student-directed. While the latter was once considered a 'gold standard' for science education (Settlage, 2007), more recent research indicates that guided inquiry, with strategic direction from the teacher, is more effective than student-directed science learning experiences (e.g., Furtak et al., 2012; Prenzel et al., 2012). However, far less is known about why these relationships may exist and little similar research has been conducted using the most recent 2015 PISA data, nor has there been any explicit focus on international comparisons of the instruction that promotes scientific literacy. More work, including the research presented here, is therefore needed, to understand better how to optimally foster scientific literacy in today's students we conducted an analysis using the US and German data to ask:

1. What relationships are observed between secondary students' scientific literacy and the instruction (inquiry-based and teacher-directed) they report (RQ#1)?
2. What are prominent instructional profiles associated with secondary students' scientific literacy (RQ#2)?

7.2.2 Methods

The study reported here was conducted as part of an international collaboration between science education researchers and assessment experts from three institutions in the United States and Germany (Forbes et al., 2020). The project involves use of statistical methods to conduct quantitative analyses on 2015 PISA science data. In its latest (2015) administration, approximately 540,000 15-year-old (secondary) students in 72 countries completed PISA. Here, we focus on data from students in the United States ($n_{US} = 5099$) and Germany ($n_{GER} = 4218$). The major domain in 2015 was science, with approximately half of the cognitive assessment devoted to science items. Collectively, these items comprise an overall measure of students' scientific literacy. Additionally, PISA includes a multitude of student questionnaire items related to science teaching and learning that comprise a number of subscales, including inquiry-based teaching (IBTEACH), teacher-directed teaching (TDTEACH), perceived feedback (PERFEED), and teachers' instructional adjustments (ADINST). For purposes of this study we focus on two subscales in the student questionnaire - IBTEACH ($n_{items} = 9$) and TDTEACH ($n_{items} = 4$) – in which students characterize the science instruction they experience.

7.2.3 Results

To address RQ#1, we conducted a one-way ANOVA to assess differences in students' scientific literacy by reported instructional practices (TDTEACH and IBTEACH). Results show a significant main effect for instruction, $F(8, 8586) = 46.8$, $p < .001$, on students' scientific literacy. Results of Tukey's HSD post-hoc tests show that students' scientific literacy is highest when they report high levels of teacher-directed instruction and moderate levels of inquiry-based teaching. In contrast, students' scientific literacy is lowest where inquiry-based teaching was reported in most to all lessons. A large (>50%) grouping of moderate levels of scientific literacy was observed which, while exhibiting varying levels of teacher-directed instruction, each exhibited limited levels of inquiry-based teaching. While nearly 40% of students reported teacher-directed instruction in many to all of their lessons, only 17% reported similar levels of inquiry-based teaching. Overall, these findings suggest students a) reported more teacher-directed instruction than inquiry-based instruction, and b) exhibited higher levels of scientific literacy in association with more teacher direction.

To address RQ #2, we conducted Latent Profile Analysis (LPA) to identify typical profiles of science instruction in each country based on instructional features (IBTEACH and TDTEACH) reported by students. LPA revealed five distinct profiles in both Germany and the United States that exhibited a particular level of consistency across countries. Three of these profiles showed similar levels of IBTEACH and TDTEACH within each profile, but differing levels across profiles (from almost never to some, over some to most, to most to all). One profile showed average levels of IBTEACH, but significantly higher levels of TDTEACH, and one profile showed very high levels of IBTEACH combined very low levels of TDTEACH. Interestingly, students reporting high levels of IBTEACH and low levels of TDTEACH showed student achievement significantly lower than the country average across both countries. However, students reporting average level of IBTEACH and high levels of TDTEACH scored significantly higher than the average in the US, and significantly lower than the average in Germany. Overall, these results show similar profiles for science instruction reported by students and similarities in students' scientific literacy associated with those profiles, but also interesting differences between both countries that warrant further exploration (see Forbes et al., 2020 for an example).

7.2.4 Discussion

Findings from this study illustrate trends in observed associations between both inquiry-based and teacher-directed science instruction and students' scientific literacy as measured in PISA, 2015. They contribute to the field's understanding of current international science education reform (GDSU, 2013; NGSS Lead States, 2013), as well as empirical perspectives on effective science teaching and learning (Furtak et al., 2012; Settlage, 2007), by building upon past and present PISA analyses (Prenzel et al., 2012). While limited by their reliance on students' self-reported data about instruction, these results lend further evidence in support of the teacher's critical role in providing guidance in effectively-designed, reform-based science classrooms and raise important questions about the nature of this relationship and reasons for observed differences between the U.S. and Germany that merit further study.

7.3 Discriminating Characteristics of Pisa Science Items According to Students' Socio-Economic-Cultural Level and Performance

Mylène Duclos, Andrée Tiberghien, and Pascale Montpied
ICAR UMR5191, UMS LLE, ENS Lyon, Lyon, France
e-mail: mylene.duclos@ens-lyon.fr; andree.tiberghien@ens-lyon.fr; pascale.montpied@ens-lyon.fr

Florence Le Hebel
ICAR UMR5191, UMS LLE, ENS Lyon, Lyon, France
University of Lyon, Lyon, France
e-mail: florence.le-hebel@ens-lyon.fr;

Valérie Fontanieu
IFE, ENS Lyon, Lyon, France
e-mail: valerie.fontanieu@ens-lyon.fr

7.3.1 Introduction to the Study

Understanding what makes a question difficult for students is crucial for both teachers and test developers (El Masri et al., 2017).

PISA assesses the students' economic social and cultural status (ESCS index) and French results from PISA science 2015 show that the influence of students' ESCS on their performance is one of the highest compared to OECD countries (OECD, 2016b). Such results offer the opportunity to investigate the links between the questions and students' difficulties in answering them according to their ESCS and their performance levels.

The aim of the study is to identify some of the main characteristics of PISA items which discriminate the students' performances, based on their ESCS and their performance level. More generally, this study aims to understand the explanatory power of these characteristics.

Numerous studies have focused on the identification of the difficulty factors of an issue and, more specifically, on the development of predictive models of question difficulty (see synthesis in Dhillon, 2003).

Several authors have proposed three main categories which distinguish the sources of difficulty (e.g. Ahmed & Pollitt, 1999; Prenzel et al., 2002). For instance, Crisp & Grayson, 2013) proposed "question attributes" (e.g. answer format) "knowledge and understanding assessed by the question", and "question processes" (undertaken to reach the question answer). Based on these studies we used the following categories: “Intrinsic characteristics of items” (linked to the meaning and to the formal aspects of the question), “Content at stake”, and “Strategies and reasoning” for our analysis.

7.3.2 *Methods*

In order to find the characteristics that may influence students’ scores, we conducted an *a priori* analysis of the items and a statistical study of students’ scores. This entailed making hypotheses on question characteristics which can be potential sources of difficulty and testing the hypotheses iteratively.

7.3.2.1 *A priori* Analysis to Characterize Possible Item Difficulties

In addition to our three categories, this *a priori* analysis also involved the PISA framework (OECD, 2016a). We identified a total of 23 characteristics: 16 main characteristics (the first seven of which are pre-established by PISA Science 2015) and 7 sub-characteristics (Table 7.1) shown in italics.

This statistical analysis aims to understand the explanatory power of these characteristics. In order to access the score differences between students according to their ESCS index, we divided the population into quartiles (equal groups of 25%). The ESCS1 (lower) group refers to the 25% most disadvantaged pupils while the ESCS4 (higher) group corresponds to the 25% most advantaged students. We proceeded approximately the same way to estimate the performance level of students in function of their performance obtained in PISA. Multiple linear regression models (STATA software) are used to identify the items’ characteristics, which influence the success rate and performance gap between the groups.

7.3.3 *Findings*

The statistical analysis shows that some characteristics have more effect according to the ESCS than the performance level while for other characteristics it is the reverse. For instance, the results show that the characteristics of the category “Strategies and reasoning” have more effect on variations in performance gaps based on students’ ESCS rather than on their level of performance. In this category, we are included the cognitive complexity (or DOK) of items (13), the answering strategy “matching” (14), the projection (characteristic 15), and the level of references to daily life reference (16). These first three characteristics are related to the ESCS of students.

Table 7.1 Characteristics of Pisa items from the a priori analysis (coded for all 183 Pisa items)

Category		Characteristic / *Sub-characteristic*	Details
Content at stake	1	Knowledge	Content knowledge, procedural knowledge, epistemic knowledge
	2	Assessed system	For the items assessing content knowledge: Physical systems, living systems, earth and space systems
	1	*Knowledge according to system*	*Crossover of content knowledge and assessed systems*
	3	Competencies	Explain phenomena scientifically; evaluate and design scientific enquiry; interpret data and evidence scientifically
	4	Context-subject	Health and disease; natural resources; environmental quality; hazards; Frontiers of science and technology
Intrinsic characteristics of item- *Formal aspects*	5	Item format	Simple multiple choice, complex multiple choice or open responses
	6	Text length	Word count
	7	Type of illustration	Photo-drawing; table; graphic; diagram; multiple illustrations; illustrations not in item
	8	Simulation	Presence or not in item
Intrinsic characteristics of item-*Aspects related to meaning*	9	Context-situation	Double context as personal-societal, personal-global, societal-global and simple context as societal, global
	10	Answer	Presence in the text and/or illustration or not.
	11	Dependence or independence	Need or not to use the information available in the item text and/or illustration.
	2	*Context-situation according to dependence*	*Crossover of context-situation and dependence*
	3	*Simple/double context according to dependence*	*Crossover of simple/double context and dependence*
	12	Link between the question and the illustration	Presence or not in item
	4	*Link between question and illustration*	*Link explicit or implicit in the item text*

(continued)

Table 7.1 (continued)

Category		Characteristic *Sub-characteristic*	Details
Strategies and reasoning	13	DOK (depth of knowledge).	Webb's DOK levels for science. This is a scale of cognitive demands (from 1 to 4) and it reflects the cognitive complexity of the question
	14	Matching possible or not	Matching between a word in the multiple choice and the same word in text of item.
	5	*Type of matching*	*Helping or disabling*
	15	"Projection" requirement present (or not)	The context of the question prompts the students to project themselves, and conceive the point of view of a community more or less close to their own life.
	6	*Type of projection*	*Close community (more or less to their own life, e.g. a student), know-how community (professions not representing a scientific community, e.g. a gardener)., knowledge community (professions representing a scientific community, e.g. a doctor)*[a]
	7	*Projection direct*	*Direct (explicit in the item text) or indirect (implicit).*
	16	Level of daily life reference in the items	High degree; moderate degree; very limited daily life reference

[a]For example, the item released "Sustainable Fish Farming" explain that the "Researchers have noticed that the water that is being returned to the ocean contains a large quantity of nutrients. Adding which of the following to the farm will reduce this problem?" and the students must choice the organism to be added to the farm to solve the problem. The statement of item involves the students trying to adopt the researcher position

"Projection" (see table above) is more beneficial to students according to their ESCS ($p < 0.05$) than as a function of their performance. Whether among low or high achievers, it is advantaged students who benefit the most from this characteristic, although among the advantaged ESCS group, those who benefit the most are high achievers ($p < 0.10$).

The presence of possible matching in an item widens the gap between low achievers according to ESCS ($p < 0.10$). Low achievers of disadvantaged ESCS seem to benefit less than other students because they don't use it to find the right answer.

Moreover, a particular interesting new result shows that the categories that distinguish students with low and high PISA performances are not the same according to their ESCS – that is it is mainly item format that distinguishes the low performing students as function of ESCS, whereas it is the DOK that distinguishes the higher performing students as function of ESCS).

7.3.4 Implications

Since we consider the characteristics projection and matching to be strategies that students can use to resolve task, we can propose as a hypothesis that, the students use certain strategies on preference to others, as a function of their ESCS and/or their performance level.

This work could help to provide a deeper understanding of how students solve the science task and allow teachers to adapt their assessments to a greater degree according to the profile of their students. Thus, it could help teachers to target these difficulties more effectively in their practice and to take them into account during assessment.

7.4 Establishing Multidimensionality: Identifying Patterns of Inquiry-Driven Science Instruction

Sara Dozier
Stanford University, Stanford, CA, USA
e-mail: osbornej@stanford.edu

7.4.1 Introduction to the Study

For over a century, teachers, researchers, curriculum developers, and policymakers have privileged inquiry as a central component of school science (e.g. Osborne, 2014). Inquiry, however, is a loosely defined concept, describing both the student knowledge and competencies that support successful engagement in scientific processes, as well as the instructional strategies to support student science learning. Broadly speaking, inquiry-based instruction seeks to allow students to develop their own understanding of scientific ideas and concepts through non-transmissive learning strategies. However, researchers and educators convey a variety of perspectives as to what counts as inquiry teaching and learning practices (e.g. Abd-El-Khalick et al., 2004). A significant body of empirical research has shown a positive relationship between different operationalized versions of inquiry-based instruction and student achievement (e.g. Furtak et al., 2012). Taken together, theoretical, empirical, and practical support for inquiry have helped maintain its status as a critical component of a high-quality science education.

The Programme for International Student Assessment (PISA) survey measures both student scientific literacy and contexts related to student learning. The student questionnaire includes an index comprising nine items that elicit frequencies of inquiry-based instructional practices in science classes (Müller et al., 2016). The OECD analysis describes a negative relationship between the *enquiry-driven instruction* index and overall student science scores (OECD, 2016b), which conflicts with the positive relationship between inquiry and student learning observed in classroom-level empirical studies (e.g. Furtak et al., 2012). Drawing on data from U.S. students, this study characterizes patterns in how students experience inquiry-based instruction in order to answer the following questions and discuss their policy implications. How does the PISA defined construct of inquiry relate to other operationalized versions that appear in the literature? Do the nine PISA questionnaire items composing the *enquiry-driven instruction* index measure a single construct of inquiry?

7.4.2 Methods

This study is a secondary analysis of data from the PISA national school sample for the United States, which was selected in accordance with the PISA 2015 technical standards (OECD, 2017).

Two models for explaining latent constructs in the *enquiry-driven instruction* index were examined. Exploratory factor analysis (EFA) was first performed using the nine items to determine the dimensionality of the index. Factors were retained if iterated principal factor analysis (IPF) showed they individually explained more than 5% of the variance between items, cumulatively accounted for more than 90% of the total variance, and had larger eigenvalues than those generated from a random data set using parallel analysis (Costello & Osborne, 2005). After promax rotation, a factor loading cutoff of 0.4 was used to determine whether an item should be associated with a given factor (Hancock & Mueller, 2010). To determine the best fit model, the unidimensional model containing all nine items was contrasted with the model that emerged from EFA using Confirmatory Factor Analysis (CFA) allowing for correlation between factors. The two models were compared for fit using Akaike's information criterion and Bayesian information criterion.

7.4.3 Results

IPF analysis extracted 3 factors, which are shown in Table 7.2. One item "Students are given opportunities to explain their ideas," did not meet the threshold for inclusion any of the three factors.

The four items loading onto the first factor described student-guided activities. The two items loading onto the second factor focus directly on practical work. In the third factor, the teacher is explaining ideas to the students. Both Akaike's and Bayesian information criteria showed that the 3-factor model that emerged from EFA better fit the data than the 1-factor model that underlies the *enquiry-driven instruction* index, as the 3-factor model had a lower criterion value for each test than

Table 7.2 Factor loading of *enquiry-driven instruction* student questionnaire items after iterated principal factor analysis of three retained factors with promax rotation

Item	Student-guided inquiry	Teacher-guided inquiry	Teacher-led traditional
There is a class debate about investigations.	0.9441	−0.1095	0.0341
Students are allowed to design their own experiments.	0.6994	0.0323	0.012
Students are required to argue about science questions.	0.5479	0.2421	−0.0109
Students are asked to do an investigation to test ideas.	0.4035	0.2602	0.1968
Students are asked to draw conclusions from an experiment they have conducted.	−0.0692	0.8548	0.0404
Students spend time in the laboratory doing practical experiments.	0.2178	0.5338	−0.0334
The teacher clearly explains the relevance of science concepts to students' lives.	0.1568	−0.1089	0.8211
The teacher explains how a science idea can be applied to a number of different phenomena (e.g., the movement of objects, substances with similar properties).	−0.1404	0.3256	0.6337
Students are given opportunities to explain their ideas.	0.1763	0.1788	0.3093

the 1-factor model. These data suggest that the 3-factor model focused on the student-guidance dimension is the best fit with these data. Based on the items loading onto each of these factors, student-reported frequencies of inquiry instructional practices vary based on the degree of student guidance. The large subject sample size at least partially mitigates concerns about low variable sampling in factors 2 and 3 (Velicer & Fava, 1998).

7.4.4 Implications

The OECD makes broad claims regarding the value of inquiry-driven instruction based on an index that covers a wide range of instructional activities. The factors emerging from this study align with the guidance dimension of inquiry-based instruction described in the Furtak et al. (2012) framework. Based on these student-reported measures of inquiry-based instruction, OECD states that "…some of the arguments against using hands-on activities in science class should not be completely disregarded" (OECD, 2016b). The multidimensionality demonstrated in the above analysis suggests that this recommendation is overly broad and that a finer grain of detail based on the 3-factor model should be considered. This study raises several questions about how to interpret these data in the service of high-quality policy recommendations. Firstly, there is the question of what, exactly, students take these questionnaire items to mean. Secondly, the assumption that if an instructional practice is effective, experiencing that practice more frequently is better should be questioned. These and other questions remain if we are to effectively use these data to recommend high quality instructional practices to policymakers and practitioners alike.

7.5 Exploiting Computer-Generated Data to Study Student Test-Taking Behavior

Davide Azzolini, Nicola Bazoli, and Loris Vergolini
FBK-IRVAPP, Trento, Italy

7.5.1 Introduction to the Study

The increased availability of computer-generated data in education allows researchers to investigate more deeply the strategies employed by students when taking a test. We used PISA 2015 log-files from European countries to study students' behavior during a low-stakes exam. Log-files information such as response time and performance at different points of the test was used to compute two novel indices of students' effort and perseverance.

PISA test items are grouped into domain-specific clusters (science, reading, or mathematics), lasting 30 min each. Two 1-h sessions are formed containing different combinations of two clusters. Each student is randomly assigned one booklet, which is made of two sessions. Hence, each student gets different combinations of test items and receives them in a different order. PISA 2015 had 66 different computer-based assessment booklets containing different cluster combinations. A booklet contains clusters of two or three different domains. This study considers only booklets containing two different domains, because students receiving three domains systematically showed lower performance than those receiving two.

7.5.2 Effort

Effort is the activation of mental power to perform a task. In the literature, effort has been measured exploiting response time information (Wise & Kong, 2005). We propose an effort index based on the difference in response time (RT, measured in milliseconds) on easy vs difficult items. More precisely, in each cluster, the RT for the five easiest items is compared to the RT for the five hardest items (OECD, 2017).[1] The main assumption underlying this index is that the more difficult an item, the more effort students have to make. Because this index is constructed at the individual level and at the cluster level, it allows overcoming "student ability bias" and "item test positioning bias". The former arises when comparing RT in easy vs difficult items located in a given point of the test but across different students. The latter occurs when comparing a given student's RT in easy vs difficult items located at different points of the test. The proposed effort index is calculated in the first cluster to avoid bias due to fatigue. In Fig. 7.1 we report the average response time on easy and difficult items for all the countries considered in the analysis. we can notice that in all the countries, on average, students use more time on difficult items than on the easy ones. On average, students spend 74 seconds on easy items and 109 on difficult ones, with a resulting difference of 35 seconds (red line in Fig. 7.1). However, the levels of effort are heterogeneous across countries.

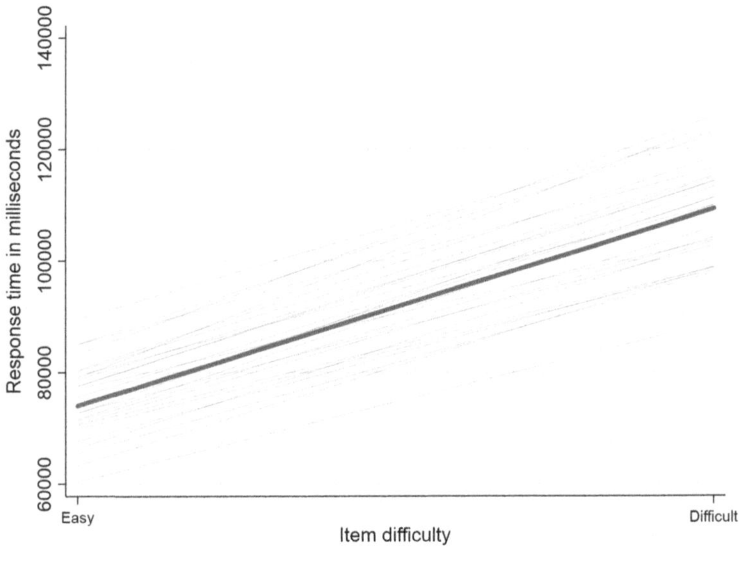

Fig. 7.1 Mean response time by item difficulty and country. (*Source*: FBK-IRVAPP analysis of PISA 2015 data. Red line: European average. Grey lines: country averages)

[1] Five is the maximum number that can be used to allow for a meaningful comparison of easy and difficult items across all clusters. The level of difficulty of an item has been estimated through Item Response Theory as illustrated in the PISA Technical report (OECD, 2017).

7.5.3 Perseverance

Perseverance is defined as "students' motivation and the impact that motivation has on self-control and the ability to withstand fatigue" (Borgonovi & Biecek, 2016, 128). We propose to measure perseverance as the difference in performance at different points of the test. This computation assumes that individual performance decreases as time goes by because of a fatigue effect. The index is computed as the difference between students' performance on cluster 1 and their performance in cluster 2. Rather than computing a simple mean score by cluster, Weighted Likelihood Estimates (WLE) of students' ability are derived for each cluster by using the international item parameters (OECD, 2017). In contrast to the simple mean score, the computation of WLE integrates items difficulty, so that WLEs are comparable from one cluster to another, independent of the cluster average difficulty (Muraki, 1992). This index can assume either positive or negative values, giving the interpretation difficult in its continuous form. Therefore, the study proposes four student profiles. *Persistently good* are students who perform above the average in the first cluster and either improve or remain constant in the second one. *Starts well but drops* are students who perform above the average in the first cluster and worsen in the second one. *Slow starters* are students who perform below the average in the first cluster and improve in the second one. Finally, *persistently weak* are students who perform below the average in the first cluster and worsen in the second one. In the overall European sample, 16.1% of students are "persistently good", while 24.0% of students are "persistently weak". However, the proportion of students belonging to each group varies substantially across countries (Fig. 7.2).

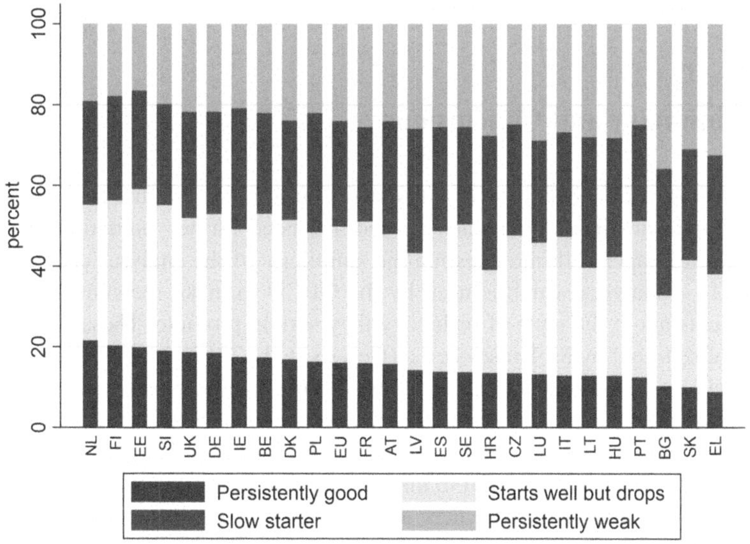

Fig. 7.2 Proportion of students belonging to the four perseverance categories in European countries. *Note*: FBK-IRVAPP analysis of PISA 2015 data

7.5.4 Concluding Remarks

Given the novelty of computer-generated data in educational research and the many challenges ahead both in terms of methodological approaches and interpretation frameworks, this study has to be considered as an exploratory one. The two proposed measures require an in-depth analysis in order to corroborate their validity and reliability. First, analyses of the response time based on item format (e.g. multiple choice or open ended) would allow us to validate the effort index. Second, the computation of these two measures should be replicated with different data. PISA is a low-stakes exam and this could imply a different students' engagement relative to a high-stakes setting. Third, the interpretation of the two measures would benefit from a thorough comparison with more traditional measurement instruments (i.e., self-reports).

Despite its exploratory nature, the study shows the potential of log-files for use in developing a deeper understanding of the cognitive and non-cognitive processes underpinning test performance. In this regard, it is important to assess the association between the two indices and overall student performance as well as designing research protocols to disentangle the causal effects of the two indices on performance. Also, the two measures could provide more insights about cross-national differences in students' learning outcomes over and beyond test performance measures. The first evidence produced thus far would suggest that top-performing countries on standardized tests are not necessarily top-performing countries in terms of student test-taking behavior (Azzolini et al., 2019). More research is needed to further explore the potential of computer-generated data to investigate student learning processes shed new light on the role played by individual, family and school characteristics for student educational outcomes.

7.6 Summary and Conclusions

What these set of papers show is really two things. First, that the analyses of the OECD should not be automatically trusted and need further validation. Both the first and third paper call into question the simplicity of the analyses conducted by the OECD – analyses which is used by the OECD to make recommendations to governments about the style of pedagogy they should promote. These findings are similar those found by McKinsey & Co (Denoël et al., 2017). In that sense, they act as a check on the OECD and challenge the narrative they promote. In short, it is important for the research community to conduct secondary analyses of the data.

The other two papers point to findings that do not emerge from the OECD analyses highlighting features that start to unpick the complexity underlying student performance. The results presented here are based mainly on European and US data thought the fourth paper looks at the data more broadly. Given that arguably the worst use of PISA data is to make cross country comparisons, the reader should be wary of making generalizations to other countries. How valuable they are to action must be left to the reader to judge but they demonstrate some of the questions that can be explored by researchers and their potential value for informing what we know about student performance.

References

Abd-El-Khalick, F., Boujaoude, S., Duschl, R. A., Lederman, N. G., Mamlok-Naaman, R., Hofstein, A., Niaz, M., Treagust, D., & Tuan, H. L. (2004). Inquiry in science education: International perspectives. *Science Education, 88*(3), 397–419. https://doi.org/10.1002/sce.10118

Ahmed, A., & Pollitt, A. (1999). *Curriculum demands and question difficulty*. Paper presented at IAEA conference, Slovenia, May.

Azzolini, D., Bazoli, N., Lievore, I., Schizzerotto, A., & Vergolini, L. (2019). *Beyond achievement. A comparative look into 15-year-olds' school engagement, effort and perseverance in the European Union*. Publications Office of the European Union.

Borgonovi, F., & Biecek, P. (2016). An international comparison of students' ability to endure fatigue and maintain motivation during a low-stakes test. *Learning and Individual Differences, 49*, 128–137.

Breakspear, S. (2012). *The policy impact of PISA: An exploration of the normative effects of international benchmarking in school system performance* (OECD education working papers, no. 71). OECD.

Costello, A. B., & Osborne, J. W. (2005). Best practices in exploratory factor analysis: Four recommendations for getting the most from your analysis. *Practical Assessment, Research & Evaluation, 10*(7), 1–9. 10.1.1.110.9154

Crisp, V., & Grayson, R. (2013). Modelling question difficulty in an a level physics examination. *Research Papers in Education, 28*(3), 346–372.

Denoël, E., Dorn, E., Goodman, A., Hiltunen, J., Krawitz, M., & Mourshed, M. (2017). *Drivers of student performance: Insights from Europe*. McKinsey & Co.

Dhillon, D. (2003). *Predictive models of question difficulty: A critical review of the literature*. AQA. https://cerp.aqa.org.uk/sites/default/files/pdf_upload/CERP-RP

El Masri, Y. H., Ferrara, S., Foltz, P. W., & Baird, J.-A. (2017). Predicting item difficulty of science national curriculum tests: The case of key stage 2 assessments. *The Curriculum Journal, 28*(1), 59–82.

Forbes, C. T., Neumann, K., & Schiepe-Tiska, A. (2020). Patterns of inquiry-based science instruction and student science achievement in PISA 2015. *International Journal of Science Education, 42*(5), 783–806.

Furtak, E. M., Seidel, T., Iverson, H., & Briggs, D. C. (2012). Experimental and quasi-experimental studies of inquiry-based science teaching: A meta-analysis. *Review of Educational Research, 82*(3), 300–329.

Gesellschaft für Didaktik des Sachunterrichts (Society for Teaching Natural and Social Sciences and Technology in Elementary School). (2013). *Perspectives framework for general studies in primary education*. Klinkhardt.

Hancock, G. R., & Mueller, R. O. (2010). *The reviewers guide to quantitative methods*. Routledge.

Hanushek, E. A., & Woessmann, L. (2012). Do better schools lead to more growth? Cognitive skills, economic outcomes, and causation. *Journal of Economic Growth, 17*(4), 267–321.

Labaree, D. F. (2014). Let's measure what no one teaches: Pisa, Nclb, and the shrinking aims of education. *Teachers College Record 116*(9), 1–14. http://www.tcrecord.org/library.

Meyer, H.-D., Tröhler, D., Labaree, D. F., & Ethan, L. (2014). Accountability: Antecedents, power, and processes. *Teachers College Record 116*(9), 1–12. http://www.tcrecord.org.

Müller, K., Prenzel, M., Seidel, T., Schiepe-tiska, A., & Kjærnsli, M. (2016). Science teaching and learning in schools: Theoretical and empirical foundations for investigating classroom-level processes. In S. Kuger, N. Jude, E. Klieme, & D. Kaplan (Eds.), *Assessing contexts of learning: An international perspective* (pp. 423–446). Springer International Publishing. https://doi.org/10.1007/978-3-319-45357-6

Muraki, E. (1992). A generalized partial credit model: Application of an EM algorithm. *ETS Research Report Series, 1*, i–30.

NGSS Lead States. (2013). *Next generation science standards. For states, by states*. The National Academies Press.

OECD. (2016a). *PISA 2015 assessment and analytical framework: Science, reading, mathematic and financial literacy*. PISA, OCDE.

OECD. (2016b). *PISA 2015 results (volume II): Policies and practices for successful schools* (Vol. II). OECD Publishing. https://doi.org/10.1787/9789264267510-en

OECD. (2017). *PISA 2015 technical report*. OECD Publishing. https://doi.org/10.1021/op8002129

Osborne, J. F. (2014). Teaching scientific practices: Meeting the challenge of change. *Journal of Science Teacher Education, 25*(2), 177–196. https://doi.org/10.1007/s10972-014-9384-1

Prenzel, M., Seidel, T., & Kobarg, M. (2012). Science teaching and learning: An international comparative perspective. In B. Fraser, K. Tobin, & C. McRobbie (Eds.), *Second international handbook of science education* (Springer International Handbooks of Education) (Vol. 24). Springer, Dordrecht.

Prenzel, M., Häußler, P., Rost, J., & Senkbeil, M. (2002). Der PISA-Naturwissenschaftstest : Lassen sich die Aufgabenschwierigkeiten vorhersagen? *Unterrichtswissenschaft, 30*(2), 120–135.

Settlage, J. (2007). Demythologizing science teacher education: Conquering the false ideal of open inquiry. *Journal of Science Teacher Education, 18*(4), 461–467.

Velicer, W. F., & Fava, J. L. (1998). Effects of variable and subject sampling on factor pattern recovery. *Psychological Methods, 3*(2), 231–251. https://doi.org/10.1037/1082-989X.3.2.231

Wise, S. L., & Kong, X. (2005). Response time effort: A new measure of examinee motivation in computer-based tests. *Applied Measurement in Education, 18*(2), 163–183.

Chapter 8
Network Analysis of Changes to an Integrated Science Course Curriculum Over Time

Jesper Bruun, Ida Viola Kalmark Andersen, and Linda Udby

8.1 Introduction

Official educational documents often reflect an agglomeration of political intentions at a given time and may change in light of educational policy changes (van den Akker et al., 2010; Schmidt et al., 1997; Dolin & Evans, 2018; Priestley & Biesta, 2013). The wording of particular curricular documents can influence the perceived possibilities for teachers charged with their implementation of new curricula. For example, Evans & Dolin (2018) found that many science teachers had little experience with reading official documents and, therefore, did not see how different notions of scientific literacy were emphasised in these documents. They found that a conceptual network tool – a tool which relied on linguistic networks (Mehler et al., 2016; Bruun et al., 2009) – could alleviate teachers' negative experiences by highlighting important themes and "less apparent interrelationships and relative emphases of various aspects of [...] scientific literacy" (Evans & Dolin, 2018). The conceptual network tool presented by Evans and Dolin represents words and connections between words in official conceptualisations of scientific literacy. The goal of their study was to help teachers identify themes and patterns in the intended outcomes for scientific literacy in different European countries in order to implement and realise these intentions in teaching. Their work highlights that even if curriculum texts can be seen to include specific intentions, teachers may not acknowledge or even be aware of these intentions. This may in part be because of a

J. Bruun (✉) · L. Udby
University of Copenhagen, Copenhagen, Denmark
e-mail: jbruun@ind.ku.dk; udby@nbi.ku.dk

I. V. K. Andersen
Roskilde Gymnasium, Roskilde, Denmark
e-mail: ivka@ind.ku.dk

O. Levrini et al. (eds.), *Engaging with Contemporary Challenges through Science Education Research*, Contributions from Science Education Research 9,
https://doi.org/10.1007/978-3-030-74490-8_8

persistence of earlier developed practices (e.g., van den Akker et al., 2010; Verbiest & Erculj, 2006) but can be alleviated by working to achieve a deeper understanding of important themes in these documents.

In general, analysing themes that emerge in official documents may play an important role in understanding the interplay between intentions, implementation, and attainment of teaching. We argue that in studying this interplay, a variety of themes might emerge in curricula over time that could be seen to reflect different stakeholders' emphases and perhaps conflicts between different views. Inspired by van den Akker et al. (2010), we argue that intentions, implementation, and (observed student) attainment may influence each other over time. Thus, teacher implementation may not reflect intentions in a curriculum (as stated by van den Akker et al., 2010). This might be ignored, be addressed in teacher professional development, or lead to a change of wording in the curriculum. Likewise, observations about student attainment might influence curriculum development to change standards or suggestions for pedagogy. In this paper, we show examples of such emerging and developing themes as they appear through the lens of a newly developed network analytical method (Bruun et al., 2019).

As an illustrative example, we use different versions of the official curriculum for a Danish integrated science course. Our aim is both to summarise emerging themes, to visualise connections between themes and development of themes through time.

8.2 Intentions in the Danish 'Basic Science Course'

In many countries, ideas of inquiry and scientific literacy have been woven into national curricula (e.g. Evans & Dolin, 2018). Often, inquiry-based science teaching, understood as science teaching in which students are motivated to engage meaningfully and autonomously with scientific content to construct knowledge and draw conclusions (e.g. Minner et al., 2010), is seen as a pedagogy that can help students achieve scientific literacy. Germany and the Nordic countries often also see scientific literacy in relation to Bildung perspectives; here, science is also important to a person's understanding and experience of their relation to the world and society (e.g. Ropohl et al., 2018).

In Denmark, ideas of scientific literacy, inquiry-based science teaching, Bildung, and interdisciplinarity were implemented officially in the largest of the country's four national upper secondary programs (called *stx*) in 2004. This was done through the introduction of the Basic Science Course (BSC, "Aftale af 28. maj", 2003). The curriculum text was then changed in 2007 and 2010.

One of the intentions of BSC is to introduce students to science through work with the basic elements of natural science. The focus should be on the commonalities and the differences within the natural science disciplines (DME, 2013). BSC teaching is meant to consist of exemplary and contemporary thematic issues where the scientific disciplines are required to work together using some degree of interdisciplinarity (Jantsch, 1972).

Another objective of the BSC is to make the students aware of the importance of knowing and understanding scientific thinking. This objective is intended to make the students able to reflect on the strengths and limitations of scientific knowledge. Moreover, the course curriculum prescribes that students have to achieve knowledge about central scientific issues, which support their curiosity in and commitment tonatural science and encourage them to learn more about science (DME, 2013). The curriculum does not specify which central scientific issues should be chosen for teaching.

The course has been the source of public debates and has been evaluated with respect to – among other things – the perspectives above (Dolin et al., 2016). As such, we may expect themes regarding, for example, scientific content, aims, teacher roles, and structure should emerge. However, we do not know the details of the themes, which tensions (if any) themes may harbour, or how themes may be connected. Thus, we argue that the BSC curriculum texts from three different years comprise a worthwhile illustrative example of our proposed way of analysing curricula.

8.3 Research Questions

This chapter reports on a larger study conducted in 2016–2017 (Andersen, 2017). For the purposes of this illustrative example, the research questions are:

1. Which interconnected themes emerge as prevalent in the Danish BSC curriculum texts for years 2004, 2007, and 2010?
2. How did selected themes evolve over the years 2004, 2007, and 2010?

8.4 Methodology

The complexity of educational systems (Evans et al., 2018) can be seen to warrant an integration of quantitative and qualitative perspectives into a mixed methods design (Johnson & Onwuegbuzie, 2004). This also holds for analysis of educational policy. Bruun et al. (2019) recently combined qualitative discourse analysis with linguistic networks in a mixed methodology for transcripts of group discussions called thematic discourse network analysis (TDNA). The analysis revealed and graphically displayed hidden themes to provide a nuanced and rich picture of the data. Bruun et al. (2019) analysed group discussions and used qualitative discourse analysis of the discussion into conversation units, where each unit "should appear to be a distinct part in the sense that it ends as the conversation is exhausted […]" (p. 325). In contrast, this study relied on official documents, which are related to different discursive and social practices (i.e. the practices of politicians, document authors, the teachers whose teaching is the document's concern, the leaders of these teachers, and other stakeholders, such as researchers and industry). Critical discourse analysis (CDA, e.g. Fairclough et al., 2011) is well suited for analysing such documents in light of different discursive and social practices. For example, Fairclough et al. (2011) reports that CDA has been used with computer-based analyses of keywords to analyse historical developments of UK political discourse. This combination of CDA and keyword analysis has demonstrated its "heuristic value in directing the analysts' gaze in unexpected and often fruitful directions" (p. 366). Here, we first show how we used critical discourse analysis, and then proceed to show, in short, how we employed TDNA. The following sections are based on Andersen's (2017) study, where interested readers may find detailed descriptions of the methodology.

8.4.1 Critical Discourse Analysis in the Present Study

We used Fairclough's three-dimensional approach (Fairclough, 1993), where a (here written) text is seen as an instance of a discursive practice, which involves the production and interpretation of texts. These, in turn, are part of social practices with social structures and power relations. Inspired by Fairclough (1992), Fig. 8.1 presents a graphical representation of how we see official Danish BSC curricular texts as embedded in discursive and social practices.

In the analysis of the curriculum text, Fairclough's approach is oriented towards linguistics. In this view, vocabulary, commonly used phrases, and passages have an impact on both discursive practices, for example, teachers' interpretations, and on social practices, like teaching practices. For us, the approach was oriented to finding themes that may characterize intentions for teaching in curricular documents. The links to discursive and social practices were then links between the text and possible processes of interpretation and use of the text itself.

8.4.2 Thematic Discourse Network Analysis of Curricular Documents

Using thematic discourse network analysis (TDNA) to find interconnected themes in the BSC curricular documents begins with using CDA to find candidate themes. In parallel, we converted each BSC curriculum document into a linguistic network (Bruun et al., 2009; Andersen, 2017). In such a network, nodes represent words/phrases, and connections (directed links) represent adjacency and order of appearance in the text. For example, the phrase *students should learn* would be represented by three nodes and two arrows: *students -> should -> learn*. Then, in each network,

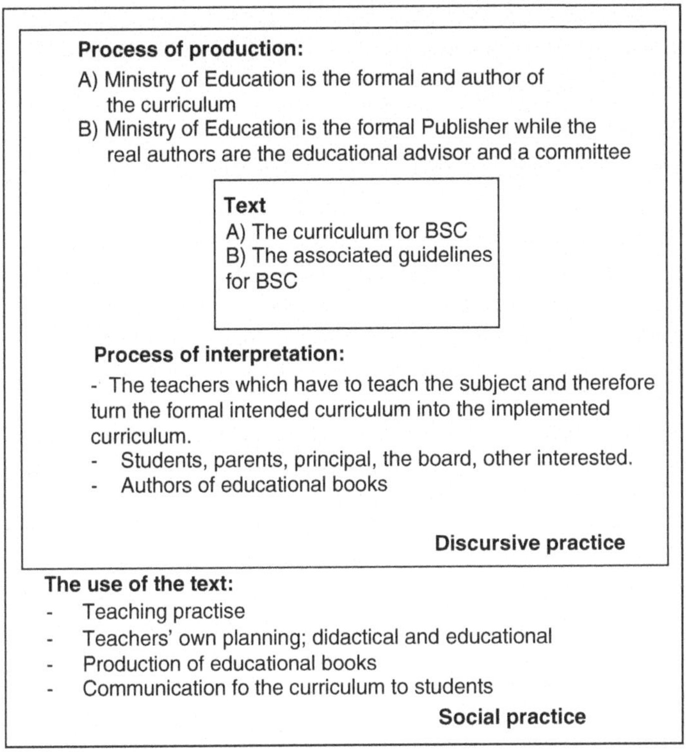

Fig. 8.1 Illustration of the curriculum for the Basic Science Course (BSC) in Fairclough's three-dimensional framework. (From Andersen, 2017)

we used a network clustering algorithm called Infomap (see e.g., Bohlin et al., 2014) to find clusters of words (Andersen, 2017; Bruun et al., 2019). In these networks, words share connections within the cluster of which they are part, but also with words outside of that cluster. On average, though, words within a cluster share more connections than they did with other clusters. Because words shared connections with other clusters, we were able to make maps of how clusters for each linguistic network were related. Thus, we were able to make candidate maps of each iteration of the BSC curriculum document. In these maps, each cluster represented a candidate theme, which could be scrutinized by analysing the connections in the cluster. This enabled a crucial step in the TDNA methodology: comparing and contrasting the candidate themes identified in the network maps with the candidate themes identified using CDA. On the one hand, we found that some CDA themes were not found as themes in the network maps, or we could only make a weak argument of correspondence. On the other, some themes emerged from the network maps, but which were not part of the initial CDA. This led to both changes to network analysis and to new interpretations of the maps in light of the CDA. Having made new networks and network maps, the process was repeated in order to reach alignment between network maps and critical interpretations. The final products were a set of networks, network maps, with clusters representing the themes, and an interpretation of these themes in light of the Danish upper secondary education.

8.4.3 Discussion of Methodological Choices

TDNA is a novel methodology, which merits a discussion so as to stress the methodological choices we made in this study. The methodology can be seen to have three modes: the qualitative analysis (here CDA), the linguistic networks, and the thematic maps. For this study, we chose CDA as the appropriate qualitative framework. For other studies, other qualitative frameworks may be more appropriate (see e.g., Bruun et al., 2019).

It is possible to integrate advances made in the field of text-mining, such as stemming, lemmatization, part of speech tagging, and tokenization (Feldman & Sanger, 2007) in TDNA. These fields rely on large sets of texts to define general rules for, for example, automatically reducing all forms of a word to a single basic form and removing prefixes and suffixes (stemming). However, we argue that in analyzing specialized texts, such as curricular texts for a scientific curriculum, the general rules developed in text-mining literature may cloud or remove aspects of meaning (Bruun et al., 2019). These aspects are important in TDNA and in the iterative development of our understanding of the texts through CDA, networks, and maps. Therefore, any use of these advancements should be monitored with care throughout the process.

We chose to use directed networks in this study, but a case could be made for using undirected networks: If the order of nodes (here, words) in general carries no meaning, then enforcing directionality could induce misleading structures. In using a directed network, we followed Masucci and Rodgers (2006), who both argue and show empirically that the order of words is an important property of human language for creating meaningful sentences. This has practical implications during the analytical process. TDNA involves continuously scrutiny of linguistic networks in relation to the original text. With directional links, it is possible to discern much of the original text in the linguistic network by following links. We argue that this reading of the text in the network provides crucial connections between network representations and original text. The directionality of links is mirrored at the thematic map-level, where it signifies the tendency of words in one module to follow words in another. Bruun et al. (2019) use this to identify how stages of argumentation in a student discussion followed each other. In this study, we do not pursue the meaning of the direction of directed links at the map-level, but refer instead to Andersen (2017) for possible interpretations.

Many network clustering algorithms are non-deterministic and may in some cases produce variations in clusters in subsequent applications of the algorithm. However, Lancichinetti and Fortunato (2009) show that Infomap is a robust choice for clustering. Even so, strategies for addressing variations exist. For example, Bruun and Evans (2020) apply Infomap 1000 times and subsequently use the most frequently appearing set of clusters. We used Bohlin et al.'s (2014) framework to apply Infomap on our linguistic networks numerous times, and did not find variations in clustering for the linguistic networks used to create the final thematic maps.

8.5 Findings

8.5.1 Central Themes in the Analysed Curricular Documents and Their Interpretations

This study revealed 13 different themes across the 3 years. Each theme is constituted by a number of connected words. Network maps show the linguistic connections between themes. Figure 8.2 shows the network map for 2010. Each circle on Fig. 8.2 represents a particular theme as produced by our TDNA. For instance, one theme is represented by the circle with the label *Importance of Science in a Bildung perspective and the BSC*. Each theme has internal network structure based on connections between words in the curriculum. The right-most network on Fig. 8.3 shows the internal network structure of the theme *Importance of Science in a*

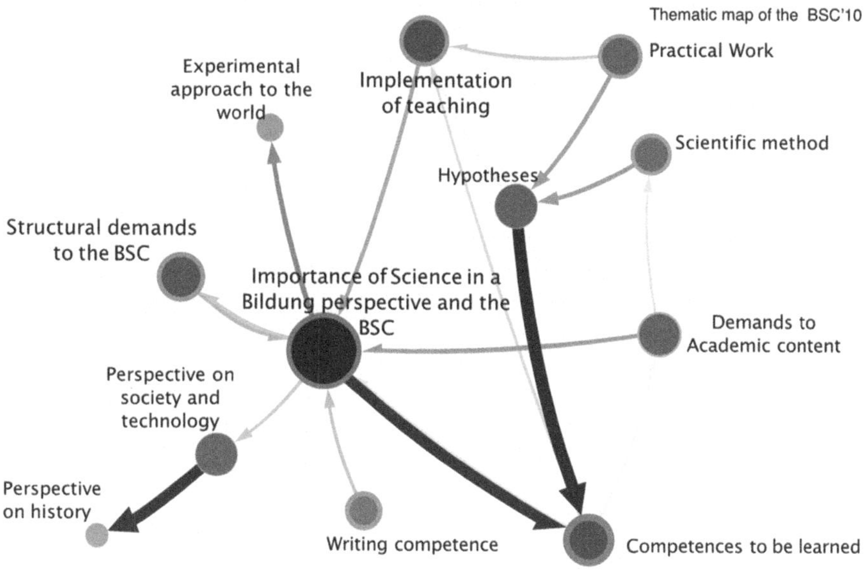

Fig. 8.2 The thematic map for BSC curricular document of 2010. Themes from Table 8.1 in bold. Note that Themes 11 (Importance of Science in a Bildung Perspective) and 13 (BSC Identity as a Course) merged to a single theme in 2010 as explained in Evolution of Themes. Thus, only three themes are in bold in this map. The sizes of circles signify the prevalence of the constituent words in the themes. The sizes of arrows reflect the tendency of words in one cluster to connect to words in another cluster. See Andersen (2017) and Bruun et al. (2019) for details. Visualisation made with MapEquation (see Bohlin et al., 2014)

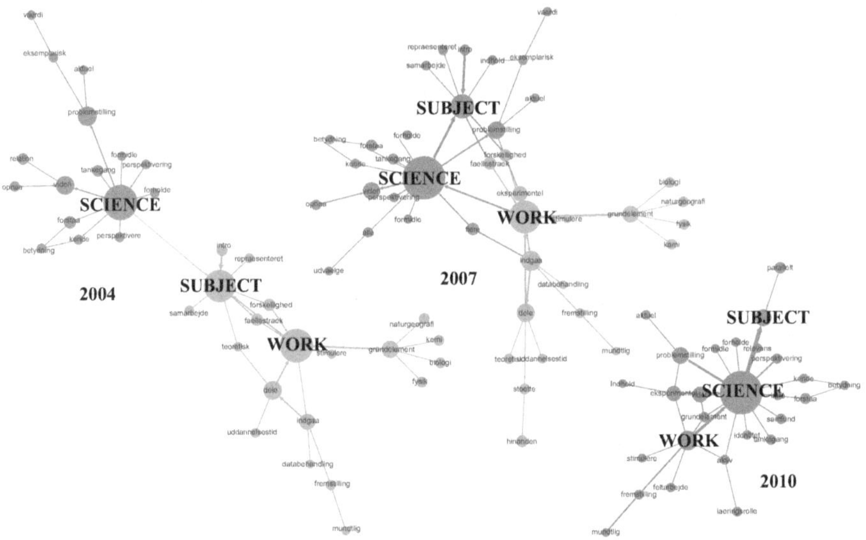

Fig. 8.3 The merger of Themes 11 and 13 over three iterations of the BSC curriculum text. Circles represent words in the curricular texts and arrows their connections. Words represented by purple circles belong to Theme 11, while words represented by green circles belong to Theme 13. Words were assigned to themes based on their connections and using the computer algorithm Infomap (see above). Over time, words in the two themes become more connected and end up in 2010 as one theme. To illustrate how this happens, we have translated the words science, subject, and work from Danish, while remaining Danish words were greyed out. Visualisation made with software Gephi (gephi.org)

Bildung perspective and the BSC, which we identified in 2010. On Fig. 8.3, the circles represent words used in the curricular text, while arrows represent connections. The arrows in Fig. 8.2 represent connections at the word-level between themes. Thus, the light blue arrow from *Implementation of teaching* to *Importance of Science in a Bildung perspective and the BSC* represents, for example, a connection between "completion" in the former theme and "experimental [activities]" in the latter.

As stressed above, the network map shown in Fig. 8.2 as well as the networks on Fig. 8.3 are *only* parts of the output of TDNA. Our interpretations and maps and networks are complementary to each other, and neither can stand alone. To illustrate, we will provide short versions of our interpretations of the overall map of 2010 and of four of the more prevalent and central themes which emerged from our TDNA of the curriculum for years 2004, 2007, and 2010. We briefly show how themes can be connected to Fairclough's discursive and social practice dimension (see Fig. 8.1) by relating them to relevant studies of policy and teacher practice. Our TDNA shows that two of the four themes, Themes 11 and 13, merge over the course of three versions of the curriculum. We illustrate this merger by showing the evolving network structure of the two themes (see Fig. 8.3), while we do not show the network structure for the remaining two themes.

Table 8.1 Themes identified in thematic maps

#	Theme	Summary of interpretations (quotes are our translations)
10	*Structural demands to the BSC*	Focus on student activity, thematic topics and pluridisciplinarity. Sample wording belonging to this theme (2004): *"the BSC implementation should be based on thematic topics which are preferably pluridisciplinary"*
11	*Importance of science in a Bildung perspective*	Focus on developing students as persons who can reflect on science as they take part in society. Sample wording belonging to this theme (2004/2007/2010): *"[students] can express a knowledge-based opinion on issues and problems with a natural sciences aspect."*
12	*Implementation of teaching*	Requirements for implementation. The theme appears in 2007 and 2010. Sample wording (2010): *"[...] observations should be integrated into teaching and choice of themes should make possible [student] completion of [student activities present in other themes]".*
13	*BSC identity as a course*	Constitutes the BSC as a subject, which introduces the natural sciences and scientific methods. Sample wording belonging to this theme (2007): *"introduction to natural science (...) through working with the basic elements of natural science with emphasis on the coherences in natural science."*

In our interpretation, the network map for 2010 shows that the BSC is intended to stage science as part of Bildung: Importance of Science in a Bildung perspective is central, with many arrows pointing to and from it. The arrows that emerge from this central theme can be interpreted as a specification of *how* science should be seen as part of Bildung which entails that students develop an *Experimental Approach to the World* as well as a *Perspective on Society and Technology*. This is then reflected in the *Competences to be Learned.* In terms of teaching, the incoming arrows can be seen as specifications to how to teach. There are *Structural Demands to the BSC*, other demands for *Implementation of Teaching*, *Demands to the Academic Content* of the course, and specified demands for how students should work with their *Writing Competencies.* Interestingly, students' work with the *Scientific Method* and *Hypotheses* are part of the overall conception of the BSC but not directly linked to the Bildung perspective.

Theme 10 (*Structural Demands to the BSC*) represents the preferred way in which the BSC should be taught. The network structure of the theme revealed that teachers are expected to place emphasis on thematic modules. In all three curricula, interdisciplinary aspects are also addressed here. In terms of linking this finding to social practice, the presence of this theme may highlight a tension between curricular intention and teacher practice. Danish teachers who work with thematically structured and interdisciplinary teaching find that, for example, time and collaboration pose obstacles to realising these demands (e.g., Elmeskov et al., 2015).

Theme 11 (*Importance of Science in a Bildung Perspective*) is prevalent in all curricular texts. The focus here is on development of students as persons who can reflect on science as they take part in society. This focus has strong links to both the Danish Bildung perspective as well as scientific literacy discussions (Evans & Dolin, 2018). Political discourse in Denmark emphasises the Bildung aspect in upper secondary schooling, and Danish upper secondary teachers find that this aspect is an important part of their practice (e.g., Dolin et al., 2016).

Theme 10 was connected to Theme 11 in all the analysed curricula. This may signify an intended connection between how the course is taught and how students should develop as human beings. Interestingly, Theme 12 (*Implementation of Teaching*) appears for the first time in 2007 and specifies in more detail, how teaching should be conducted (e.g., making observations, using themes as a basis for experimentation). As such, it could be seen as an elaboration of Theme 10, although only few linguistic connections appeared between the two. Thus, the curriculums seemed open to teacher interpretations of how different student activities were meant to support, for example, interdisciplinarity and the use of thematic topics.

Theme 13 (*BSC Identity as a Course*) conveys how the BSC should be construed as a school subject in Danish upper secondary schools. It should be seen as an introduction to the natural sciences and how these and their methods are related to each other. Linking natural sciences and different methods and models used in natural sciences may have a profound impact on teachers' practice if teachers are used to focus only on one scientific school discipline at time. For example, Andersen (2017) also observed teachers' implementation of the BSC and found that most of the observed teachers predominantly addressed one scientific school discipline in their teaching. Table 8.1 summarises these themes.

8.5.2 Evolutions of Themes

Figure 8.3 shows the merger of Themes 11 and 13 as illustrated by the network structure of the two themes. In 2004 and 2007 they appear as two distinct but connected themes: *Importance of Science in a Bildung Perspective* (Theme 11) and *BSC Identity as a Course* (Theme 13). Notice that there is only one link between the two Themes in 2004 (a green arrow from *science* to *subject*). This is the reason the Infomap algorithm identifies these themes as separate. In 2010, the word *science* is central and connects (among other words) *work* and *subject*. The structure is denser than in 2004 and Infomap identifies the words as one theme. We interpret 2007 as a middle position: there are more connections between words in each theme, but they are still distinct. We have highlighted *science, subject,* and *work*, because these words are highly connected in the network and prevalent in the curricular texts. The merger could be seen as a shift in intentions: Instead of working with different science subjects, science is seen as the overarching subject. This may indicate that the intentions for the BSC develop from an agglomeration of different disciplines to a discipline in its own right.

As seen in Table 8.1, Theme 10 mentions pluridisciplinarity (Jantsch, 1972, p. 15–16). The framing of pluridisciplinarity changes over time. This change is mainly visible as a substitution of and later removal of words: From teaching topics being "*preferably pluridisciplinary*" (2004), to being "*normally pluridisciplinary*" *(2007)* to being "*thematic plurisciplinary topics*" (2010). This can be seen as a move from pluridisciplinarity as a suggestion to becoming the norm to becoming taken for granted.

Theme 12, which first appeared in 2007 gained clearer connections to Practical Work, Hypotheses, Competences to be Learned, and Theme 11+13 in 2010 (See Fig. 8.2). The theme contains national instructions to how teaching should be implemented and the connections to Practical Work and Hypotheses provides further indications of this instruction. For example, practical work should be integrated into teaching as opposed to an appendix to teaching. Also, observations should in some cases be used by students to generate hypotheses. The Danish system relies heavily on teacher autonomy (Evans et al., 2018), such instructions may be seen as decreasing that autonomy.

8.6 Discussion

This study has potential implications for our understanding of the interplay between curriculum texts, practice, and political trends. For instance, one could ask if the merger between Themes 11 and 13 in 2010 aligns with public discussions about the relationship between Bildung and Science in preceding years. In terms of practice, an investigation of teachers' perception of the purpose of the BSC aligns well with the merger between Themes 11 and 13; the BSC is often seen as a Bildung course with emphasis on interdisciplinarity (Dolin et al., 2016). However, we emphasise that the type of interdisciplinarity advocated in the BSC curriculums is pluridisciplinarity which does not necessitate that subjects coordinate with respect to a joint problem (Jantsch, 1972, p. 15–16).

It is important to emphasise that the final products of our TDNA consisted of all three parts: our thematic network maps, linguistic networks, and an interpretation. A complete picture will include taking all these parts into consideration. For example, thematic maps did not show every grain of detail of clusters and their interconnections. Therefore, it is also interesting to ask which themes were not shown in the thematic maps. In our case, one theme was consistently present for all years, but was never prevalent enough to show up on the thematic maps. The theme represented a string of words: *inductive –> teaching principle –> prioritise –> autonomous –> work processes* (translated from Danish). This string of words was connected to the rest of the network only through its connection with the word *teaching (teaching* and *teaching principle* are different words in Danish*).* Since *teaching* is part of Theme 12, this small cluster is connected to *Implementation of Teaching*. In fact, the initial CDA identified this as a subtheme of Theme 10. However, the theme consistently showed up as a separate theme, weakly connected to the rest of the network. The appearance of Theme 12 in 2007 and 2010 may signify that a discursive practice that specifies in more detail what students should do in teaching has gained ground.

8.6.1 Appropriateness of Thematic Maps

Just as was the case with group discussions (Bruun et al., 2019) and keywords analysis (Fairclough et al., 2011), we found that using TDNA revealed themes that were not part of the initial analysis. Furthermore, we made connections between these themes (and their development) and other discursive markers - in this case, other articles, and reports. However, we have not shown that our thematic maps correspond to either the intentions of the curriculum authors or explicit interpretations of teachers. We may speculate that curriculum texts are formulated in official bodies through a process that involves both political intentions as well as educational expertise. This process has to the best of our knowledge not been documented.

Instead, in order to gauge the appropriateness of thematic maps with regards to political intentions, future research could further integrate the methodology with critical discourse analysis to compare their development to political debates about the curriculum or investigate further links with survey data. For example, one could look for evidence that the emergence of *Implementation of Teaching* in 2007 was correlated with confusion about how to implement teaching in this new school subject. In connection to teacher interpretations, maps like these have been used to analyse other curriculums (Elmeskov et al., 2015) and, as mentioned, in teacher professional development (Evans & Dolin, 2018). To gauge whether the maps in this study capture teacher interpretations or could help bridge the gap between official intentions and current teacher practices, future action research could strive to use these maps in trial teacher professional development.

8.7 Conclusion

In this work, we have illustrated how thematic discourse network analysis (TDNA) integrated with critical discourse analysis (CDA) can be used to identify and interpret interconnected themes in three different iterations of a curriculum text. We have also shown how themes may change internally and externally signifying potential differences in political intentions and possible interpretations. Further research may link these potential differences to accounts of political decisions and to teacher implementation.

Acknowledgement This work is supported by Novo Nordisk Foundation grant NNF18OC0052291.

References

Aftale af 28. maj 2003 mellem Regeringen […] om reform af de gymnasiale uddannelser. (2003). http://www.gl.org/uddannelse/udd.politik/Documents/Politisk%20aftale%20-%20reform.doc. Accessed 6 Sept 2020.

Andersen, I. V. K. (2017). *Interdisciplinarity in the basic science course*. Master's thesis, University of Copenhagen, Copenhagen, Denmark. Available from https://www.ind.ku.dk/publikationer/studenterserien/interdiciplinarity-in-the-basic-science-course/

Bohlin, L., Edler, D., Lancichinetti, A., & Rosvall, M. (2014). Community detection and visualization of networks with the map equation framework. In R. Ding & Wolfram (Eds.), *Measuring scholarly impact* (pp. 3–34). Springer.

Bruun, J., & Evans, R. H. (2020). Network analysis of survey data to identify non-homogeneous teacher self-efficacy development in using formative assessment strategies. *Education Sciences, 10*(3), 54. https://doi.org/10.3390/educsci10030054

Bruun, J., Evans, R. H., & Dolin, J. (2009). Investigating scientific literacy documents with linguistic network analysis. In *ESERA; European Science Education Research Association: 2009 Conference* (p. 239). European Science Education Research Association (ESERA).

Bruun, J., Lindahl, M., & Linder, C. (2019). Network analysis and qualitative discourse analysis of a classroom group discussion. *International Journal of Research and Method in Education, 42*(3), 317–339. https://doi.org/10.1080/1743727X.2018.1496414

DME. (2013). *Danish Ministry of Education: Læreplan: Naturvidenskabeligt grundforløb – stx*, juni 2013. https://www.retsinformation.dk/Forms/R0710.aspx?id=152507#Bil45. Accessed 6 Sept 2020.

Dolin, J., & Evans, R. H. (Eds.). (2018). *Transforming assessment: Through an interplay between practice, research and policy*. Springer.

Dolin, J., Jacobsen, L. B., Jensen, S. B., & Johannsen, B. F. (2016). Evaluering af naturvidenskabelig almendannelse i stx-og hf-uddannelserne. *MONA Forskningsrapportserie for matematik- og naturfagsdidaktik, 3*, 1–144.

Elmeskov, D. C., Bruun, J., & Nielsen, J. A. (2015). Evaluering af bioteknologi A som forsøgsfag i stx og htx. *MONA Forskningsrapportserie for matematik- og naturfagsdidaktik, 1*, 1–99.

Evans, R. H., Cross, D., Grangeat, M., Lima, L., Nakhili, N., Rached, E., … Ronnebeck, S. (2018). European educational systems and assessment practice. In J. Dolin & R. Evans (Eds.), *Transforming assessment: Through an interplay between practice, research and policy* (pp. 211–226). Springer.

Evans, R. H., & Dolin, J. (2018). Taking advantage of the synergy between scientific literacy goals, inquiry-based methods and self-efficacy to change science teaching. In O. E. Tsivitanidou, P. Gray, E. Rybska, L. Louca, & C. Constantinou (Eds.), *Professional development for inquiry-based science teaching and learning* (pp. 105–120). Springer.

Fairclough, N. (1992). *Discourse and social change* (Vol. 10). Polity press.

Fairclough, N. (1993). Critical discourse analysis and the marketization of public discourse: The universities. *Discourse & Society, 4*(2), 133–168.

Fairclough, N., Muldering, J., & Wodak, R. (2011). Critical discourse analysis. In T. A. Van Diejk (Ed.), *Discourse studies: A multidisciplinary introduction* (pp. 258–284). Sage.

Feldman, R., & Sanger, J. (2007). *The text mining handbook: Advanced approaches in analyzing unstructured data*. Cambridge university press.

Jantsch, E. (1972). Inter- and transdisciplinary university: A systems approach to education and innovation. *Higher Education Quarterly, 1*(1), 7–37.

Johnson, R. B., & Onwuegbuzie, A. J. (2004). Mixed methods research: A research paradigm whose time has come. *Educational Researcher, 33*(7), 14–26.

Lancichinetti, A., & Fortunato, S. (2009). Community detection algorithms: A comparative analysis. *Physical Review E, 80*(5), 056117.

Masucci, A. P., & Rodgers, G. J. (2006). Network properties of written human language. *Physical Review E, 74*(2), 026102.

Mehler, A., Lücking, A., Banisch, S., Blanchard, P., & Job, B. (Eds.). (2016). *Towards a theoretical framework for analyzing complex linguistic networks*. Springer.

Minner, D. D., Levy, A. J., & Century, J. (2010). Inquiry-based science instruction—What is it and does it matter? Results from a research synthesis years 1984 to 2002. *Journal of Research in Science Teaching: The Official Journal of the National Association for Research in Science Teaching, 47*(4), 474–496.

Priestley, M., & Biesta, G. (Eds.). (2013). *Reinventing the curriculum: New trends in curriculum policy and practice*. Bloomsbury.

Ropohl, M., Nielsen, J. A., Olley, C., Rönnebeck, S., & Stables, K. (2018). The concept of competence and its relevance for science, technology and mathematics education. In J. Dolin & R. Evans (Eds.), *Transforming assessment* (pp. 3–25). Springer.

Schmidt, W. H., McKnight, C. C., Valverde, G., Houang, R. T., & Wiley, D. E. (Eds.). (1997). *Many visions, many aims: A cross-national investigation of curricular intentions in school mathematics* (Vol. 1). Kluwer.

Akker, J. van den, Fasoglio, D., & Mulder, H. (2010). *A curriculum perspective on plurilingual education*. Online resources by the Council of Europe. http://www.oerp.ir/sites/default/files/parvande/SLO_persp2010_EN.pdf. Accessed 6 Sept 2020.

Verbiest, E., & Erèulj, J. (2006). Building capacity in schools—dealing with diversity between schools. In M. Pol (Ed.), *Dealing with diversity—a key issue for educational management* (pp. 65–80). Brno, Czech Republic: Masaryk University.

Part II
Expanding the Evidence Base

Chapter 9
Developmental Patterns of Students' Understanding of Core Concepts in Secondary School Chemistry

Sascha Bernholt and Lars Höft

9.1 Introduction

The acquisition of a sound scientific understanding is a central goal of science education in school. In recent years, researchers have recommended to focus teaching on core concepts that are fundamental to the discipline, in contrast to teaching a broad range of different topics and content aspects (Bransford et al., 2000). This recommendation addresses a long-standing criticism that science teaching often only provides students with rather superficial knowledge of a broad range of discrete facts and distinct topics whose interconnections might not be obvious for the students (Duschl et al., 2011).

Educational reforms in many countries have adopted this focus and have established educational standards that are increasingly structured by means of central subject-specific concepts or scientific practices, which are to form the focus of teaching activities in class (cf. Bernholt et al., 2012). However, empirical research on students' understanding repeatedly indicates that students struggle to acquire a sophisticated understanding even of the most fundamental science concepts, for instance the structure and composition of matter (Hadenfeldt et al., 2016), chemical reactions (Yan & Talanquer, 2016), or energy (Neumann et al., 2012).

Only few studies have followed students over longer time periods, so that little is known about the development of students' conceptual understanding across secondary school (Emden et al., 2018; Löfgren & Helldén, 2009). However, numerous studies have shown that the learning of (normative) scientific concepts is highly dependent on the pupils' existing ideas (Duit & Treagust, 1998); and these ideas

S. Bernholt (✉) · L. Höft
Department of Chemistry Education, Leibniz Institute for Science and Mathematics Education, Kiel, Germany
e-mail: bernholt@leibniz-ipn.de

O. Levrini et al. (eds.), *Engaging with Contemporary Challenges through Science Education Research*, Contributions from Science Education Research 9,
https://doi.org/10.1007/978-3-030-74490-8_9

prove to be extremely robust in relation to teaching interventions (Allen, 2010; Garnett et al., 1995).

In recent years, research on the interplay between student conceptions, instruction, and the learning of scientific concepts and practices has attracted increasing attention under the label of *learning progressions* (Bernholt & Sevian, 2018; Duschl et al., 2007; Ford, 2015). Learning progressions describe models of teaching and learning in a specific domain over a longer period of time—ideally several school years (Alonzo & Gotwals, 2012). These models aim to describe increasingly complex levels of understanding in a domain and how students can be best supported to acquire these higher levels (Duschl et al., 2007).

In order to investigate how students progress in their understanding of a particular concept, items are needed that can elicit students' level of understanding of this concept. With regard to assessment, artifacts from students' learning in the classroom (e.g., students' drawings; Krajcik et al., 2012), interviews (Yan & Talanquer, 2016), or open-ended tasks (Chen et al., 2016) are considered to provide valid information about students' understanding. However, these approaches are costly both with respect to time and money when it comes to investigating larger samples of students. Here, such studies often use multiple-choice items—sometimes at the cost of validity.

Briggs, Alonzo, Schwab, and Wilson (2006) suggested to combine the efficiency of classical multiple-choice items with the validity of open-ended items by means of so called ordered multiple-choice (OMC) items. These OMC items provide several response options, of which one is correct and corresponds to the highest level of understanding that can be assessed with this item. The remaining answer options represent either a scientifically inappropriate or an incomplete understanding (e.g. by addressing common alternative conceptions) and correspond to specific lower levels of understanding, based on theoretical or empirically validated models of understanding a particular concept. Hence, two fundamental assumptions of this item format are that students systematically employ one predominant theory or mental model across related problem contexts (Steedle & Shavelson, 2009) and that these theories or mental models can be ranked in terms of their sophistication (Alonzo & Gotwals, 2012; Briggs et al., 2006). There are some empirical findings that OMC items have the potential to assess the level of conceptual understanding of students almost as well as interviews and that learning gains can be made visible by students' shifts in applying theories or mental models that are increasingly sophisticated (Alonzo & Steedle, 2009; Chen et al., 2016). However, most studies are based on cross-sectional or pre-post data, thus providing limited insights into patterns on the individual level over longer timespans (Duncan & Gotwals, 2015; Taber, 2017).

In the present study, we scrutinize students' conceptual understanding of three fundamental concepts of chemistry education: the structure and composition of matter (covering both substance-particle and structure-property relationships), chemical reaction, and energy. These so-called basic concepts are integral part of the curriculum (for the participating schools) and are intended to guide teachers in imparting increasing levels of understanding and thus to foster the progression of cumulative knowledge acquisition over the course of secondary school chemistry (KMK, 2004). In this study, the progression of students' conceptual understanding is assumed to follow subsequent levels of sophistication, ranging from (0) everyday ideas, (1) hybrid ideas (particles in a continuum etc.), (2) simple, Daltonian particle ideas, (3) differentiated particle ideas, to (4) systemic particle ideas (intra- and inter-molecular interactions) (in case of the concepts matter and chemical reaction; for details cf. the systematic review by Hadenfeldt et al., 2014) or (0) everyday ideas, (1) forms and sources of energy, (2) energy transfer, (3) energy degradation, and (4) energy conservation (in case of the energy concept; for details cf. the literature review by Neumann et al., 2012). For instance, students are expected to start without the use of a particle concept when explaining a specific phenomenon like evaporation. Later, students learn about particles in school, but sometimes integrate the idea of particles and the perception of matter as continuous (e.g., in the form of hybrid models of particles as entities embedded in matter), before they understand particles as the building blocks of matter (with nothing between the particles). This understanding becomes more sophisticated by using a differentiated atom model and by its application to explain interactions in a system of particles (cf. Andersson, 1990; Liu & Lesniak, 2006; Stevens et al., 2010). Building up upon these theoretical assumptions, the research question of the present study aims at examining to which extend students' understanding of the three concepts (structure and composition of matter, chemical reaction, and energy) actually increases over the course of secondary school and which developmental patterns can be identified on the individual level across this timespan.

9.2 Methods

In order to investigate the conceptual understanding of a diverse sample, we combined classical multiple-choice items and ordered multiple-choice items (OMC; Briggs et al., 2006) in a ratio of 1:2 per test form. While multiple-choice items were included to cover factual knowledge of chemical terms, definitions, and principles (that are considered to be an integral part of students' conceptual understanding), the answer options of the OMC items were designed to differentiate between five levels of understanding a particular concept (cf. the description provided above; Hadenfeldt et al., 2014; Neumann et al., 2012). Based on a multi-step item development procedure, questions and answer options were selected, reviewed by an expert panel of chemistry education researchers and teachers, piloted (both quantitatively and by think-aloud interviews), and compiled into test forms by a common item anchor design with both grade-specific and common (i.e., across-grade) items. Each participating student answered 30 items (ten for each of the three concepts) addressing varying chemical phenomena out of a pool of 114 items. Students' answers to multiple-choice items were coded as incorrect (0) or correct (1), or, in case of OMC items, answers were assigned partial credits from 0 to 4 in correspondence to the theoretical model. The highest level of understanding assessed varies across items, with a focus on levels 1 to 3 (i.e., aiming to discriminate between two to four levels of understanding), as we expected most students to have a medium level of understanding (Alonzo & Steedle, 2009; Hadenfeldt et al., 2016). The highest level is addressed only by two to three items per concept and test form, as prior studies indicate that most students struggle to acquire this level of sophistication in understanding a particular concept (Hadenfeldt et al., 2016; Neumann et al., 2012), but the full range of levels is covered by each set of items presented to students of a particular grade.

9.2.1 Sample

In total, $N = 3299$ students attending grades 5 to 12 in Germany participated in the project. Students' in grades 5 (denoted as cohort LS5; $n = 724$) and 9 (denoted as US9; $n = 769$) were then followed for 3 years (i.e., four measurement waves) and are in focus of the present analyses. All students were attending the Gymnasium (the highest track of secondary schools in Germany, preparing students for higher studies). The five schools that were willing to participate in this study represent a convenience sample of schools. Within these schools, all classes were included in the data collection.

9.2.2 Data Analysis

To account for the matrix design of the test (i.e., not all items having been administered to all students), we utilized Item Response Theory (IRT) to analyse the data. A three-dimensional multi-group generalized partial credit model (MG-GPCM) was used for analysing the coded results of student answers at each measurement occasion, with students' grade (from 5 to 12) as grouping variable. As it is possible for students to opt-out from chemistry courses after grade 10, data from students who did not attend chemistry courses in a particular grade were excluded from the analysis. The expected a posteriori based on plausible values reliability (EAP/PV) was calculated and found to range from .68 to .84 for the different concepts and measurement waves, indicating a sufficiently high reliability of the measured construct. Also, estimates of item fit (Weighted Mean Square estimates and standardized mean-square fit statistics) were considered and found to meet typical cut-off values (WMNSQ from 0.7 to 1.3; Wright & Linacre, 1994). Finally, weighted likelihood estimates (WLE) were calculated as a measure of students' understanding of each concept, as measured by the test. IRT parameter estimates of the four measurement waves were then linked based on a generalisation of the log-mean-mean linking in accordance with Haberman (2009) to obtain the same metric across grade levels. While this approach allows test results to be compared between students and over time, the metric at the first measurement point is arbitrarily centred at mean of

0, so that an interpretation of the values in terms of the assumed theoretical levels of understanding on which the item design was based and, thus, of individual or medium levels of understanding is not possible.

With regard to analysing trends over time, two separate lines of analysis were conducted. First, we were interested in the development of students' understanding of the three concepts over time. Based on students' WLE scores for each of the concepts at each measurement wave, unconditional True Individual Change Models (TICM; Steyer et al., 2014) and Latent Change Score Models (LCSM; Little et al., 2014) were analysed (single-indicator approach). In both models, a difference score is modelled as a latent variable that indicates, for each student, the difference between his/her understanding of a particular concept at one measurement occasion and the previous occasion. The models thus allow the analysis of the latent mean course (TICM, focusing on comparing students' obtained *scores* from grade to grade) as well as the analysis of the stability of the change of students' understanding (LCSM, focusing on the degree of *changes* in students' scores from grade to grade) over time.

Second, we then analysed the obtained item difficulty parameters of the OMC items as a function of the levels of understanding, based on the theoretical models mentioned above. The aim of this analysis was to see if the levels of understanding (which were used to construct the different answer options in the OMC items) indeed represent a hierarchy as hypothesized by the theoretical underpinning (Hadenfeldt et al., 2013; Steedle & Shavelson, 2009). For this purpose, so-called Thurstonian thresholds were calculated for the different answer options in the OMC items. These thresholds indicate the difficulty values for preferring an answer option of a higher level of understanding (with a probability of >50%) over the next lower level for each item. In addition, the longitudinal data were used to investigate whether students prefer answers on a particular conceptual level across the items at a particular measurement wave and whether this preference shifts towards higher levels over time.

9.3 Results

To investigate the development of students' understanding of the three concepts over time, TICMs and LCSMs were used. Fit indices indicate a good fit of these models to the data (CFIs > .94; SRMRs < .034; RMSEAs < .075; Hu & Bentler, 1999). The mean values of students' obtained scores (based on TICMs) indicate substantial learning gains in each concept for both cohorts (cf. Fig. 9.1). Here, learning gains are generally lower in the lower grades (cohort LS5), as expectable. Across grades 5 to 8, students show the highest progress in understanding the structure and composition of matter, while gains in understanding the concepts of chemical reaction and energy are about half the size (cf. Table 9.1). In grades 9 to 12, the TICMs indicate a continuation in students' learning from year to year, especially

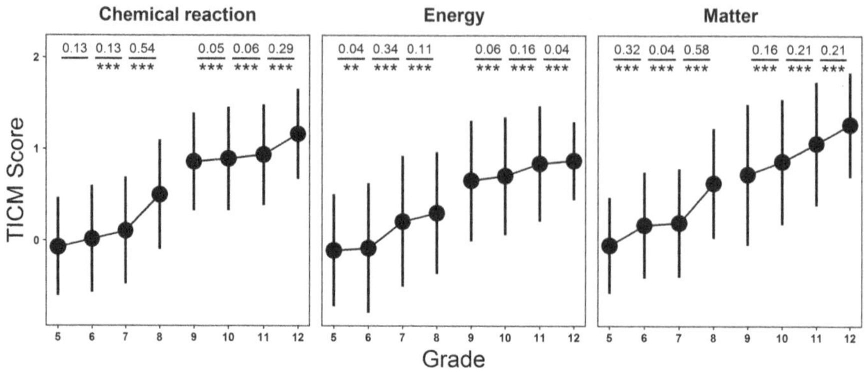

Fig. 9.1 Mean values and standard deviations of students' understanding of the concepts chemical reaction, energy, and matter across the grades 5 to 8 (cohort LS5) and 9 to 12 (cohort US9), respectively, and effect sizes (Cohen's *d*) of year-to-year learning gains, calculated using True Individual Change Score models; . $p > .05$, $^{*}p < .05$, $^{**}p < .01$, $^{***}p < .001$

Table 9.1 Mean values (M) of the True Individual Change Score Models (TICM) and mean difference scores (ΔM) with corresponding *p*-value level of the Latent Change Score Models (LCSM) for the two cohorts LS5 (grades 5 to 8) and US9 (grades 9 to 12), the three concepts (*CR* chemical reaction, *EN* energy, *MA* matter), and the four measurement waves (T1 to T4)

	Cohort	Lower Secondary (LS5; grades 5 to 8)			Upper Secondary (US9; grades 9 to 12)		
	Concept	CR	EN	MA	CR	EN	MA
TICM	*M* (T1)	−0,07	−0,109	−0,048	0,858	0,655	0,724
	M (T2)	0,016	−0,082	0,167	0,893	0,707	0,863
	M (T3)	0,106	0,210	0,195	0,935	0,842	1062
	M (T4)	0,500	0,301	0,627	1165	0,871	1268
LCSM	*ΔM* (T1-T2)	0,086	0,025*	0,216***	0,037***	0,049***	0,135***
	ΔM (T2-T3)	0,090***	0,293***	0,027***	0,042***	0,135***	0,198***
	ΔM (T3-T4)	0,391***	0,094***	0,431***	0,242***	0,035***	0,214***

$^{*}p < .05$, $^{**}p < .01$, $^{***}p < .001$

with regard to the concept of matter, for which students show the highest mean scores across the three concepts.

Based on the LCSMs, difference scores were modelled as a latent variable to capture the year-by-year changes in students' understanding of a particular concept between two consecutive measurement points on the individual level. *P*-values for the mean difference scores indicate that these mean value differences are significantly different from zero in both cohorts for all three concepts and all four measurement waves, with the only exception of students' understanding of the concept chemical reaction when transitioning from grade 5 to grade 6 (cf. Table 9.1).

While the results in Table 9.1 indicate significant positive learning gains for students' understandings of the three concepts in both cohorts, it is important to note that this analysis focuses on the mean values of the two cohorts within each grade.

When inspecting the range of students' individual difference scores for each concept and cohort, a substantial portion of students actually have negative latent change scores (i.e., <0), which indicates a regress in their understanding of particular concepts when transitioning to higher grades (Fig. 9.2).

When categorizing students' difference scores, whether these are positive (i.e., indicating positive learning gains) or negative (i.e., indicating a regress), it can be seen that about 16 to 46% of students in cohort LS5 ($M = 33\%$) and about 18 to 43% of students in cohort US9 ($M = 31\%$) show negative learning gains between two consecutive grade levels in one of the three concepts. When further considering students' individual trajectories across the four measurement waves, only 20 to 30% of students reveal positive learning gains in all three grade transitions, within a particular concept (cf. Fig. 9.3 for cohort US9). The numbers differ slightly between concepts and cohorts, but the overall pattern of the distribution of students according to when and how often positive learning gains are detected when transitioning from grade to grade in this analysis is strikingly similar, across cohorts and concepts.

The second line of analysis focuses on the OMC items included in the tests. Here, we intended to analyse the obtained item difficulty parameters as a function of the levels of understanding, based on the theoretical models that were used in the item design. By analysing the Thurstonian thresholds of the MG-GPCM model (i.e., the difficulty values for preferring an answer option of a higher level of understanding (with >50%) over the next lower level for each item), 47% (Chemical reaction), 18% (Energy), and 47% (Matter) of the variance in the thresholds can be accounted for by the respective level of understanding, depending on the chemical concept (Fig. 9.4).

Accordingly, the Thurstonian thresholds follow on average a pattern from easier to harder levels, in correspondence to the assumed theoretical levels of understanding (cf. Hadenfeldt et al., 2013; Neumann et al., 2012). However, when analysing

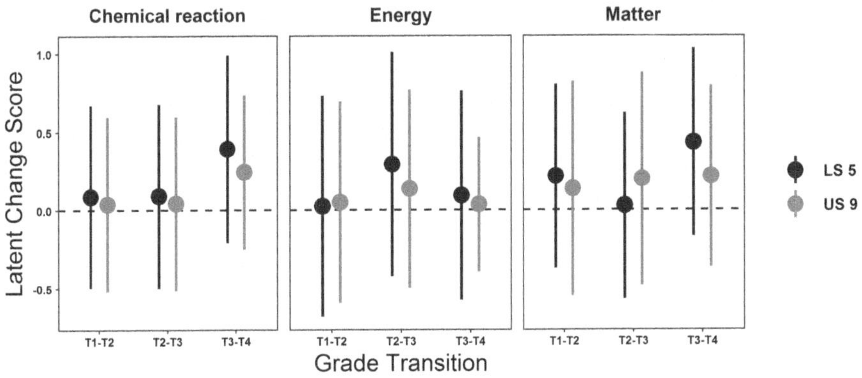

Fig. 9.2 Mean values and standard deviations of students' individual latent difference scores between subsequent grades for the concepts chemical reaction, energy, and matter across the grades 5 to 8 (LS5) and 9 to 12 (cohort US9), respectively, calculated using Latent Change Score models

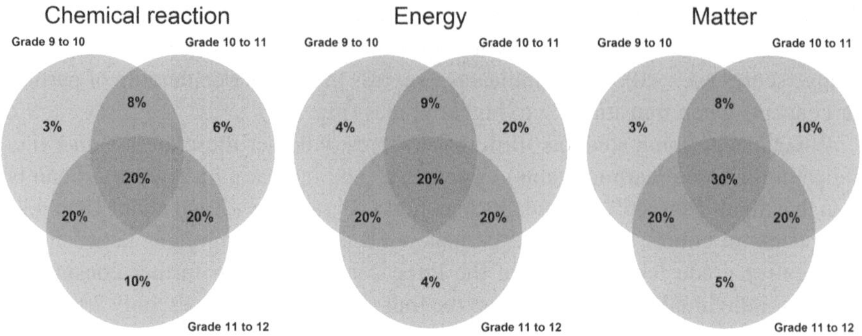

Fig. 9.3 Venn diagrams illustrating the distribution of students in cohort US9 according to whether their LCSM difference score indicates positive learning gains in their understanding of the concepts chemical reaction (left), energy (middle), and matter (right) only in one transition (numbers in one of the circles), in two transitions (numbers in the intersection of two circles), or in all three transitions (numbers in the intersection of all three circles) between the grades 9 to 12. Numbers for students in cohort LS5 are comparable

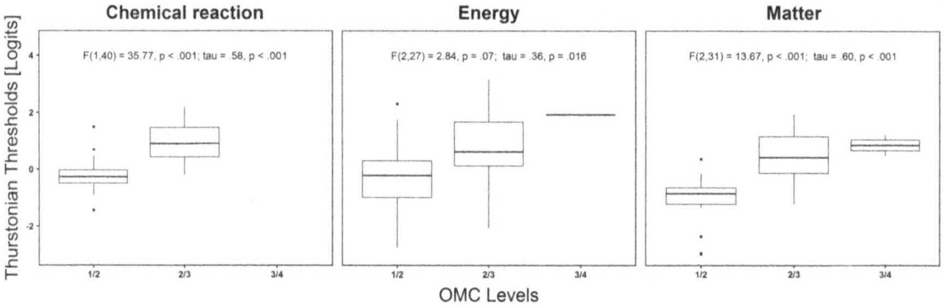

Fig. 9.4 Distribution of the Thurstonian thresholds over level of understanding required to select a particular answer option of the OMC items, within each of the three concepts

students' answer options across items within each measurement wave, in most cases no stable preference for a particular level of understanding was detectable. Even with a rather liberal criterion of selecting an answer option at a particular level of understanding in at least 2/3 of all items, only about 33% (ranging from 21% to 48% depending on grade and concept) of the students indicated a preference for a particular level of understanding (Steedle & Shavelson, 2009).

9.4 Discussion

The analysis of students' overall performance indicates substantial learning gains from grade to grade, with small to medium effect sizes. The magnitude of these annual learning gains are comparable to other longitudinal findings (cf. Bloom et al., 2008). However, the differentiated analysis of three fundamental chemistry concepts indicates that students' progress in understanding these three concepts does not follow a linear trajectory, but reveals both stagnation and gains in different years over the course of the 4 years covered by this study. These years of stagnation and gains can often be traced back to emphases in the timetable or the curriculum of the participating schools (Bernholt et al., 2020). The years of stagnation, however, raise the question whether the current curriculum and how it is enacted in the schools actually support students' cumulative learning or whether this finding rather indicates substantial room for improvement.

This concern is reinforced by the analysis on the individual level. While the aggregated mean values in the two cohorts at group level indicate at least a constant or even increasing understanding, a significant proportion of the students actually shows regress in their performance development. In both cohorts, i.e. in grades 5 to 8 and grades 9 to 12, only about 20 to 30% of the students follow the often assumed learning path of continuous annual learning gains. The majority of students, however, show a regress in their conceptual understanding at least once during their participation in this study. Although students participated voluntarily in this study and were repeatedly informed that they could dropout at any time without consequences, the motivation to repeatedly take additional (no-stakes) chemistry tests was certainly limited for some students. However, a continuous decrease in all three concepts was only observed for 3% (LS5) and 8% (US9) of the students, respectively.

Hence, while general positive trends are observable in the broader sample, the trajectories of individual students show much more variation than commonly expected. Aside from factors related to the study context (e.g., test taking motivation) that might account for some of this variation, further factors seem to influence students' performance over time. Motivation, triggers in the items, or context features that correspond to a student's learning history might play a role (Sevian & Couture, 2018) and need further attention. In addition, students' learning paths are certainly influenced by the sequences and logic of concepts in instruction, which was not taken into account in this study.

Overall, this study provides little evidence that students' trajectories of their conceptual understanding reflect repeated and continuous cumulative learning opportunities. When further taking into account students' answer patterns in the OMC items, it seems that the assumption of stability in performance is also challenged within measurement waves. With regard to the assumption underlying the design of the OMC items, the results at group level support the presumption here: Students' progression across grades indeed follows, on average, the hierarchy of ordered levels as postulated by the theoretical models (cf., Hadenfeldt et al., 2016). When conducting a more fine-grained analysis based on longitudinal data, however, it does not seem that students actually progress through these levels step-by-step in developing an understanding of the three concepts covered in the test. Also, students' answer patterns often do not indicate the stable preference for answer options of a certain level of sophistication, as it is generally and often implicitly assumed (Steedle & Shavelson, 2009). Rather, the majority of students does not perform comparably across item contexts (diSessa & Wagner, 2006; Fischer & Bidell, 2006). However, more fine-grained analyses are needed to identify and characterize the intermediate states students' actually pass over time (cf. Zabel & Gropengiesser, 2011).

A major constraint of the analysis presented here is the absence of proximal information about teaching and learning processes in the classrooms of the participating students. Information of the quantity and quality of learning opportunities related to the three concepts would provide valuable insights into how these opportunities affect the development of students' conceptual understanding over time and to better explain deviations of students' trajectories from the assumed theoretical progression, as documented in this study. Incorporating both student and class-level variables would also require a multilevel approach (as students in the same class, in most cases, would have access to the same learning opportunities). Due to the limited number of classes participating in this study and, especially, due to the varying class composition in the upper secondary cohort (US9), a multilevel analysis was not feasible to implement in the present analysis so that we were not able to take the impact of this data structure into account.

With regard to identifying students' individual trajectories of learning a particular concept, annual measurement waves are certainly too coarse-grained to identify nuanced patterns. While the variation in students' performance across time and contexts has been addressed both theoretically (e.g., diSessa & Wagner, 2006) and empirically (e.g., Amaral et al., 2018), even the quantitative approach taken in this study indicates a substantial amount of fluctuation, particularly when taking the individual level into account.

Acknowledgement This research was funded by the Swedish Research Council under grant number 721-2013-2180.

References

Allen, M. (2010). Learner error, affectual stimulation, and conceptual change. *Journal of Research in Science Teaching, 47*(1), 151–173. https://doi.org/10.1002/tea.20302

Alonzo, A. C., & Gotwals, A. W. (Eds.). (2012). *Learning progressions in science: Current challenges and future directions*. Sense Publishers.

Alonzo, A. C., & Steedle, J. T. (2009). Developing and assessing a force and motion learning progression. *Science Education, 93*(3), 389–421. https://doi.org/10.1002/sce.20303

Amaral, E. M. R. D., Ratis Tenório da Silva, J. R., & Sabino, J. D. (2018). Analysing processes of conceptualization for students in lessons on substance from the emergence of conceptual profile zones. *Chemistry Education Research and Practice, 19*(4), 1010–1028. https://doi.org/10.1039/C8RP00050F

Andersson, B. (1990). Pupils' conceptions of matter and its transformations (age 12–16). *Studies in Science Education, 18*(1), 53–85. https://doi.org/10.1080/03057269008559981

Bernholt, S., & Sevian, H. (2018). Learning progressions and teaching sequences – Old wine in new skins? *Chemistry Education Research and Practice, 19*(4), 989–997. https://doi.org/10.1039/C8RP90009D

Bernholt, S., Neumann, K., & Nentwig, P. (Eds.). (2012). *Making it tangible – Learning outcomes in science education*. Waxmann.

Bernholt, S., Höft, L., & Parchmann, I. (2020). Die Entwicklung fachlicher Basiskonzepte im Chemieunterricht – Findet ein kumulativer Aufbau im Kompetenzbereich Fachwissen statt? *Unterrichtswissenschaft, 48*(1), 35–59. https://doi.org/10.1007/s42010-019-00065-4

Bloom, H. S., Hill, C. J., Black, A. R., & Lipsey, M. W. (2008). Performance trajectories and performance gaps as achievement effect-size benchmarks for educational interventions. *Journal of Research on Educational Effectiveness, 1*(4), 289–328. https://doi.org/10.1080/19345740802400072

Bransford, J. D., Brown, A. L., & Cocking, R. R. (2000). *How people learn: Brain, mind, experience and school*. National Academy Press.

Briggs, D. C., Alonzo, A. C., Schwab, C., & Wilson, M. (2006). Diagnostic assessment with ordered multiple-choice items. *Educational Assessment, 11*(1), 33–63. https://doi.org/10.1207/s15326977ea1101_2

Chen, J., Gotwals, A. W., Anderson, C. W., & Reckase, M. D. (2016). The influence of item formats when locating a student on a learning progression in science. *International Journal of Assessment Tools in Education, 3*(2), 101–122. https://doi.org/10.21449/ijate.245196

DiSessa, A. A., & Wagner, J. F. (2006). What coordination has to say about transfer. In J. P. Mestre (Ed.), *Current perspectives on cognition, learning, and instruction. Transfer of learning from a modern multidisciplinary perspective* (pp. 121–154). IAP.

Duit, R., & Treagust, D. (1998). Learning in science: From behaviourism towards social constructivism and beyond. In B. J. Fraser & K. G. Tobin (Eds.), *International handbook of science education* (pp. 3–25). Kluwer.

Duncan, R. G., & Gotwals, A. W. (2015). A tale of two progressions: On the benefits of careful comparisons. *Science Education, 99*(3), 410–416. https://doi.org/10.1002/sce.21167

Duschl, R. A., Schweingruber, H. A., & Shouse, A. W. (Eds.). (2007). *Taking science to school: Learning and teaching science in Grades K-8*. Nation Academic Press.

Duschl, R. A., Maeng, S., & Sezen, A. (2011). Learning progressions and teaching sequences: A review and analysis. *Studies in Science Education, 47*(2), 123–182. https://doi.org/10.1080/03057267.2011.604476

Emden, M., Weber, K., & Sumfleth, E. (2018). Evaluating a learning progression on 'Transformation of Matter' on the lower secondary level. *Chemistry Education Research and Practice, 19*(4), 1096–1116. https://doi.org/10.1039/C8RP00137E

Fischer, K. W., & Bidell, T. R. (2006). Dynamic development of action, thought, and emotion. In W. Damon & R. M. Lerner (Eds.), *Theoretical models of human development. Handbook of child psychology* (6th ed., pp. 313–399). Wiley.

Ford, M. J. (2015). Learning progressions and Progress: An introduction to our focus on learning progressions. *Science Education, 99*(3), 407–409. https://doi.org/10.1002/sce.21169

Garnett, P. J., Garnett, P. J., & Hackling, M. W. (1995). Students' alternative conceptions in chemistry: A review of research and implications for teaching and learning. *Studies in Science Education, 25*(1), 69–96. https://doi.org/10.1080/03057269508560050

Haberman, S. J. (2009). *Linking parameter estimates derived from an item response model through separate calibrations* (ETS Research Report: ETS RR-09-40). ETS.

Hadenfeldt, J. C., Bernholt, S., Liu, X., Neumann, K., & Parchmann, I. (2013). Using ordered multiple-choice items to assess students' understanding of the structure and composition of matter. *Journal of Chemical Education, 90*(12), 1602–1608. https://doi.org/10.1021/ed3006192

Hadenfeldt, J. C., Liu, X., & Neumann, K. (2014). Framing students' progression in understanding matter: A review of previous research. *Studies in Science Education, 50*(2), 181–208. https://doi.org/10.1080/03057267.2014.945829

Hadenfeldt, J. C., Neumann, K., Bernholt, S., Liu, X., & Parchmann, I. (2016). Students' progression in understanding the matter concept. *Journal of Research in Science Teaching, 53*(5), 683–708. https://doi.org/10.1002/tea.21312

Hu, L.-t., & Bentler, P. M. (1999). Cutoff criteria for fit indexes in covariance structure analysis: Conventional criteria versus new alternatives. *Structural Equation Modeling: A Multidisciplinary Journal, 6*(1), 1–55. https://doi.org/10.1080/10705519909540118

KMK. (2004). *Bildungsstandards im Fach Chemie für den mittleren Schulabschluss*. Sekretariat der Ständigen Konferenz der Kultusminister der Länder in der Bundesrepublik Deutschland.

Krajcik, J. S., Sutherland, L. M., Drago, K., & Merritt, J. (2012). The promise and value of learning progression research. In S. Bernholt, K. Neumann, & P. Nentwig (Eds.), *Making it tangible – Learning outcomes in science education* (pp. 261–283). Waxmann.

Little, T. D., Schnabel, K. U., & Baumert, J. (Eds.). (2014). *Modeling longitudinal and multilevel data: Practical issues, applied approaches, and specific examples*. Psychology Press.

Liu, X., & Lesniak, K. (2006). Progression in children's understanding of the matter concept from elementary to high school. *Journal of Research in Science Teaching, 43*(3), 320–347. https://doi.org/10.1002/tea.20114

Löfgren, L., & Helldén, G. (2009). A longitudinal study showing how students use a molecule concept when explaining everyday situations. *International Journal of Science Education, 31*(12), 1631–1655. https://doi.org/10.1080/09500690802154850

Neumann, K., Viering, T., Boone, W. J., & Fischer, H. E. (2012). Towards a learning progression of energy. *Journal of Research in Science Teaching, 50*, 162–188. https://doi.org/10.1002/tea.21061

Sevian, H., & Couture, S. (2018). Epistemic games in substance characterization. *Chemistry Education Research and Practice, 19*(4), 1029–1054. https://doi.org/10.1039/C8RP00047F

Steedle, J. T., & Shavelson, R. J. (2009). Supporting valid interpretations of learning progression level diagnoses. *Journal of Research in Science Teaching, 46*(6), 699–715. https://doi.org/10.1002/tea.20308

Stevens, S. Y., Delgado, C., & Krajcik, J. S. (2010). Developing a hypothetical multi-dimensional learning progression for the nature of matter. *Journal of Research in Science Teaching, 47*(6), 687–715. https://doi.org/10.1002/tea.20324

Steyer, R., Partchev, I., & Shanahan, M. J. (2014). Modeling true intraindividual change in structural equation models: The case of poverty and children's psychological adjustment. In T. D. Little, K. U. Schnabel, & J. Baumert (Eds.), *Modeling longitudinal and multilevel data: Practical issues, applied approaches, and specific examples*. Psychology Press.

Taber, K. S. (2017). Researching moving targets: Studying learning progressions and teaching sequences. *Chemistry Education Research and Practice, 18*(2), 283–287. https://doi.org/10.1039/C7RP90003A

Wright, B. D., & Linacre, M. J. (1994). Reasonable mean-square fit values. *Rasch Measurement Transactions, 8*(3), 370.

Yan, F., & Talanquer, V. (2016). Students' ideas about how and why chemical reactions happen: Mapping the conceptual landscape. *International Journal of Science Education, 37*(18), 3066–3092. https://doi.org/10.1080/09500693.2015.1121414

Zabel, J., & Gropengiesser, H. (2011). Learning progress in evolution theory: Climbing a ladder or roaming a landscape? *Journal of Biological Education, 45*(3), 143–149. https://doi.org/10.1080/00219266.2011.586714

Chapter 10
Learning Evolution – A Longterm Case-Study with a Focus on Variation and Change

Martin Scheuch, Jaqueline Scheibstock, Heidemarie Amon, Gerald Fuchs, and Christine Heidinger

10.1 Introduction

Evolution is the most important theory in biology as it integrates nearly all biological phenomena. Only through the theory of evolution can the uniqueness and diversity of living beings and natural phenomena be described and explained. Thus, the teaching and learning of biology as a theory-based science subject needs the theory of evolution as its central theoretical basis (Harms & Reiss, 2019). However, in the Austrian state curriculum, the theory of evolution plays only a minor role: It is only mentioned twice — first, in grade 7[1] (at lower secondary level in Austria) and, second, in grade 12 (last year of upper secondary level), very close to the final examination. Therefore, as part of this project, we developed a series of task-based teaching sequences (for grades 5, 8, 9, and 10) to enable learning of evolution in secondary education to be extended over a broader period of time and more continuous (Scheuch et al., 2019). In our research on this topic, we conducted a longitudinal study focusing on the conceptual development of students who took part in the newly developed evolution curriculum. In Scheuch et al. (2019), we presented the

[1] We are using the Anglo-American grade system for better understanding. In Austria, there are four years of primary school (age 6–10), four years of lower secondary school (age 10–14) and four years of upper secondary school (age 14–18).

M. Scheuch (✉)
University College for Agricultural and Environmental Education, Vienna, Austria

Austrian Educational Competence Centre for Biology, University of Vienna, Vienna, Austria
e-mail: martin.scheuch@haup.ac.at

J. Scheibstock · H. Amon · G. Fuchs · C. Heidinger
Austrian Educational Competence Centre for Biology, University of Vienna, Vienna, Austria
e-mail: jaqueline.scheibstock@sagasser.com; heidemarie.amon@univie.ac.at

O. Levrini et al. (eds.), *Engaging with Contemporary Challenges through Science Education Research*, Contributions from Science Education Research 9,
https://doi.org/10.1007/978-3-030-74490-8_10

first results from this research, which dealt with the conceptual development of the concept of 'selection' from interviews in grade 8 and grade 10; in this contribution, we focus on the development of learners' understanding of the concepts of 'variation' and 'change' from grade 8 until grade 12.

10.1.1 Development of Students' Conceptions in Evolution

It is well known that students use their personal everyday experiences when trying to understand natural phenomena (Coley & Tanner, 2012; Gropengießer, 2007; Kampourakis, 2014). Conceptual learning in science involves a gradual reconstruction or growth of everyday conceptions as scientific concepts are encountered through instruction (Krüger, 2007). Students' pre-conceptions are, in general, very resistant to change, however (Strike & Posner, 1992). So, what do we know about students' conceptions in evolution and their conceptual development?

Students tend to think that organisms are responsible for their own adaptation to environmental changes. In other words, it is the individual itself that initiates change in its body in order to survive changes in the environment (Baalmann et al., 2004; Gregory, 2009; Harms & Reiss, 2019). Such conceptions of goal-directed adaptation are related to another common naïve pattern, teleological thinking – i.e. a phenomenon is explained in terms of its goal; most problematic is explaining evolution in terms of purposeful design (Kampourakis, 2020), rather than mechanistic and stochastic processes.

Students, typically, do not consider variation between individuals within a population when trying to understand evolutionary change (e.g. Lehrer & Schauble, 2012). This can be explained in part by the prevalence of students' essentialist thinking about species (all individuals of a species are the same and united by a shared essence) which ignores variation between individuals, or within a population (Coley & Tanner, 2012; Hammann & Asshoff, 2015; Harms & Reiss, 2019; Kampourakis, 2014; Kattmann, 2015). However, variation is a precondition for understanding all other evolutionary concepts – e.g. selection (natural, sexual, artificial) - and is therefore crucial for learning about evolutionary processes (Scheuch et al. 2019). Alred et al. (2019) found that students had difficulty in recognizing variation at all in living beings: More than 90% of the students in middle school, more than 80% in high school, still over 50% at undergraduate level did not have proper understanding of variation. This is a huge proportion of students who are still stuck in essentialist thinking and will, therefore, have difficulties with the role of variation in evolutionary processes. These results contrast starkly with the correct professional/technical description of variation and change in populations, as expressed by Ernst Mayr (2004, p. 29):

> In a biopopulation […] every individual is unique, while the statistical mean value of a population is an abstraction. No two of the six billion humans are the same … and …The properties of populations change from generation to generation in a gradual manner.

This concise statement of a central biological idea by Ernst Mayr is, in a sense, so simple and yet so hard to get, if students apply goal-oriented thinking on an individual level and neglect variation within populations as well as thinking in generations.

Helping students to transform their personal pre-conceptions into a more scientific understanding of evolution often takes years (Wandersee et al., 1995). In a cross-sectional study by To et al. (2017), for example, the comparison of the reasoning about evolution of students aged 12, 14 and 16 years showed that only a small group of students of all ages demonstrated more scientific reasoning; all others applied mostly teleological and essentialist ideas in the context of evolutionary phenomena.

One reason for the difficulties students face in understanding evolutionary concepts is the underlying structure of their intuitive understanding of, and reasoning about, biological phenomena. As already stated, everyday experiences with biological phenomena lead to pre-conceptions and intuitive knowledge about the living world. This intuitive knowledge is shaped by three central underlying modes of thinking: anthropocentric thinking, teleological thinking and essentialist thinking (Alred et al., 2019; Coley & Tanner, 2012; McLure et al., 2020). These modes of thinking are highly relevant for biology teaching, since they make mechanisms of evolutionary processes come across as counter intuitive to laypersons. Southerland et al. (2001) used the concept of "phenomenological primitives" (p-prims) to describe how basic patterns of thought (e.g. "*need as a rational for change*") have profound implications for understanding the concepts of variation, selection, reproduction, the genetic background, and evolutionary change itself.

10.1.2 Teaching Sequences on Evolution

Teaching and learning science should always include working with the students' pre-instruction everyday conceptions and thinking (Taber, 2014). Jelemenská et al. (2010) tackled this challenge and developed research-based teaching sequences, which were based on student conceptions of evolution extracted from literature. The procedure of planning and refining by the biology education researcher with the teachers was published as well (Jelemenská, 2012). The results of the planning process were two teaching sequences on evolution for secondary education (grade 8 and 10; students from the age of 14 to 16); these completed a teaching sequence for grade 12 that already existed based on the Austrian state curriculum. The main ideas introduced to the students in each sequence are described briefly here (for more details see Scheuch et al., 2019):

Grade 8: "Variation and Selection"

Within a period of three lessons, the teacher introduced the students to the basic evolutionary terms "variation" and "selection." A computer simulation of dog breeding was used as an example of artificial selection. This was designed to first focus students' attention on the existence of variation within a population and then to help clarify its role in the breeding process in the population over a number of reproductive cycles. In addition, natural selection was introduced via a film and illustrated through the evolution of 14 species of finches on the Galapagos Islands, which emerged out of a little founder population of one species, originating from the South American coast. Additionally, students worked on assignments on the studies conducted in the 1970s by the Grants about those finches and the extreme weather events, illustrating the role of environmental changes (e.g. Grant & Grant, 1989). Finally, the students, assisted by the teacher, compared the differences between artificial and natural selection: *selection* with an aim and a *breeder* as an active selector compared to environmental conditions and differences in reproductive success.

Grade 10: "Natural and Sexual Selection"

Sexuality and behavioral research were the main concepts of this teaching module. At the beginning of the five lessons that made up this module, an internet game was used to revisit the concepts of natural selection and variation. The game reconstructed the phenomenon of industrial melanism in peppered moths in the nineteenth century. Then, the formation of traits due to sexual selection, such as the length of tail feathers, was our focus to highlight the concept of variation. Students got pictures of specimens with long tail feathers and were asked to hypothesize about this phenomenon based on guiding questions (e.g. advantages and disadvantages of the trait). Students learned that Darwin noticed that certain traits, such as conspicuous feathers or overly large antlers, could not be explained by natural selection. The instructional goal was for the students to understand that sexual selection explains why organisms with apparently disadvantageous but attractive characteristics survived, and that evolution does not pursue a certain aim – and certainly does not strive for a perfect, ideal adaptation to the environment.

Grade 12: "Evolution and Genetics"

The state curriculum prescribes genetics and evolution in grade 12. During this year molecular mechanisms of heredity, basic biochemical pathways from DNA to proteins, gene regulation, and human genetics are linked with the chemical and biological evolution at the molecular level. Moreover, the mechanisms of evolution, the descent of man, as well as evolution as a basis for biodiversity and change round off the biological education in the last year of schooling.

10.2 Research Questions

The aim of this long-term study is to investigate the conceptual development in students who attended the biology lessons of the teaching and learning sequence on evolution just introduced. We investigated students' conceptions of 'variation' and 'change' from the age of 14 (grade 8) to the age of 18 (grade 12). Thus, this study is guided by the following research questions: What student conceptions of 'variation' and 'change' are formed from grade 8 to grade 12, and how do these conceptions develop over time?

10.3 Methods

This study adopted a qualitative, descriptive, and exploratory research design and followed a case study approach (Yin, 2009) which aimed to reveal the development of the conceptions of three students over five years. We collected data via problem-centered, open and guideline-based interviews (Baalmann et al., 2004; Fenner, 2013; Johannsen & Krüger, 2005) as shown in Fig. 10.1.

We used a pre-post design to study the same group of students in grade 8 (age 13–14), grade 10 (age 15–16), and finally in grade 12 (age 17–18). We initially selected four students in a Viennese Gymnasium (three girls and a boy) thought to vary in their interest, knowledge, attitudes, and marks in biology. The selections were made based on group discussion about evolution with the whole class (24 students) and based on the students' biology teacher's judgement. The boy refused to take part in the interviews in grade 10 and in the post interview of grade 12; therefore, the results of the analysis of his interviews are excluded from the analysis presented here. The remaining three students (pseudonyms: Anna, Bianca & Claudia) were interviewed about their conceptions of 'variation' and 'natural selection' in grade 8 and 'sexual selection' in grade 10 (pre- and post to the teaching sequence). In grade 12, the students underwent a pre-post interview about the role of genetics in evolution. All previous concepts of evolution were dealt with again in later interviews with a special focus on problems of understanding revealed in earlier interviews.

The interviews were semi-structured and included activities to elicit students' thinking about different biological phenomena (e.g. reasoning about the evolution of the giraffe's long neck: Johannsen & Krüger, 2005). In the post-interview in grade 12, all previous concepts were recapitulated and students were asked to

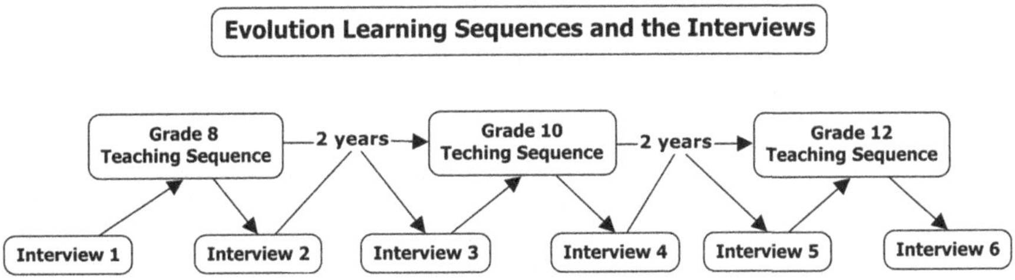

Fig. 10.1 Scheme of data collection and evolution teaching and learning (lower and upper secondary level)

construct a concept map while arranging key concepts in the topic of evolution on cards. Students were then asked to explain their concept map.

Data sources consisted of verbatim transcripts of all the interviews and photos of the different concept maps (at different stages of their construction) produced in the grade 12 post interviews. In order to identify student conceptions, we analyzed this data using qualitative content analysis according to Gropengießer (2005) and Mayring & Fenzl (2014). We then coded with deductive as well as inductive codes, then reduced and explicated the material. The deductive codes were derived from the literature on students' pre-conceptions about evolution (e.g. Baalmann et al., 2004; Gregory, 2009; Kattmann, 2015). The results from all interviews were then compared across all grades in order to trace the development of students' conceptions of 'variation' and 'change'. Data taken out of the interviews in the following section are marked with the capital letter of the pseudonym of the participant (**A**nna, **B**ianca, **C**laudia), the number of the interview (1–6) and the respective transcript lines.

10.4 Results

In general, the analysis shows that all students have developed their concepts regarding 'variation' over the five years of study and have integrated knowledge from the biology lessons. However, underlying modes of thinking proved to be very stable. Different patterns of conceptual change were identified across the three cases.

In the following sections, we show the development of the students' conceptions related to 'variation' and 'change', highlighting three common themes: dichotomous thinking, difficulties with linking variation to genetics, and teleological thinking.

10.4.1 Variation

The fact that variation within a population is present and that it is a prerequisite for selection is anchored in the thinking of the students in grade 12, after all the teaching sequences. They assumed some kind of variation right from the beginning, in grade 8. As seen in her first interview, Claudia starts with anthropocentric thinking about variation: "*there must be some kind of variation in animal populations because humans also do not look all the same*" (C1/145–149). Similarly, in grade 8, Anna states that individual humans, animals and plants are different (A1/212–258), but she is unsure about microorganisms (A1/259–271). Interestingly, even in her grade 12 interview, Bianca's concept of variation is only used for humans and animals and is not generalized to plants; she states that individual humans and animals "*all have all a bit different DNA, but not plants, those are only copies*" (B5/239–297). She explains the fact that plants of the same population sometimes look different by appealing to differences in environmental conditions of individual plants. It is noteworthy that, overall, the students speak more of 'differences' rather than of 'variation'.

10.4.2 Dichotomous Thinking Regarding Changes in the Population

Despite this awareness of variation, the assumption that essentialist thinking regarding variation has been overcome by all three students would be premature. When considering variation in relation to selection, all three students use a very simplistic version of variation that reflects essentialist thinking. Very often, the students group the animals of a population into two distinct groups: one group with (identical) advantageous features and one group without these advantageous features (A2/1–22, B2/63–99, B5/336–391, C1/160–179, C2/1–18). That is, the students display dichotomous thinking regarding the different attributes of individuals in a population instead of conceptualizing variation with the help of normal distributions. As a consequence, survival and reproductive success are conceptualized to be the same for all members of a group. Examples of this mode of thinking can be shown in students' arguments involving individuals of a population having a bright or dark appearance (mice, peppered moth), having long or short necks or legs (giraffes, horses): *"only those with longer necks have survived and developed"* (A2/1–22), surviving or dying due to more or less camouflaged fur colors or due to enhanced or reduced ability to find food (e.g. *"an albino mouse will not survive that easy like a normal one [...] it is more visible and therefore an easy prey"* (B5/336–391). While there is some awareness of variation, this is essentialist thinking and evolution cannot be explained continuously within this mode of thought. In this mode of thought, variation is only there initially and then disappears after one reproductive cycle, and the outcome is a homogenous group with identical favored traits. This finding contradicts students' explanations of variability of individuals in terms of genetics, which is addressed in the next section.

10.4.3 Genetics

The idea that there is a genetic basis of variation appears first in Anna's second interview (A2/92–117: "*Genes make the difference because they come from the father and the mother*") and in the third interviews of Bianca (B3/322–332) and Claudia (C3/14–28). But genetic processes appear to be a black box for all three students and their explanations of variability are vague and seldom linked to other evolutionary concepts like selection. Somehow, the students know that variation is a product of sexual reproduction, but it largely remains a mystery for them how exactly this works and so is understood only superficially: "*only the next generation can change due to mixture of genes*" (C3/242–256) and "*Genes are the basis for inheritance. In mating of two individuals the genes produce change.*" (B4/59–127). The prevalent concept that accounted for the production of variability was "*mixing of genes*" (B3/485–501, B4/1–18, B5/239–289, 289–311, B6/670–689, A2/92–117, C3/242–256, C5/63–91, 345–360). None of the students was able to describe production of genetic variability in an appropriate way, even if they used correct terms (e.g., recombination): "*Mixing means that half of the genes come from the father and half from the mother. Recombination itself is the mixing*" (B6/1479–1508) and "*The moment of recombination is unclear to me.*" (B6/1509–1530). Hardly any evidence can be found across all 18 interviews where attributes of individuals or populations (phenotypic variability) are directly linked to genetic traits (genetic variability). Therefore, it seems that the students view the phenotypic and genetic variability as two separate types of variability: One type of variability linked to evolution via selection of favorable attributes due to external selective pressure (environment, nature, and habitat); and the other is mating and the mixing of genes. Claudia explicitly mentioned these two types together: "*The environment is more decisive […]. Looking for mating partners is on the individual level; (but) the environment affects the entire population or even the entire species.*" (C5/494–533). The students do not understand that phenotypic and genetic variability are different levels of conceptualization closely linked to one another.

10.4.4 Teleological Reasoning – A Basis for Thinking

It is interesting to note that all three students continue to stick to teleological arguments in explaining change in a population up until the final interview in grade 12. All three students started with the conception that *individuals act conscious of their adaptation* (A1, B1, C1) in describing the growth of the giraffes' neck in the first interview. In all cases, this conception evolved over the years to the more scientific conception that *nature/the environment/the habitat selects*. But there is evidence that this conceptual development is only partial and that teleological thinking and looking for final causes still prevails. The students still seem to be convinced that species evolve for a certain reason. Examples of this thinking are reflected in the following interview excerpts: "*Certain attributes are <u>needed</u> in certain situations or certain habitats*" (A4/1–6); "*First earth changes, then the species, because <u>they want</u> to survive*" (C1/150–159) or "*<u>they needed</u> something for survival, therefore they adapted*" (C4/266–288); "*natural selection is perhaps the change of environmental factors, that <u>nature can influence</u> evolution somehow*" (C6/485–491). In the students' conceptions, organisms, individuals, populations or the whole species are always reacting to a previous change in the environment; in the last example, nature is acting in a goal-directed fashion.

Bianca's case is special, because she usually focused on environmental factors that influence variability during the lifetime of organisms in order to explain the changes in the appearance of an organism. Therefore, she demonstrated highly anthropomorphic thinking connected with teleological reasoning: "*mice choose to live near humans, because this is an advantage compared to those living in the woods, they get more to eat and are therefore stronger*" (B5/392–415) or "*Due to environmental changes horses needed other attributes*" (B5/416–451). This aspect remains stable in Bianca's interviews until grade 12. She also applies teleological thoughts to reasoning about mating choice: "*the female bird chooses the partner with the best attributes for the optimal survival of the offspring*" (B4/59–127). The appeal to intentional mating choices as a basis for adaptation and change is also prevalent in Claudia's third interview (C3/195–219) and fourth interview "*should the next generation need longer wings, this can only happen through mating*" and "*female choice is about looking for the partner with the exactly needed attributes*" (C4/34–44/10–33). In later interviews, she modifies her conceptions insofar as individuals are seen no longer as aiming at improving their offspring on purpose: "*reproduction: it is not totally conscious, not that they are born and know 'this is my aim', more probably it is about instincts and drive*" (C5/331–344 & 694–721).

10.5 Discussion and Outlook

Students include more and more scientific conceptions as their understanding develops, but some initial, intuitive understanding remains. It is very interesting how the newly acquired knowledge is - or is not - attached to this thinking.

Variation among individuals of a population of animals or other living beings is difficult to recognize in everyday life. The conceptual development of two of the three students focused on here can be understood as an extension of corresponding concepts from humans, where variation can be easily recognized in animals and then extended to plants and other living beings (with uncertainty about microorganisms). Bianca did not complete this sequence of gradual extension. She states in grade 12 that poppy plants are copies and only look different due to differences of current environmental conditions. This is understandable against the background that students have problems recognizing plants as living beings. This finding links to other research on students' conceptions about plant blindness (e.g. Lampert et al., 2020). Another finding is that students tend to think about variation in terms of differences in appearance. This conception seems to be easier for students to apply on natural phenomena because it focuses on concrete, observable features of organisms instead of on the much more abstract concepts of 'variation' or 'variability'. Variation includes the genotype as well as the phenotype of an organism and the links between those levels, which is – as we saw – very difficult for the students to grasp.

The **dichotomous thinking** about variation of attributes is interesting. During the teaching sequence, we introduced at least two examples in which we extensively discussed normal distribution-based descriptions of attributes (bill size of Darwin finches and tail feather length of barn swallows with graphs of the distribution).

Despite these examples, it seems easier for students to simplify the complexity of a normal distribution into two categories (big vs. small and long vs. short). However, reducing this complexity results in a highly simplified view of evolution, retaining the problems of essentialist thinking. Alred et al. (2019) report a similar conception of 'dichotomous variation'. They considered 'dichotomous variation' to be a 'low level of understanding' of variation in a large cohort of middle school to undergraduate students. They argue that this represents a form of naïve understanding and far short of the canonical, scientific understanding of variation. Incorporating attention to normal distributions in the teaching modules did not produce a significant conceptual development towards a more scientific account of variation. There is a need to revise or expand the approach taken to instruction about normal distributions.

Another big concern is the missing integration of **genetics** and inheritance as a basis for evolutionary processes. Genetics is covered in detail very late in Austrian schools (grade 12), although some relevant individual concepts (Mendelian genetics, mitosis, and meiosis) are addressed earlier. Therefore, the students did express ideas about genetics in the interviews starting in grade 8, but by grade 12, they are still having great difficulty connecting this knowledge to evolutionary concepts like variation and change. All these concepts (genetic variability, recombination, mutation, variation, selection and change) exist but without meaningful connections in students' thinking. It is well-known in biology education that linking concepts on different levels – genetics and evolutionary processes – is very difficult (Jördens et al., 2016). Evolutionary change happens on different levels: the individual (ontogenetic) level, the population (phylogenetic) level as well as the ecological level – and all three levels are interrelated. The above-mentioned contradiction in students' conceptions highlights this separation: essentialist or dichotomous thinking and genetic argumentation about unique individuals. Examples of evolutionary phenomena where links between these levels can be easily apprehended have to be used in teaching to help the students establish the relationships between these different levels (e.g. Jördens et al., 2016).

Another great hurdle for the students remains **teleological thinking**. In our study, we could detect a shift in teleological reasoning from individual behavior for adaptation to several other conceptions (e.g. from the individual to the population or even nature as an active selector; mating for purpose of adaptation). It seems that teaching about evolution allowed students to relate the acquired concepts and facts to their preconceived teleological reasoning which remains stable. Southerland et al. (2001) work also contributes to the interpretation of these findings. They argue that the underlying thinking with "*need as a rational for change*" is much harder to deal with because of its unconscious and intuitive use in explaining phenomena. This supports the idea that this is an underlying mode of thinking or a persistent cognitive bias (Nehm et al. 2012; Pobiner et al., 2019), and all newly learned ideas are interpreted accordingly. Sorting the results into newly developed conceptions and underlying modes of thought (Southerland et al. 2001) may help in interpreting the gains and pitfalls in the learning process, as these modes seem to have different consequences for learning in the three students. A follow up investigation could be a fine-grained analysis of different modes of teleological reasoning. Such a follow up investigation would seem promising in light of Kampourakis' (2020) recent review in which he found several different types of teleological argumentation.

These further analyses of our data will help us gain an in-depth understanding of how these underlying modes of thinking – teleological, anthropomorphic and essentialist thinking – influence the development of students' conceptions of evolutionary phenomena. Only when evolution teaching also becomes effective at this very basic level of understanding, can a scientific understanding of the theory of evolution be achieved.

Our longitudinal study is one of the first where well-known student conceptions about evolution are traced over five years of learning; beginning in the lower secondary level and including the whole upper secondary levels. The students' learning about evolution was found to be far from a linear development in understanding, from everyday knowledge towards scientific conceptions. Instead, the students show difficulties in overcoming underlying modes of thought, such as teleological, essentialist or anthropocentric thinking. Even the design of learning sequences, which was based on student conceptions research, did not help the learners overcome their naïve ideas; moreover, the result was an idiosyncratic amalgam of previous thinking and scientific ideas. Many more longitudinal studies at all levels and with the different evolutionary topics are needed to learn more about pathways of developing understanding, on the one hand, and the needed instructional support to help learners overcome difficulties, on the other.

Acknowledgement This project was supported by the Hochschuljubiläumsstiftung der Stadt Wien (Grant Nr. H-316715/2017).

References

Alred, A. R., Doherty, J. H., Harttley, L. M., Harris, C. B., & Dauer, J. M. (2019). Exploring student ideas about biological variation. *International Journal of Science Education, 41*(12), 1682–1700.

Baalmann, W., Frerichs, V., Weitzel, H., Gropengießer, H., & Kattmann, U. (2004). Students' conceptions about processes for adaptation – Results of an interview study. *Zeitschrift für Didaktik der Naturwissenschaften, 10*, 7–28.

Coley, J. D., & Tanner, K. D. (2012). Common origins of diverse misconceptions: Cognitive principles and the development of biology thinking. *CBE-Life Sciences Education, 11*, 209–215.

Fenner, A. (2013). *Schülervorstellungen zur Evolutionstheorie, Konzeption und Evaluation von Unterricht zur Anpassung durch Selektion*. Justus-Liebig-Universität.

Grant, B. R., & Grant, P. R. (1989). Natural selection in a population of Darwin's finches. *The American Naturalist, 133*(3), 377–393.

Gregory, T. R. (2009). Understanding natural selection: Essential concepts and common misconceptions. *Evolution: Education and Outreach, 2*(2), 156–175.

Gropengießer, H. (2005). Qualitative Inhaltsanalyse in der fachdidaktischen Lehr- und Lernforschung. In P. Mayring & M. Gläser-Zikuda (Eds.), *Die Praxis der qualitativen Inhaltsanalyse* (pp. 172–189). Beltz.

Gropengießer, H. (2007). Theorie des erfahrungsbasierten Verstehens. In D. Krüger & H. Vogt (Eds.), *Theorien in der biologiedidaktischen Forschung* (pp. 105–116). Springer.
Hammann, M., & Asshoff, R. (2015). *Schülervorstellungen im Biologieunterricht – Ursachen für Lernschwierigkeiten*. Klett-Kallmeyer.
Harms, U., & Reiss, M. (2019). The present status of evolution education. In U. Harms & M. Reiss (Eds.), *Evolution education re-considered.: Understanding what works* (pp. 1–20). Switzerland.
Jelemenská, P., Amon, H., Wenzl, I. (2010). Was verstehen Sie bitte unter Evolution? – Unterrichtsplanung unter Berücksichtigung von Schülervorstellungen. Report. IMST. https://www.imst.ac.at/imst-wiki/images/d/d5/1847-Langfassung-Jelemenska.pdf. Accessed 12 Jan 2020.
Jelemenská, P. (2012). Lehrervorstellungen zum Lehren und Lernen von Evolution – eine Fallstudie zum fachdidaktischen Coaching. *Zeitschrift für Didaktik der Naturwissenschaften, 18*, 229–259.
Johannsen, M., & Krüger, D. (2005). Schülervorstellungen zur Evolution – eine quantitative Studie. *Berichte des Institutes für Didaktik der Biologie, 14*, 23–48.
Jördens, J., Asshoff, R., Kullmann, H., & Hammann, M. (2016). Providing vertical coherence in explanations and promoting reasoning across levels of biological organization when teaching evolution. *International Journal of Science Education, 38*(6), 960–992.
Kampourakis, K. (2014). *Understanding evolution*. Cambridge University Press.
Kampourakis, K. (2020). Students' "teleological misconceptions" in evolution education: Why the underlying design stance, not teleology per se, is the problem. *Evolution: Education and Outreach, 13*(1), 1–12.
Kattmann, U. (2015). *Schüler besser verstehen – Alltagsvorstellungen im Biologieunterricht*. Aulis Verlag.
Krüger, D. (2007). Die conceptual change-Theorie. In D. Krüger & H. Vogt (Eds.), *Theorien in der biologiedidaktischen Forschung* (pp. 81–90). Springer.
Lampert, P., Müllner, B., Pany, P., Scheuch, M., & Kiehn, M. (2020). Students' conceptions of plant reproduction processes. *Journal of Biological Education, 54*(2), 213–223.
Lehrer, R., & Schauble, L. (2012). Seeding evolutionary thinking by engaging children in modelling its foundations. *Science Education, 96*(4), 701–724.
Mayr, E. (2004). *What makes biology unique? – Considerations on the autonomy of a discipline*. Cambridge University Press.
Mayring, P., & Fenzl, T. (2014). Qualitative Inhaltsanalyse. In N. Baur & J. Blasius (Eds.), *Handbuch Methoden der empirischen Sozialforschung* (pp. 543–556). Springer Fachmedien.
McLure, F., Won, M., & Treagust, D. (2020). Students' understanding of the emergent processes of natural selection: The need for ontological conceptual change. *International Journal of Science Education*. https://doi.org/10.1080/09500693.2020.1767315
Nehm, R. H., Beggrow, E. P., Opfer, J. E., & Ha, M. (2012). Reasoning about natural selection: Diagnosing contextual competency using the ACORNS instrument. *American Biology Teacher, 74*, 92–98.
Pobiner, B., Watson, W. A., Beardsley, P. M., & Bertka, C. M. (2019). Using human examples to teach evolution to high school students: Increasing understanding and decreasing cognitive biases and misconceptions. In U. Harms & M. Reiss (Eds.), *Evolution education re-considered: Understanding what works* (pp. 185–205). Springer Nature.
Scheuch, M., Scheibstock, J., Amon, H., & Bauer, H. (2019). Teaching evolution along a learning progression: An Austrian attempt with a focus on selection. In U. Harms & M. Reiss (Eds.), *Evolution education re-considered: Understanding what works* (pp. 81–100). Springer Nature.
Taber, K. S. (2014). *Student thinking and learning in science: Perspectives on the nature and development of learners' ideas*. Routledge.
To, C., Tenenbaum, H. R., & Hogh, H. (2017). Secondary school students' reasoning about evolution. *Journal of Research in Science Teaching, 54*(2), 247–273.

Southerland, S. A., Abrams, E., Cummins, C. L., & Anzelmo, J. (2001). Understanding students' explanations of biological phenomena: Conceptual frameworks or P-prims? *Science Education, 85*, 328–348.

Strike, K. A., & Posner, G. J. (1992). A revisionist theory of conceptual change. In R. A. Duschl & R. J. Hamilton (Eds.), *Philosophy of science, cognitive psychology, and educational theory and practice* (pp. 147–176). State University of New York Press.

Wandersee, J. H., Good, R. G., & Demastes, S. S. (1995). Research on biology education about evolution: An inventory. *Zeitschrift für Didaktik der Naturwissenschaften, 1*, 43–54.

Yin, R. K. (2009). *Case study research: Design and methods* (4th ed.). SAGE.

Chapter 11
What Is City Air Made of? An Analysis of Pupils' Conceptions of Clean and Polluted Air

Èlia Tena and Digna Couso

11.1 Introduction

Air pollution is one of the most important environmental problems in cities. This phenomenon has important effects on human, animal, and vegetable life. Changing this situation needs the active involvement of citizens, as traffic is a very important source of air pollution. As a consequence, air pollution is considered a hot topic in the promotion of scientific literacy for responsible citizenship (OCDE, 2020).

To be able to act in an informed and empowered way about air pollution, pupils need an understanding of the model of matter applied to gases and ideas on the structure and nature of air pollutants. This implies the mastery of an appropriate particle model of matter starting in early years. From a constructivist viewpoint, this can only be done by considering ideas on this topic and designing and guiding learning so that these ideas can evolve appropriately, as pupils construct improved versions of their models of matter applied to air and pollution.

The aim of this paper is to share our initial research results, obtained in the context of the implementation of an evidence-based teaching and learning sequence (TLS) about children' initial and final models regarding air pollution.

È. Tena (✉) · D. Couso
CRECIM and Departament de Didàctica de la Matemàtica i les Ciències Experimentals-UAB, Barcelona, Spain
e-mail: elia.tena@uab.cat; digna.couso@uab.cat

O. Levrini et al. (eds.), *Engaging with Contemporary Challenges through Science Education Research*, Contributions from Science Education Research 9,
https://doi.org/10.1007/978-3-030-74490-8_11

11.2 Teaching and Learning the Particle Model of Matter

The particle model of matter, and especially the idea that matter is discrete, is a recognized core idea of scientific literacy (Harlen, 2010). From a (socio)constructivist perspective that view learning as the evolution of ones' ideas to make better sense of experience and teaching as the optimal guided instruction to support this evolution (Taber, 2011), knowing children's beliefs about the model of matter and how their ideas progress with schooling is essential (Karata et al. 2013).

Research on pupils' ideas about matter show us that children usually do not have a scientifically adequate understanding of the model of matter (Driver et al., 1994; Hadenfeldt et al., 2014; Merritt & Krajcik, 2013; Talanquer, 2009). On the contrary, pupils display significant alternative conceptions about matter, particularly when they talk about gases like air (Thornber et al., 2016). Some of the most important ones are the ideas of gases as a massless transient; air and oxygen as synonymous; air as a single substance rather than a mixture of gases (Driver et al., 1994) and also the use of properties of matter at the macro level (e.g. colour) to explain atomic or molecular features (Meijer et al., 2013). Moreover, some cross-age studies show that schooling has a moderate, often very slow, impact in the improvement of pupils' ideas on matter (Karata et al., 2013).

There are different aspects related to the teaching of this topic that can explain pupils' difficulty to build an adequate model of matter along schooling, such as the focus on the historical account of the construction of the atomic idea instead of into pupils' own initial ideas, which are strongly based on their macroscopic and qualitative experience with matter (Merritt & Krajcik, 2013). In addition, teaching about matter do not usually take into account the different scales necessary to study and interpret it, including the macroscopic scale (between 0.1 and 1 m), the mesoscopic scale (between 10^{-7} and 10^{-1} m) and the submicroscopic (between 10^{-10} and 10^{-9} m) (Meijer et al., 2013).

11.3 Air Pollution as a Relevant Phenomenon

Existing research shows us that people have a limited understanding and significant alternative conceptions on air pollution, their effects and possible strategies to reduce it (Mandrikas et al., 2017). Some of these alternative ideas are that: pollution only exists when we can see, feel, or taste it with our senses (Thornber et al., 2016); pollutants can only be in gas form and the confusion between pollution and other environmental problems like the Green House Effect or the Ozone Layer degradation. The increase in the prevalence of some of these alternative ideas by the end of formal education would suggest that schooling does not help children to overcome them (Thornber et al., 2016).

Focusing on air pollution due to suspended particle matter (PM), the most relevant regarding pollutants in cities due to traffic, to understand this phenomenon in an adequate way, implies mastering a sophisticated model of matter (Solé et al.,

2020). On the one hand, it implies overcoming pupils' interpretations based on their direct observations of the macro scale to focus on entities at the meso scale (particulate solid matter suspended in air) and the atomic- molecular scale (the actual particles air is made of). Additionally, it implies being able to differentiate the meso and the atomic-molecular scales, overcoming the polysemy of the word particle which is usually used for both (for instance for dust particles and for oxygen particles).

11.4 Models and Modelling in Primary School

Mastering a sophisticated model of matter that allows us to describe, predict and explain adequately the air pollution phenomena requires instruction that addresses the above mentioned challenges. One approach that is consistent with a (socio)constructivist view of learning and focuses on helping students express and confront their initial models in order to develop them is model and modelling-based instruction (Chinn & Buckland, 2012). In school settings, this approach involves the construction of models by pupils and implies imagining new theoretical entities and building explanations about the world that are consistent with the available evidence (Duschl et al., 2011). The promotion of modelling in classrooms require helping pupils to imagine and express their initial ideas about a phenomena, put their ideas to test, make changes in their models to make them more coherent with their observations and finally, to compare these ideas with the consensus knowledge in (school) science (Couso & Garrido-Espeja, 2017; Schwarz et al., 2009).

In agreement with Lehrer & Schauble (2019) the goal of introducing modelling and models in the science classroom is twofold, both epistemic and epistemological. First, we involve pupils in modelling scientific phenomena to help them learn conceptual knowledge in the form of school scientific models. These models are educationally reconstructed versions of the models of science that are targeted in school because they have potential to explain many phenomena (Couso & Garrido-Espeja, 2017). Second, we aim to develop both procedural and epistemic knowledge and competence, by involving students in a modelling practice that helps them to participate in school science in a way that is coherent with how real science develops. This helps them to understand how individuals and communities generate scientific knowledge and to systematize this kind of practice (Lehrer & Schauble, 2019). From a learning perspective, some authors have argued how the learning of epistemic practices related to model and modelling-based instruction is important not only in terms of learning about science, but also for aiding concept development by promoting a school culture where the use of evidence for changing our ideas is valued (Chinn & Buckland, 2012).

Despite the potential and interest that literature attaches to modelling and models as a crucial part of the process of sense-making (Oh & Oh, 2011), modelling practice is rarely incorporated into elementary schools (Schwarz et al., 2009). Often, models are considered highly abstract entities inadequate for young children (Schwarz et al., 2009). However, as research has shown, children are capable of thinking in both concrete and abstract ways, generating verbal and graphical representations about different everyday phenomena if the task has an appropriate cognitive level. When the modelling process is well-directed, children' ideas are quite sophisticated and closer than expected to the scientific ideas (Lehrer & Schauble, 2019).

11.5 Research Aims

The aim of this research is to analyse the ideas of pupils aged between 10 and 12 years-old about air pollution when they are involved in a model/modelling-based TLS specifically designed to promote the construction of an adequate model of matter applied to the air pollution phenomenon. Particularly, we focus on: (1) What are pupils' initial and final ideas about the structure and nature of clean air? and (2) What are pupils' initial and final ideas about the structure and nature of polluted air?

11.6 Context and Methods

The present study has been developed in the context of an educational project in which more than 12 schools and 647 pupils ages 10–12 have participated.

In the project, a research-based TLS of 12 classroom hours was iteratively developed, implemented in real classrooms and evidence-based modified during the 2018–2019 and 2019–2020 school years. To do so, we followed a Design-Based Research methodology (DBR Collective, 2003) from a participatory perspective that took into account the participating teachers' views (Couso, 2017). The TLS followed a didactical approach based on principles of model/modelling-based instruction. Specifically, we used the modelling cycle (Couso & Garrido-Espeja, 2017) to organize the sequence so that each targeted scientific idea could be progressively developed by involving children in the process of expressing, using, evaluating and revising their models.

11.6.1 Data Collection

To analyse pupils' ideas on clean and polluted air, during the implementation of the final, refined version of the TLS[1] (course 2019–2020) we have collected individual productions of pupils at the beginning and at the end. In order to promote the use of

[1] Information about the project ParticipAire and the TLS is available at: https://ddd.uab.cat/record/225073?ln=ca

their model of matter at different scales (macro, meso and atomic-molecular levels), we asked participating pupils to draw and describe in written form two hypothetical samples (bags) of clean and polluted air as seen both with the naked eye and with something that allowed them to look at its insides. Pupils' multimodal productions were gathered to be able to analyse their own initial models of matter applied to the phenomenon.

For the analysis in this paper, data from 24 pupils were collected in a public urban school with different socio-economical and ethnic profiles, including children from a range of ability groups.

11.6.2 Analysis

The constant comparative method (Miles et al., 2014) has been used to analyse pupils' ideas. The system of coding categories was developed using a twofold top-down and bottom-up approach, by using categories identified in previous research or literature on the topic and considering emerging categories from the data. The final system of categories included in Table 11.1 analyses children' ideas in terms of both the structure and nature of clean and polluted air.

The dimension of structure uses categories inspired in previous literature on children' ideas of gas matter. As different authors have pointed out (Hadenfeldt et al., 2014; Merritt & Krajcik, 2013; Talanquer, 2009), one of the most important aspect to analyse pupils' progress toward a deeper understanding of matter is that of continuity, including matter seen as continuous, semi-continuous or discreet. Another important aspect to understand pupils' views of matter is that of scale (macro, meso and atomic-molecular), as scale is essential to interpret the differences between clean and polluted air. Following Acher et al. (2007) and Meijer R. et al. (2013), in primary school education "particles" can be understood as small parts in a mesoscopic scale, which can be interpreted as an initial step for the construction of actual idea of atomic/molecular particles. In the description of our categories (Table 11.1) we have used the word "particle" between commas to refer to any meso/micro entity the pupils might refer to.

The analysis of pupils' ideas for the dimension of nature has also been done based on previous literature on ideas about clean air (Driver et al., 1994) and polluted air (Pruneau et al., 2005; Thornber et al., 2016). Based on those, one of the most important ideas to be built in primary school education is that some everyday substances (e.g. air) are formed by a mixture of different components (e.g. Oxygen, Nitrogen...) (Driver et al., 1994). Therefore, the identification of one or more than one component as part of clean and/or polluted air has been categorised. On the other hand, we also included pupils' ideas about what are the concrete clean and polluted air components they refer to.

Table 11.1 Categories used to codify pupils' ideas on the structure and nature of clean and polluted air

		Category	Definition
Structure of clean and polluted air	**Idea of continuity**	**Continuous (Cont)**	Pupils describe air or pollution structure as a continuum substance, with no underlying structure.
		Semi-continuous (Semicont)	Pupils describe air or pollution using an initial concept of "particle". They understand these "particles" as granules or small parts of one substance embedded in the same or other continuous substance, that is, with some supporting material between them
		Discontinuous (Discont)	Pupils describe air or pollution structure using a concept of "particle". They understand these "particles" as granules or small meso/micro (non-macroscopic) entities without any supporting material between them.
	Idea of scale	**Macro scale (Macro)**	Pupils refer to clean and/or polluted air as materials or substances. These substances can be perceptible by the senses and closely connected to human scale (seen with the naked eye).
		Meso scale (Meso)	Pupils refer to clean and/or polluted air as structures made by "particles" in terms of smalls parts of the substance that have the same properties of the substance. These structures are small but could be seen with a magnifier.
		Atomic-Molecularscale (Micro)	Pupils refer to clean and/or polluted air as structures made by "particles" in terms of atoms or molecules, that is, particles that do not share the properties of the substance. These particles could not be seen with any magnifier.
Nature of air and pollution	**Idea of components**	**Do not represent Earth's atmosphere components (No Rep)**	Pupils do not represent or describe the nature of air and/or pollution. They focus on sources, express tautological answers…
		Natural Earth's *a*tmosphere *c*omponents(NEAC)	Pupils refer to air and/or pollution as made by natural Earth's atmosphere components like O_2, CO_2, N_2, noble gases…
		No natural Earth's atmosphere components (NoNEAC)	Pupils refer to air and/or pollution as made by no natural Earth's atmosphere components like virus and bacteria, dust, fumes,…
	Idea of mixture	**1 component**	Pupils represent air or pollution as made up by a single component.
		+1 component	Pupils represent air or pollution as made by a mixture of components.

11.7 Results and Discussion

11.7.1 Pupils' Ideas Regarding Clean and Polluted Air as Seen with the Naked Eye

The analysis of pupils' ideas on clean and polluted air as seen with the naked eye shows that their ideas are quite homogeneous: most of the children (83%) initially identify clean air as a continuous transparent substance. However, when pupils refer to polluted air in initial productions, they express diversity of ideas. The number of children who represent polluted air as transparent is only 30%. In other 30% of the cases, pupils represent air pollution as a visible grey substance (Figs. 11.1 and 11.2). A small number of children (13%) relate air pollution with the emergence of "particles" that we can see without any magnifier. A smaller percentage of them (8%) associate air pollution to macro objects such as trash on (e.g. plastics, bags). These results are in line with results obtained by Pruneau et al. (2005).

After the TLS the percentage of pupils' that identify clean and polluted air both as transparent substances without introducing any other macroscopic entity is 83% This means that most students after the TLS maintain their adequate macroscopic view of clean air and substantially modify in an adequate way their macroscopic view of polluted air. As such, at the end of the TLS pupils' have overcome the important alternative idea that pollution only exist when we can feel it (Thornber et al., 2016).

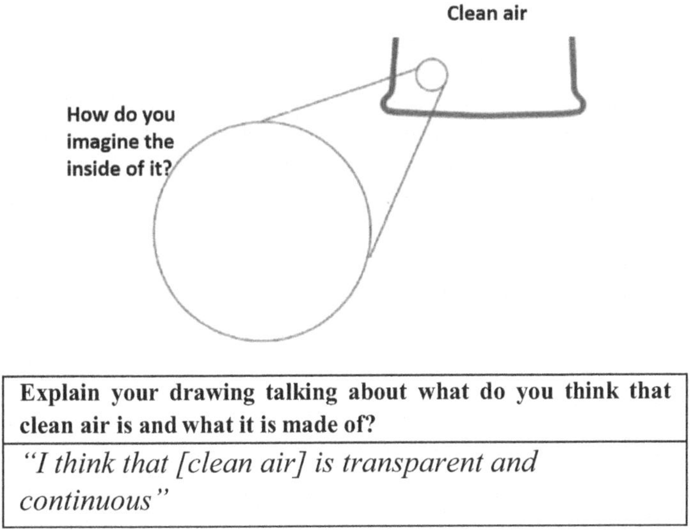

Fig. 11.1 Initial production of P26 about how they imagine clean air if you look inside of it. They imagine clean air as continuous matter in a macro scale. Also, they do not represent the Earth's atmosphere components (NoRep)

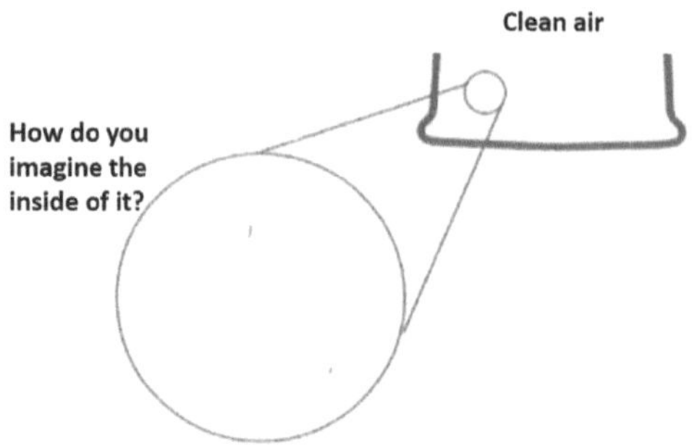

Explain your drawing talking about what do you think that clean air is and what it is made of?
"I think that [clean air] is made of by clean particles and O2, H2O, Nitrogen and CO2"

Fig. 11.2 Final production of P26 about how they imagine clean air if you look inside of it. Clean air is imagined as semi-continuous matter in a meso scale (there are little dots in their drawing that are not included in the naked eye representation which have macro properties such as being clean). Different components of the natural atmosphere are also identified (NEAC)

11.7.2 Pupils' Ideas Regarding Clean Air When Looking Inside of It

Analysis of children's ideas on clean air when looking inside of it with an imaginary magnifier was expected to show pupils' ideas on continuity and scale compatible with an initial particle model of matter. However, a large range of understandings of continuity and use of diverse scales is found in pupils' drawings and explanations.

Regarding ideas on structure, most of the pupils (92%) conceived air as matter, with more than half of the pupils (58%) expressing a continuous and macroscopic view of clean air at the beginning (Fig. 11.1 for an exemplary answer). This is coherent with previous research which identifies the idea of discontinuous matter and also the idea of scale as some important challenges for primary school education (Driver et al., 1994; Hadenfeldt et al., 2014; Meijer et al., 2013). In their final productions, however, an idea of gas as semi-continuous matter (including an initial concept of "particles" floating in a continuous supporting material) is expressed by 72% of the children (Fig. 11.2 for an exemplary answer). In agreement with previous research we consider this semi-continuous model of air as an adequate enough idea on the structure of gas matter for 11 and 12 years-old pupils (Karata et al., 2013), as it is a necessary conceptual step towards a more sophisticated model of gas matter (Talanquer, 2009).

Additionally, a change in the scale in which children represent clean air is highlighted. As mentioned, in the pre-productions children usually use the idea of a macro scale. However, in the post-productions the most usual representation of the inside of clean air is in the meso scale (91%) (Fig. 11.2). Despite not being the correct scale to represent the air molecules and atoms that made clean air, in agreement with Meijer et al. (2013), we consider this meso scale useful for pupils' gradual learning about atomic-molecular particles.

Regarding pupils' ideas on the nature of clean air, it is important to highlight that there are many differences between children's ideas about what clean air is made of independently on their quite homogeneous pre-ideas of structure. Initially 79% of them see clean air as made of a unique substance (1 component). In most of the cases, they refer to it as made by air or oxygen. As Driver et al. (1994) pointed out, pupils often use the word oxygen and other air gases as synonymous to "air." At the end of the TLS, about 78% of children understand air as made of more than one component. If we focus on the type of components that pupils talk about, we can see that initially most of their productions (79%) do not specify that (See Fig. 11.1 for an exemplary answer). They use tautological expressions (like "air is made of air") or focus their attention on the sources (like "clean air appears in the forest"). At the end of the TLS there is an increase in the number of children that mention that air is made of different gases commonly present in the atmosphere or NEAC (47%) (Fig. 11.2). However, only in few cases pupils mention all natural earth's atmosphere gases.

Interestingly, in their final productions some pupils' answers include representation of pollution or pollutants on their clean air drawings (NoNEAC 28%) (Fig. 11.3). We attribute it to the fact that in the TLS the "particles" of some common materials

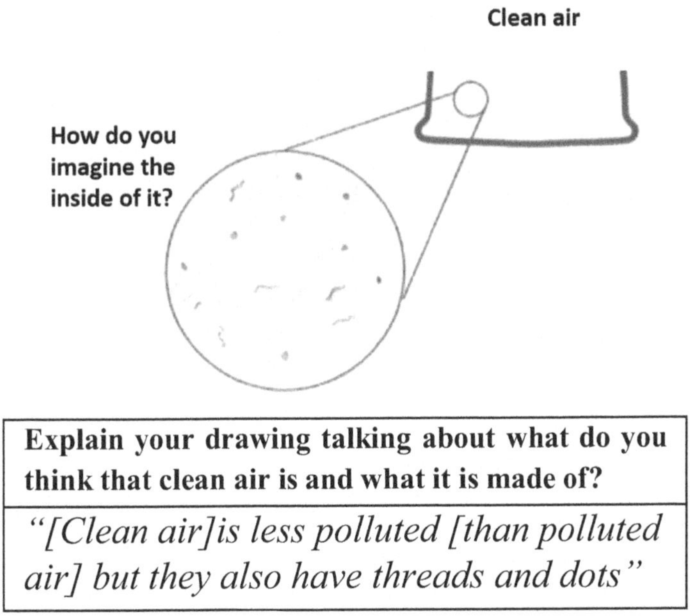

Fig. 11.3 Final production of P16 about how they imagine clean air if you look at its inside. They also include some pollutants (threads...) on their drawing and writing of clean air

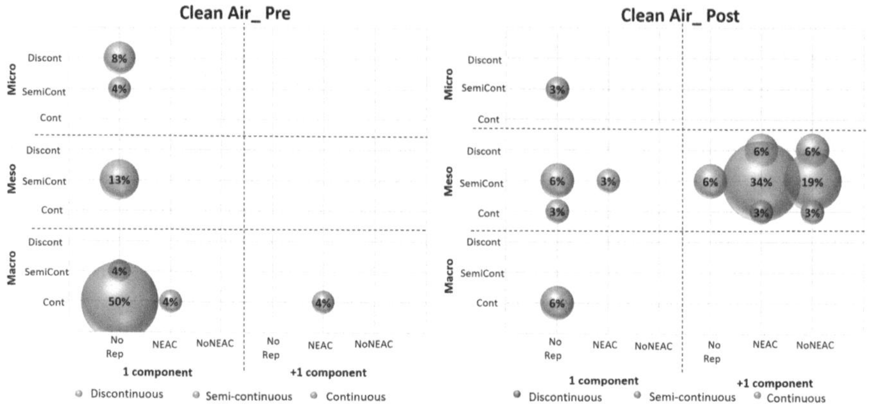

Fig. 11.4 Co-occurrence analysis between structure and nature of clean air in pupils' initial and final productions

such as dust, sand or pollen are introduced as natural pollutants. This can influence their representations of clean air, as rarely pupils identify "natural" sources with pollution (Thornber et al., 2016).

The percentages mentioned above for each of the categories that describe structure and nature of clean air are included in Fig. 11.4, which represents a co-occurrence analysis between both dimensions at the pre and post moment. One relation that can be identified is that children whose ideas of clean air involve a macroscopic view of matter usually think that clean air is made of a unique substance. However, in the case of children whose ideas of clean air involve the meso scale we cannot identify any direct relationship between structure and nature.

11.7.3 Pupils' Ideas About Pollution When Looking Inside of It

Our analysis of children's ideas about polluted air when looking inside of it expected to show pupils' ideas of matter at a meso scale level. However, as we can see in clean air, pupil's use a large range of scales specially in their initial productions.

Regarding the discontinuity of matter of polluted air in initial productions, 22% of students identify polluted air as continuous. However, even at an initial stance most students' express the idea of a certain semi-continuity in the polluted air structure (74%), which implies that pollution is understood by children as something added to the air (Fig. 11.5). Discontinuous ideas about matter applied to polluted air are very scarcely mentioned (4%). The percentage of these more sophisticated ideas of semi-continuity increase in students' final productions (97%) (Fig. 11.6).

The most important change between pre and post pupils' ideas regarding the structure of air pollution are the scale in which they represent the phenomena. In a

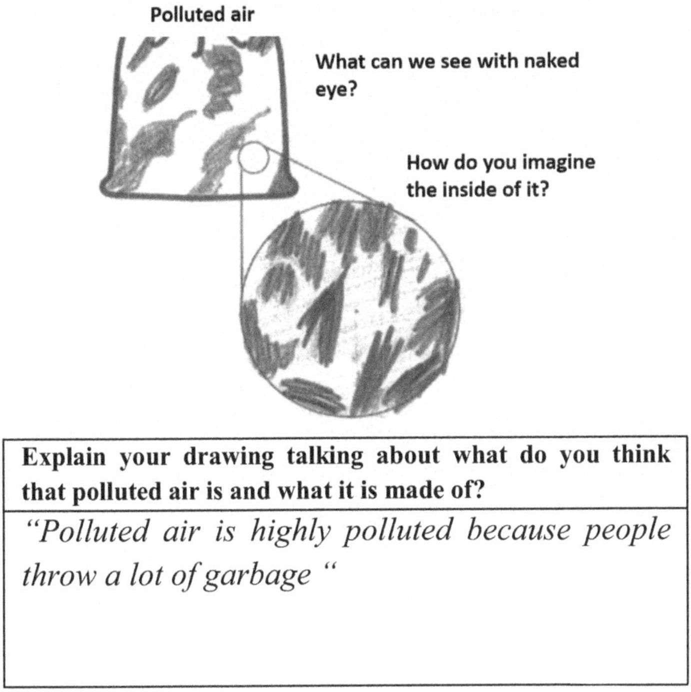

Fig. 11.5 Example of the most common initial production about how they imagine polluted air when you look at the inside of it. Pupil 04 imagines polluted air as a semi-continuous matter in a macro scale. Also, they represent pollution as a single NoNEAC (garbage)

similar way as with clean air, at the beginning, more than a half of the pupils (59%) represent air pollution in a macro scale. At the end of de TLS, on the contrary, 97% of pupils' answers are in a more adequate meso scale.

The pupils' ideas on both dimensions of structure (continuity and scale) of polluted air show us that there is a more sophisticated model of matter regarding structure in the context of air pollution than in the context of clean air. This can be related to the difference of scale of both phenomena, which allows to have direct experience of the existence of pollutants but not of atomic particles. In fact, when we talk scientifically about pollution, we are necessarily in the meso scale, as "particles" of pollutants are from 10^{-7} y 10^{-5} m in suspension in the air. This makes possible to trap pollutants (for instance with a filter or sticky surface) that cannot be seen with the naked eye but can be observed with a magnifier, as pupils do in the TLS. In this sense, students in the TLS have a direct experience with the meso scale that they cannot obviously have with the atomic-molecular scale, as the particles of air (atomic and molecular particles) are of 10^{-10} y 10^{-9} m and can only be conceived in our imagination.

Regarding the nature of pollution, our results diverge strongly from previous research, such as that of Thornber et al. (2016), whose focus was mostly on gas pollutants as CFCs and Ozone in the context of the greenhouse effect. In our research

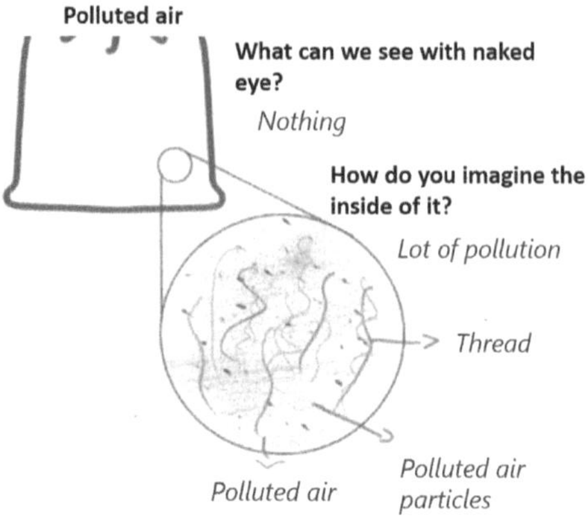

Explain your drawing talking about what do you think that polluted air is and what it is made of?
"[Polluted air] is made of by carbon dioxide, threads and polluted particles. I also draw black on the back because of threads and carbon dioxide"

Fig. 11.6 Example of the most common final production about how they imagine polluted air when you look at the inside of it. Pupil 05 imagines polluted air as a semi-continuous matter in a meso scale (we can only see pollution in the non-naked eye representation). Also, they identify different NoNEAC (threads, polluted air particles…)

in the context of polluted air in cities, we have found two different ways of understanding pollution: (1) pollution as an emergence or change in NEAC, like CO_2; and (2) pollution as an emergence or change in NoNEAC, like particles exhausted by cars, microorganisms, dust, etc. Almost all the children's final productions include the idea of pollution as an emergence or change in NoNEAC (63%) (Fig. 11.7). This idea is closer to the scientific idea of pollution in the TLS.

A deeper analysis about the NoNEAC shows us that the most important ideas that have emerged in pupils' representations after the TLS are the idea of "particles" coming from cars or industry (pre 27% and post 71%) and the idea of threads or hairs (pre 0% and post 79%) as a pollutants (See Fig. 11.7 for exemplary answers). Also, in a coherent way with Pruneau et al., (2005), thrash is one of the ideas present in the initial drawings and explanations of pupils in a quite representative percentage (21%). These naïve ideas on the nature of polluted air have almost disappeared in the final productions.

On the other hand, a "contamination" view is also an important alternative idea in children's productions. As Dimitriou and Christidou (2007, p. 26) explained: "contamination refers to the presence of pathogenic microorganisms in the air", while "pollution" refers to the "presence of gaseous, solid or liquid substances". In Spanish, the words "contaminación" and "polución" are frequently used as

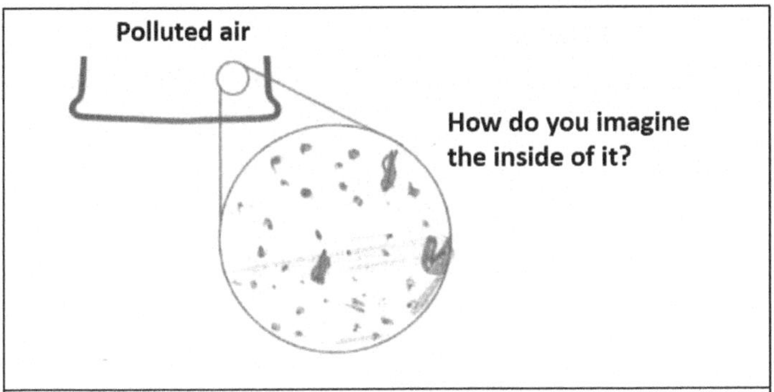

Explain your drawing talking about what do you think that polluted air is and what it is made of?

"Polluted air is grey on the back and it has lot of black dots that are soot and asphalt. Polluted air is also made of by O2, H2O, CO2 and Nitrogen."

Fig. 11.7 Example of pupil's final production that identify car and/or industry particles (soot and asphalt) as pollutants

equivalent. This can explain the 8% of pupils who identify microorganisms as a pollutant (Fig. 11.8). After the CoVID-19 situation, it is possible that this view is even more present in students' answers.

In a similar way as with clean air, most of the children at the beginning think about pollution as made of one single component (70%). However, at the end of the TLS 93% of pupils talk at least about two components when referring to pollutants.

The co-occurrence analysis crossing students' views on both the structure and nature of polluted air as shown in the pre and post questionnaire is included in Fig. 11.9. We can see that in the case of polluted air there is a very important improvement between pre and post pupils' views, shifting from very diverse pre-ideas that combine all the scales, continuity, and nature views possible, to a majority view of pollution as a semi-continuous phenomenon in the meso scale that involves more than 1 component, mostly NoNEAC.

11.8 Conclusions and Implications

Our analysis about children preconceptions shows that the two main challenges for 10–12-year-old students when modelling clean and polluted air are: (1) Overcoming the idea of air as a continuous substance and (2) Appreciating air as a mixture made of different components. These results are in agreement with previous research on this topic (Driver et al., 1994; Hadenfeldt et al., 2014; Talanquer, 2009).

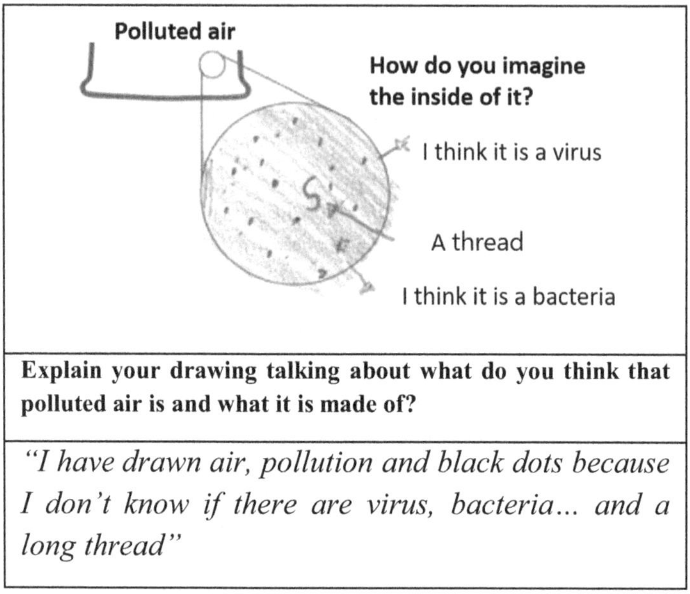

Fig. 11.8 Example of pupil's final production that identify infectious particles (virus and bacteria) and thread as pollutants

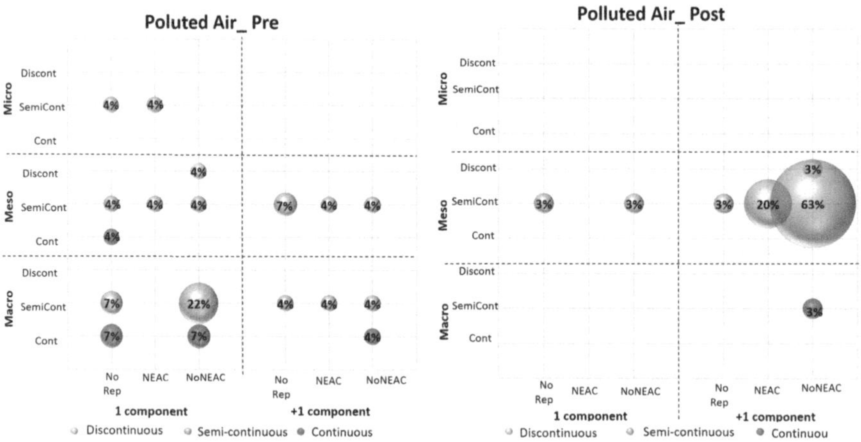

Fig. 11.9 Co-occurrence analysis between structure and particle nature of air pollution pupils' initial and final ideas

Additionally, this research has pointed out that involving children in a model/modelling-based TLS about air pollution is a promising context for improving pupils' ideas on matter and for overcoming challenges. Our results show it is easier for children to build the idea of semi-continuity regarding "particles" in the meso scale (particles as parts or particulate matter) than in the atomic-molecular scale (particles as atoms and molecules), as Meijer et al.(2013) have pointed out. This signals the adequacy of the context of air pollution as one that triggers students' initial ideas closer to the scientific ones. In addition, we consider that the promising results identified are related to the modelling nature of the TLS, that allows confronting initial ideas in order to promote conceptual progression and model-evolution. For instance, having pupils express their initial models and test them experimentally, observing that even when they do not see anything in the air with their naked eye there can be different entities in it that can be trapped and observed with a magnifier glass, shows to impact strongly their final views, in which they add a lot of the observed elements (e.g. threads, sand).

Despite the improvement in children's ideas about matter in the context of clean and polluted air, the final productions of pupils still show some important alternative ideas, like the idea of air particles embedded in some supporting material or that particles have the same properties as the whole substance (semi-continuity idea). As some previous researchers have pointed out, building these ideas requires a deeper understanding of the atomic-molecular scale (Hadenfeldt et al., 2014; Talanquer, 2009). However, from a learning progression perspective, the ideas that pupils have built in the TLS can be understood as a necessary step to move forward to more adequate ideas regarding the particle model of matter.

In this sense, our results suggest that in primary school education we need to explore deeper the potential for children's understanding of matter of building ideas about different phenomena from a macro and meso scale perspective, as a previous step to the introduction of the atomic-molecular and subatomic scale.

Acknowledgement This research was done in PhD Education programme at UAB and partially funded by Spanish Government (PGC2018-096581-B-C21), ACELEC group (2017SGR1399) and FECYT (FCT-17-11955). Tena was supported by AGAUR under FI programme.

References

Acher, A., Arcà, M., & Sanmartí, N. (2007). Modelling as a teacher learning process for understanding materials: A case study in primary education. *Science Education, 91*(1), 398–418.

Chinn, C., & Buckland, L. (2012). *Model-based instruction: Fostering change in evolutionary conceptions and in epistemic practices* (Issue October 2016).

Couso, D. (2017). Participatory approaches to curriculum design from a design research perspective. In *Iterative design of teaching-learning sequences: Introducing the science of materials in European SCHOOLS*. Springer.

Couso, D., & Garrido-Espeja, A. (2017). Models and modelling in pre-service teacher education: Why we need both. In *Cognitive and afective aspects in science education* (Vol. 3, pp. 245–261).

DBR Collective. (2003). Design-based research: An emerging paradigm for educational inquiry. *Educational Researcher, 32*(1), 5–8.

Dimitriou, A., & Christidou, V. (2007). Pupils' understanding of air pollution. *Journal of Biological Education, 42*(1), 24–29.

Driver, R., Squires, A., Rushworth, P., & Wood-Robinson, V. (1994). Part III: Materials and their properties. In *Making sense of secondary science: Research into children's ideas* (pp. 137–229). Routledge.

Duschl, R. A., Maeng, S., & Sezen, A. (2011). Learning progressions and teaching sequences: A review and analysis. *Studies in Science Education, 47*(2), 123–182.

Hadenfeldt, J. C., Liu, X., & Neumann, K. (2014). Framing students' progression in understanding matter: A review of previous research. *Studies in Science Education, 50*(2), 181–208.

Harlen, W. (2010). *Principles and big ideas of science education*. Association for Science Education.

Karata, F. Ö., Ünal, S., Durland, G., & Bodner, G. (2013). What do we know about students' beliefs? Changes in students' conceptions of the particulate nature of matter from pre- instruction to college. In *Concepts of matter in science education* (pp. 231–247). Springer.

Lehrer, R., & Schauble, L. (2019). Learning to play the modelling game. In *Towards a competence-based view on models and modeling in science education* (Vol. 145, Issue 3, pp. 221–236). Springer.

Mandrikas, A., Stavrou, D., & Skordoulis, C. (2017). Teaching air pollution in an authentic context. *Journal of Science Education and Technology, 26*(2), 238–251. https://doi.org/10.1007/s10956-016-9675-8

Meijer M., Bulte A. M. W., & Pilot, A. (2013). *Macro-micro thinking with structure-property relations: Integrtatin "meso-levels" in secondary education. 19*, 485–520.

Merritt, J., & Krajcik, J. (2013). Learning progression developed to support students in building a particle model of matter. In G. Tsaparlis & H. Sevian (Eds.), *Concepts of matter in science education* (Vol. 19, pp. 11–45). Springer.

Miles, M. B., Huberman, A. M., & Saldaña, J. (2014). Qualitative Data Analysis (3rd Editio). Arizona State Univeristy.

OCDE. (2020). *PISA 2024 strategic vision and direction for science* (Issue March).

Oh, P. S., & Oh, S. J. (2011). What teachers of science need to know about models: An overview. *International Journal of Science Education, 33*(8), 1109–1130.

Pruneau, D., Richard, J. F., Langis, J., Albert, G., & Cormier, M. (2005). The evolution of children's ideas on pollution in the framework of experiential and socioconstructivist activities [1]. *International Journal of Environment and Sustainable Development, 4*(1), 17–34.

Schwarz, C. V., Reiser, B. J., Davis, E. A., Kenyon, L., Achér, A., Fortus, D., Shwartz, Y., Hug, B., & Krajcik, J. (2009). Developing a learning progression for scientific modeling: Making scientific modeling accessible and meaningful for learners. *Journal of Research in Science Teaching, 46*(6), 632–654.

Solé, C., Tena, È., & Couso, D. (2020). ¿Qué modelo de materia explica la contaminación? Explorando el modelo corpuscular en la mesoescala. *Alambique: Didáctica de Las Ciencias Experimentales, 101*, 30–36.

Taber, K. S. (2011). Constructivism as educational theory: Contingency in learning, and optimally guide instruction. In *Education theory* (pp. 89–100). Nova Science Publishers.

Talanquer, V. (2009). On cognitive constraints and learning progressions: The case of "structure of matter". *International Journal of Science Education, 31*(15), 2123–2136.

Thornber, J., Stanisstreet, M., & Boyes, E. (2016). School students' ideas about air pollution: Hindrance or help for learning ? *Journal of Science Education and Technology, 8*(1), 67–73.

Chapter 12
Undergraduates' Grasp of Evidence for Evaluating Scientific Knowledge Claims Associated with Socioscientific Issues

Won Jung Kim and Alicia C. Alonzo

12.1 Introduction

There is widespread agreement that science education should produce scientifically literate people who can relate and apply science to their lived experiences (Feinstein, 2011; National Research Council, 2012)—e.g., making decisions about socioscientific issues (SSIs). SSIs are personally- and socially-meaningful science-related issues (Rudolph & Horibe, 2016; Zeidler et al., 2005). Making informed decisions about SSIs requires people to evaluate the trustworthiness of associated scientific knowledge claims. Otherwise, people are vulnerable to being persuaded to take positions that are not in their own best interest (or in the interest of their communities). However, SSI-related knowledge claims are particularly challenging to evaluate because they are often uncertain, due to inherent uncertainty in scientific claims, and conflicting, due to different stakeholders marshalling evidence to support different viewpoints (Kolstø, 2001).

There are many examples of people making decisions about SSIs without critically evaluating associated knowledge claims. For example, in deciding not to vaccinate their children, parents may be convinced by personal testimonials—rather than critically evaluating testimonials in light of scientific evidence refuting a link between vaccines and autism. In order to support informed decision-making about SSIs, it is crucial to describe, in detail, the practice of critically evaluating uncertain and conflicting scientific claims. This practice has been considered within the larger context of research on epistemic cognition (e.g., Chinn et al., 2011; Lombardi et al., 2016; Sinatra et al., 2014).

W. J. Kim (✉) · A. C. Alonzo
Michigan State University, East Lansing, MI, USA
e-mail: kimwon10@msu.edu; alonzo@msu.edu

O. Levrini et al. (eds.), *Engaging with Contemporary Challenges through Science Education Research*, Contributions from Science Education Research 9,
https://doi.org/10.1007/978-3-030-74490-8_12

Within the tradition of epistemic cognition, Duncan et al. (2018) recently proposed Grasp of Evidence (GOE) as a theoretical framework to describe and support understandings of evidentiary reasoning necessary for engaging with science as a 'competent outsider' (Feinstein, 2011). Duncan et al. argue that, although science education standards, such as the US *Next Generation Science Standards* (NGSS Lead States, 2013), highlight the importance of evidence, they do not explicate 'the epistemic features and roles of evidence' (p. 909) necessary for sophisticated and complex engagement with evidence. The GOE framework seeks to differentiate among different forms of evidence and among different ways of engaging with evidence. GOE is particularly relevant for our interest in SSIs because uncertain and conflicting claims require 'evaluation in a framework of alternatives and evidence' (Kuhn et al., 2017, p. 233).

In this chapter, we take up Duncan et al.'s (2018) call to 'explore the utility of [the GOE] framework as an analytic tool' (p. 933), using this framework as the lens for examining undergraduate students' critical evaluation of SSI-related knowledge claims. In particular, we ask: *How do students' responses to socioscientific scenarios reveal their GOE?* By answering this question, we explore use of the GOE to examine evaluation of SSI-related scientific knowledge claims and use our data to provide empirical illustration of the GOE framework.

12.2 Grasp of Evidence: Laypeople's Understanding of Evidence

Duncan et al. (2018) argue that 'laypeople need to grasp two distinct, yet interrelated, aspects of evidence' (p. 910): experts' use of evidence (i.e., how scientific claims are generated) and laypeople's use of evidence (i.e., how non-experts can use evidence to engage with science). Different ways of engaging with evidence are represented as five dimensions of the GOE framework. Four dimensions reflect experts' evidentiary practices: *analysis* (identifying and comprehending components of scientific studies), *evaluation* (examining the quality of evidence), *interpretation* (examining the strength of evidence), and *integration* (identifying and weighing relevant evidence). The fifth dimension reflects laypeople's use of secondhand reports of evidence (Sharon & Baram-Tsabari, 2020).

Within each of these dimensions, Duncan et al. (2018) use the AIR model of epistemic cognition (Chinn et al., 2014) to specify epistemic components of each practice. *Epistemic Aims and Values* (EAs) 'are the kinds of epistemic products' people 'set to achieve (aims)… and the importance of those products (values)'; *Epistemic Ideals* (EIs) 'are the criteria used to evaluate whether epistemic aims have been achieved' and *Reliable Epistemic Process* (REPs) 'are the diverse processes' used 'to achieve epistemic aims' (p. 914).

Duncan et al. (2018) provide the EA, along with examples of EIs and REPs, for each dimension of the GOE framework; we referred to these to inform our analysis of students' responses to the socioscientific scenarios. For example, the EA of the evidence evaluation dimension is 'determining if evidence is of high quality and whether conclusions can be trusted'. An EI example is '*Conclusiveness* (i.e., ruling out confounds and alternative explanations for the findings), and an REP example is 'Evaluating the appropriateness of study design (e.g., appropriate samples and comparisons)' (p. 915).

12.3 Method

Data collection for this study took place in the context of a semester-long interdisciplinary course at a large public university in the western United States. The course aimed to help undergraduate students apply scientific-style critical thinking to make decisions about scientific and non-scientific issues by introducing concepts and principles that scientists have developed to generate and evaluate knowledge claims.

12.3.1 Data Sources

The primary data source for this study was transcripts of interviews conducted with 15 of the 95 students enrolled in the course. Students were selected by stratified random sampling, considering academic year and major, from the 72 students who agreed to be interviewed. Each student participated in a total of 3–5 one-hour video-recorded interviews, once every 2–3 weeks after the first one-third of the semester. In each interview, participants were prompted to respond to questions about scenarios designed to mimic everyday decision-making and to share the reasoning behind their responses.

Across the interviews, students responded to a total of 36 scenarios. Initially, we selected 10 scenarios (with a total of 96 responses) that (1) prompted students to evaluate claims and make decisions and (2) effectively elicited the reasoning underlying students' evaluations and decision-making.

As described below, in this chapter, we focus on two of these scenarios: CFC and Chocolate (see Appendix). The CFC scenario asks students to discuss how a legislator would go about deciding whether to ban a type of chemical (CFCs). Students consider arguments provided by scientists (for a ban) and industry representatives (against a ban). The Chocolate scenario asks students to decide whether they would change their dietary habits based on a news report of a study claiming chocolate causes weight loss.

12.3.2 Data Analysis

In order to understand students' critical evaluation of the socioscientific scenarios, we conducted an initial grounded theory analysis of students' transcripts (Charmaz & Belgrave, 2012), along with a concurrent review of relevant literature, using two iterative steps. First, we open-coded idea units for features that appeared relevant to the cautious and informed evaluation of scientific knowledge claims, incorporating these features into an evolving definition of a construct we called 'epistemic critique'. Second, we connected these features to relevant concepts from the literature. Towards the end of our iterative process, we recognized that Duncan et al.'s (2018) GOE framework captured much of what we sought to describe using 'epistemic critique', prompting our current investigation.

The two frameworks seemed to identify the same aspects of people's evaluation of scientific knowledge claims; however, there was not a one-to-one correspondence between the GOE framework and our features of *epistemic critique*. Therefore, we conducted a new analysis of the CFC and Chocolate scenarios using the GOE framework in order to explore its utility. We chose these two scenarios for several reasons: (1) in our initial, iterative analyses, we had coded these two scenarios using the AIR framework (Chinn et al., 2014); (2) as compared to the other eight scenarios, they more explicitly elicited students' understandings of evidence; and (3) features identified in responses to these scenarios corresponded to a range of components of the GOE framework.

We conducted a content analysis of responses to the CFC and Chocolate scenarios (13 responses to each) using Duncan et al.'s (2018) GOE framework: five evidentiary practices and three epistemic components, along with specific examples of the epistemic components within each of the five dimensions.

The two authors independently coded idea units in the interview transcripts using the GOE framework and then discussed to develop a shared understanding of the GOE framework in relation to our data, as well as consensus as to the applied codes. This process was aided by previous discussions of the interview transcripts as part of our initial analyses.

During the coding process, we recognized that the GOE framework did not fully capture students' engagement with the scenarios. In particular, epistemic concepts—those required for evaluating the trustworthiness of knowledge claims—were not explicitly included, yet seemed important for describing how students evaluated SSI-related scientific knowledge claims. Thus, we added another grounded theory analysis. Similar to our initial analysis, we articulated epistemic concepts through an iterative process of coding and consultation of relevant literature.

12.4 Findings

In our data we identified evidence of students' understandings of: (1) the two types of evidentiary practices (experts' and laypeople's) and (2) three of four dimensions of experts' use of evidence. We further identified epistemic concepts that seemed to underlie students' GOE and to account for meaningful differences in their evaluations of SSI-related scientific knowledge claims.

12.4.1 Understanding of Experts' and Laypeople's Use of Evidence

The GOE framework helps to illuminate whether students drew on understandings of experts' or of laypeople's evidentiary practices. We illustrate this difference with Evan's and Tyler's responses to the CFC scenario. While both students considered the scientists' claims to be more trustworthy than the aerosol companies' claims, they focused on different types of evidentiary practices. Evan attended to experts' evidentiary practice:

> If a large portion of the [scientific] community had independent studies, like if a lot of studies found the same result, I would be more inclined to believe it. then it's much harder to deny or just to step aside and say we don't know yet.

In this excerpt, Evan exhibited understanding of EI 'Replicated evidence' (Duncan et al., 2018, p. 916) from the evidence integration dimension by indicating that he would tend to trust claims that draw on replicated evidence from multiple studies.

In contrast, Tyler attended to laypeople's evidentiary practice:

> The aerosol industry, like they'd be so biased on like ... I would definitely like lean towards the side of the scientists because like they're more experts; they have like a better opinion. ... This [what causes Ozone depletion] was never like a polarized issue. They [scientists] basically like discovered it. It's not like they were polarized before and like were trying to figure out more about it.

Tyler exhibited understanding of EI 'Source trustworthiness' (degree of expertise, integrity, lack of bias, etc.)' (Duncan et al., 2018, p. 916). In this excerpt, he considered both scientists' lack of bias and status as experts. First, while identifying the aerosol industry as biased, he seemed to absolve scientists of similar bias, suggesting that—because scientists discovered Ozone depletion before it was a polarizing issue—their findings would not have been biased by the controversy presented in the scenario. Second, he identified scientists as having more expertise ('they're more experts', they have a 'better opinion').

12.4.2 Understanding of Practices Within Scientists' Use of Evidence

When attending to understandings of experts' evidentiary practices, students focused on three of the four dimensions in the GOE framework: *evaluation*, *interpretation*, and *integration*. Overall, the Chocolate scenario prompted students to examine the study using understandings of the *evaluation* dimension, since the accompanying questions (see Appendix) focused students' attention on the study design. However, as illustrated below, students also exhibited understandings of the *integration* and *interpretation* dimensions.

We use Asra's example as representative of how all 13 students exhibited GOE in the *evaluation* dimension:

> [T]hey have a control group; they have three groups actually in this case. And it seems like they did it right ...[However], I'm still sceptical... A period of 21 days ... that's not enough... so, I would want to see this study replicated and possibly redone in different ways before I'd be ready to completely change my diet because of it. There's only 16 adults in this, age what? 19–67, so I mean that's a pretty good range age wise, but... you've got 16 people divided into three groups, ... They haven't even taken a random sample here. ... You need to have... a larger scope; you can't just be testing five people and assume that it's representative of the population.

Asra's concerns are indicative of the REP 'Evaluating the appropriateness of study design (e.g., appropriate samples and comparisons)' (Duncan et al., 2018, p. 915). Although Asra evaluated the study design positively in terms of the inclusion of a control group and the wide range of ages represented by study participants, she expressed scepticism due to the study's duration (21 days) and the small sample size, noting that 5 people in a given treatment group would not be representative of the broader population.

Asra's excerpt also illustrates the *integration* dimension. Like Evan (responding to the CFC scenario), Asra demonstrated understanding of the EI *Replicated evidence* by explicitly calling for the study to be replicated. In addition, her call for the study to be 'redone in different ways' suggests attention to the EI '*Variety* of evidence (i.e., multiple types/lines of evidence)' and/or the EI '*Consistency of support* (i.e., lack of contradictory evidence)' (Duncan et al., 2018, p. 916). It is unclear what Asra means by 'redone'; however, her call for the study to be done 'in different ways' suggests understanding of the value of additional confirmatory evidence.

In contrast to Asra, James exhibited understanding of the *interpretation* dimension. James attended to the REP 'Developing arguments that systematically connect evidence to models' (Duncan et al., 2018, p. 915):

> If they gave like a complete explanation of what the chocolate does to your body to make you lose weight, and maybe young people, I'd be less sceptical. ... I would just be curious about what kind of chocolate, and what in the chocolate is actually making you lose weight. Even if I knew that the study was valid, I'd want to know why, biologically, like how that works... like a deeper explanation.

We interpreted James's expressed desire to understand the mechanism behind evidence of chocolate's effect on weight loss as related to an understanding of scientists' work to connect evidence to explanatory models.

12.4.3 Epistemic Concepts Underlying Students' GOE

In addition to illustrating dimensions of the GOE framework, we also unpacked students' reasoning to identify epistemic concepts that seemed to underlie their GOE. Here, we describe two sets of epistemic concepts that appeared particularly important in students' responses to the CFC and Chocolate scenarios: inherent uncertainty of scientific claims; and randomized controlled trial (RCT).

Two lines of work were especially relevant to our efforts to capture understandings of these two epistemic concepts: dimensions of reliability in science (Allchin, 2011), a framework for understanding the nature of science, and the concepts of evidence framework (Gott et al., 2015; Roberts & Johnson, 2015), which further specifies relevant concepts from Allchin's framework by describing knowledge underlying understandings of scientific evidence. Inherent uncertainty of scientific claims is reflected in Allchin's concept 'error and uncertainty' (p. 525); concepts of evidence further unpack uncertainty by describing how scientists present 'confidence limits' to 'indicate the degree of confidence that can be placed on the datum' and explaining what specific confidence limits mean (Gott et al., 2015, p. 7). Similarly, Allchin's framework includes the concept 'controlled experiment', and the concepts of evidence framework explicitly defines 'randomised controlled trial (RCT)' as random assignment of a large sample to treatment groups, such that 'confounding variables will even out', leaving only the difference due to the treatment (Roberts & Johnson, 2015, p. 356).

12.4.4 Inherent Uncertainty of Scientific Claims

The concept of the inherent uncertainty of scientific claims appeared to underlie some students' understanding of laypeople's evidentiary practice, particularly regarding the REP 'identify who the experts are, including level and relevance of expertise'. This could be seen, for example, in Brooke's response to the CFC scenario:

> One thing I liked about scientists for instance was the fact that they did admit, 'Okay, there is this hole in the ozone layer, we don't 100% know what it is, but we kind of think this could be one of the reasons'. While it's like the other one [claim from the CFC companies] seems to be a lot more confident…, they can't be that sure.

Her consideration of scientists as more trustworthy than the aerosol industry seems to reflect understanding that scientific claims, particularly predictive ones as in this scenario, are inherently uncertain and that reporting levels of confidence in such claims can be a strength of the scientific process.

In contrast, the scientists' uncertainty made it difficult for Matt to trust their claim:

> Well, … if they can't exactly explain it because then it is really uncertain and it's just hard to take …side with the scientists just because they're saying, 'We think this but we can't say why we think this'.

In this excerpt, he did not exhibit the same understanding of the inherent uncertainty of scientific claims that Brooke exhibited.

12.4.5 Randomized Controlled Trial (RCT)

Concepts regarding RCT appeared to underlie some students' understanding of *evidence evaluation*, particularly the REP *Evaluating appropriateness of study design*, as demonstrated in responses to the Chocolate scenario. For example, Brooke and Caren both identified the use of control and experimental groups as an important criterion for determining the effect of one variable (chocolate) on another (weight loss). Using this concept, both evaluated the design of the chocolate study to be sound.

However, they differed in their understanding of the other important criterion for RCT: random assignment to treatment groups, requiring a sufficiently large sample. As illustrated below, Brooke demonstrated understanding of this concept, and Caren did not. Brooke considered a large sample crucial for randomization and, thus, for determining the effect of the target variable (chocolate diet):

> By randomizing it in like a bigger group, the odds are that we're going to get people with the specific genetic conditions and some that don't, some with these specific personal habits, some that don't. So in the overall like larger scale, these things are probably going to cancel out through randomizing … just like keeping everything intact and just changing one variable.

In contrast, Caren considered randomization into control and treatment groups to be sufficient (despite the small sample size):

> They did try to randomize the groups. And they did intervene actively on the groups … So, I mean that's a good thing they did there …The sample size is a little small … Maybe it would be good to at least start out with … It may be an advantage. … it may be good to start… Have a smaller group.

As these examples illustrate, specific epistemic concepts could be identified underlying the GOE that students exhibited. In particular, Brooke and Matt illustrate how the epistemic concept of inherent uncertainty of scientific claims may underlie the *layperson's* REP *Identify who the experts are*, while Brooke and Caren illustrate how epistemic concepts related to RCT may underlie the *evaluation* REP *Evaluating the appropriateness of study design*. In both cases, Brooke demonstrated understanding of the epistemic concept, while her counterpart did not. The contrasts between Brooke and Matt and Caren, respectively, provide some indication of how understanding of epistemic concepts may affect students' evaluations of SSI-related scientific knowledge claims.

12.5 Discussion

Drawing on our findings, we discuss both how the GOE framework was useful in analysing our data and how it could be further unpacked to increase its utility. First, the GOE framework allowed us to make important distinctions among students' understandings of different evidentiary practices. By distinguishing between experts' and laypeople's use of evidence, the GOE framework brings attention to the lay use of evidence. In our data, students engaged with both types of evidentiary practices when considering socioscientific scenarios. Although students' engagement in the practices of scientists (as advocated by current science education reforms, e.g., NRC, 2012) is vital to their understanding of scientists' use of evidence, engagement with SSIs may be important for developing students' understanding of laypeople's use of evidence and, thus, for empowering students as competent outsiders who use science wisely in their daily lives.

In addition, the GOE framework allowed us to distinguish the practice of *evidence evaluation* from other expert evidentiary practices: *evidence interpretation* and *evidence integration*. GOE associated with *evidence evaluation* can be seen in traditional images of students' engagement in science inquiry, which emphasise experimental investigation and, thus, understandings related to 'controlling variables' and 'identifying sources of error' (NRC, 2012, p. 43). By calling attention to other evidentiary practices, the GOE supports attention to other aspects of scientists' work—such as modelling (*evidence interpretation*) and working with evidence collected by others (*evidence integration*)—and, thus, to the importance of providing opportunities for students to engage with a range of scientific practices. As demonstrated by the different understandings students used to evaluate socioscientific scenarios, understandings related to these practices are important for engaging with scientific claims, not only as scientists, but also as citizens.

Second, although we found the GOE framework useful for making distinctions among different uses of evidence, unpacking the epistemic concepts underlying the framework may increase its utility. Brooke used the same components of the GOE framework as did her counterparts Matt and Caren, but Brooke drew on specific epistemic concepts to support her more well-informed evaluations of the socioscientific scenarios. Her understanding of the uncertainty of scientific claims allowed her to resist the aerosol industry's critiques of the scientists' claims in the CFC scenario, and her understanding of criteria for RCT allowed her to recognize a fatal flaw in the design of the chocolate study. In both cases, these epistemic concepts would allow Brooke, as a competent layperson, to avoid being fooled by those attempting to use science to persuade her. Although epistemic concepts may be inferred from the GOE framework, these concepts need to be unpacked so that students' opportunity to develop GOE is not unduly dependent on teachers' ability to make these inferences.

In our study, we unpacked only some epistemic concepts—and illustrated even fewer in this chapter. Our engagement with relevant literature suggests that others' frameworks may be useful for further articulating epistemic concepts underlying the GOE framework. For example, Gott et al.'s (2015) concepts of evidence describe 'a body of knowledge which underlies an understanding of scientific evidence' (p. 1), but at a much smaller grain size than that of the GOE framework. By articulating, in more detail, understandings associated with specific components of evidentiary practices, the concepts of evidence framework could be used to fill in some of the epistemic concepts underlying the GOE framework.

12.6 Conclusion

This study provides an empirical exploration of the utility of the GOE framework. As Duncan et al. (2018) suggested, we see implications of the framework (as well as our proposed unpacking of the framework) for researchers and science educators. First, the GOE framework seems useful for describing and distinguishing among understandings of different evidentiary practices, calling attention to practices that receive less emphasis in current educational settings. Focus on laypeople's use of evidence and on experts' interpretation and integration of evidence has the potential to allow researchers to learn more about and teachers to provide more support for students' understanding of these practices. Second, we suggest that epistemic concepts should be explicitly included in studies of and efforts to support students' GOE. Future studies could more systematically investigate how epistemic concepts relate to students' GOE. Such studies could consider a wide range of epistemic concepts, useful to engage with a variety of different SSIs. We hope that this empirical study will contribute to further investigations of, and support for, students' informed engagement with SSIs and, ultimately, decisions that are personally and societally beneficial.

Acknowledgement This research is funded by the Gordon and Betty Moore Foundation through Grant GMBF 4945 to Michigan State University (PI, Alicia Alonzo).

Appendix: Interview Scenarios

CFC

In the 1970s, there was a debate over chlorofluorocarbons (CFCs), commonly used in aerosols. On the basis of available information, the US government had to decide whether to institute a ban on CFCs. On one side, scientists reported an ozone hole over Antarctica, which was probably attributable to CFCs, although they could not explain exactly how it had formed. On the other side of the debate, the aerosol industry argued that (a) decisions should not be made on such uncertain science (e.g., natural causes – such as volcanoes – could also explain increased levels of chlorine in the stratosphere); (b) the scientists are biased by self-interest, playing up the risk of CFCs in order to obtain funding for their research; and (c) any substitutes that would be developed to replace CFCs would definitely be expensive, not to mention dangerous. If you were a congressperson tasked with making a decision about this issue, how would you go about weighing the information provided on the two sides of the debate?

Chocolate

1. Around Easter last year, headlines appeared, touting the benefits of chocolate for weight loss. What questions would you want to ask before you would be willing to change your diet on the basis of this study?
2. A study was carried out with 16 healthy German adults (aged 19–67) over a 21-day period. Subjects were randomly assigned to one of three treatment groups: a control group (instructed to make no changes to their current diet), one group that followed a low-carb diet, and one group that followed a low-carb diet supplemented with a daily 1.5 oz bar of dark chocolate. Based on a large number of pre- and post-diet blood tests and other health measures, the researchers reported that people on the low-carb diet plus chocolate regime lost weight 10% faster than did people on the low-carb only regime. Would you use the results of this study as the basis for a change in your diet? Why or why not?
3. Suppose another group of researchers is interested in a follow-up study to explore whether chocolate causes weight loss. They recruit 1000 participants and plan to randomly assign them to treatment (4 oz. of dark chocolate per day) and control (no change to daily diet) groups, following each group for 2 years.

However, they are concerned that other variables (such as genetics, personal eating habits, and overall feelings of wellbeing) may have a greater influence on participants' weight as compared to their consumption of chocolate. Will the researchers' plans to randomly assign participants to control and treatment groups be sufficient to address the researchers' concerns? Why or why not? If not, what would you recommend the researchers do to improve their study?

References

Allchin, D. (2011). Evaluating knowledge of the nature of (whole) science. *Science Education, 95*(3), 518–542. https://doi.org/10.1002/sce.20432

Charmaz, K., & Belgrave, L. (2012). Qualitative interviewing and grounded theory analysis. In *The SAGE handbook of interview research: The complexity of the craft* (2nd ed., pp. 347–366). Sage Publications, Inc.

Chinn, C. A., Buckland, L. A., & Samarapungavan, A. (2011). Expanding the dimensions of epistemic cognition: Arguments from philosophy and psychology. *Educational Psychologist, 46*(3), 141–167. https://doi.org/10.1080/00461520.2011.587722

Chinn, C. A., Rinehart, R. W., & Buckland, L. A. (2014). Epistemic cognition and evaluating information: Applying the AIR model of epistemic cognition. In D. Rapp & J. Braasch (Eds.), *Processing inaccurate information* (pp. 425–453). MIT Press.

Duncan, R. G., Chinn, C. A., & Barzilai, S. (2018). Grasp of evidence: Problematizing and expanding the next generation science standards' conceptualization of evidence. *Journal of Research in Science Teaching, 55*(7), 907–937. https://doi.org/10.1002/tea.21468

Feinstein, N. (2011). Salvaging science literacy. *Science Education, 95*(1), 168–185. https://doi.org/10.1002/sce.20414

Gott, R., Duggan, S., Roberts, R., & Hussain, A. (2015). *Research into understanding scientific evidence*. http://community.dur.ac.uk/rosalyn.roberts/Evidence/CofEv_Gott%20et%20al.pdf

Kolstø, S. D. (2001). Scientific literacy for citizenship: Tools for dealing with the science dimension of controversial socioscientific issues. *Science Education, 85*(3), 291–310. https://doi.org/10.1002/sce.1011

Kuhn, D., Arvidsson, T. S., Lesperance, R., & Corprew, R. (2017). Can engaging in science practices promote deep understanding of them? *Science Education, 101*(2), 232–250. https://doi.org/10.1002/sce.21263

Lombardi, D., Nussbaum, E. M., & Sinatra, G. M. (2016). Plausibility judgments in conceptual change and epistemic cognition. *Educational Psychologist, 51*(1), 35–56. https://doi.org/10.1080/00461520.2015.1113134

National Research Council (U.S.). (2012). *A framework for K-12 science education: Practices, crosscutting concepts, and core ideas*. The National Academies Press.

NGSS Lead States. (2013). *Next generation science standards: For states, by states*. The National Academies Press.

Roberts, R., & Johnson, P. (2015). Understanding the quality of data: A concept map for 'the thinking behind the doing' in scientific practice. *The Curriculum Journal, 26*(3), 345–369. https://doi.org/10.1080/09585176.2015.1044459

Rudolph, J. L., & Horibe, S. (2016). What do we mean by science education for civic engagement? *Journal of Research in Science Teaching, 53*(6), 805–820. https://doi.org/10.1002/tea.21303

Sharon, A. J., & Baram-Tsabari, A. (2020). Can science literacy help individuals identify misinformation in everyday life? *Science Education*. Advance online publication.

Sinatra, G. M., Kienhues, D., & Hofer, B. K. (2014). Addressing challenges to public understanding of science: Epistemic cognition, motivated reasoning, and conceptual change. *Educational Psychologist, 22*(2), 123–138. https://doi.org/10.1080/00461520.2014.916216

Zeidler, D. L., Sadler, T. D., Simmons, M. L., & Howes, E. V. (2005). Beyond STS: A research-based framework for socioscientific issues education. *Science Education, 89*(3), 357–377. https://doi.org/10.1002/sce.20048

Chapter 13
Psychological Patterns in Chemistry Self-Concept: Relations with Gender and Culture

Lilith Rüschenpöhler and Silvija Markic

13.1 Introduction

Although some studies about the chemistry self-concepts of secondary school students exist (e.g. Nielsen & Yezierski, 2016; Nieswandt, 2007), knowledge in this field is still limited. For instance, the distribution of chemistry self-concept regarding gender and cultural background remains unknown. This would be important to understand because research has shown that self-concepts impact career decisions (e.g. Eccles & Wang, 2016).

Regarding the field of science, much more information is available, both regarding the job market and the students' self-concepts. In Germany, more men than women work in science, which represents a gender gap that is present in many Western countries (OECD, 2009a). This is reflected in students' science self-concepts: in many countries, young women tend to have lower self-concepts in science than young men (e.g. Germany: Jurik et al., 2013; U.S.: Riegle-Crumb et al., 2011). Besides, students belonging to a population's dominant ethnic group tend to show stronger science self-concepts than those belonging to non-dominant ethnic groups (Rüschenpöhler & Markic, 2019a).

This article aims to provide more knowledge about chemistry self-concepts of secondary school students which is described in more detail in Rüschenpöhler and Markic (2020). It investigates the relation of secondary school students' chemistry self-concept with (i) the students' gender and cultural backgrounds, (ii) their learning goal orientations, and (iii) their perception of social support and the perception of their linguistic abilities in chemistry.

L. Rüschenpöhler (✉) · S. Markic
Ludwigsburg University of Education, Ludwigsburg, Germany
e-mail: rueschenpoehler@ph-ludwigsburg.de; markic@ph-ludwigsburg.de

O. Levrini et al. (eds.), *Engaging with Contemporary Challenges through Science Education Research*, Contributions from Science Education Research 9,
https://doi.org/10.1007/978-3-030-74490-8_13

13.2 Theoretical Background

In the following, selected key findings about secondary students' science self-concept will be summarized in order to locate findings on chemistry self-concept in the science education literature. The literature on chemistry self-concept is considered in a separate section. We follow the definition of academic self-concepts as the perception of one's abilities in specific academic fields (Shavelson et al., 1976).

13.2.1 The Relation of Science Self-Concept with Achievement

Self-concept has received much attention due to its correlation with achievement (e.g. Jansen et al., 2014). This is interesting because it shows how mental phenomena can affect one of the central outcomes in school. In the literature, a reciprocal relationship between self-concept and achievement is assumed: achievement enhances self-concepts and positive self-concepts enhance achievement (Marsh, 1990; Marsh & Craven, 2006). The impact of self-concept on academic achievement could be mediated by learning goal orientations that are positively correlated with self-concept (Dishon-Berkovits, 2014). Positive chemistry self-concepts could thus positively influence a learning goal orientation in chemistry and thereby impact learning behaviour and achievement in chemistry.

13.2.2 The Impact of Gender and the Students' Cultural Backgrounds

Science self-concepts are not equally distributed. In many countries, boys tend to have stronger science self-concepts than girls (e.g. Germany: Jurik et al., 2013; U.S.: Riegle-Crumb et al., 2011). This association of science with gender has been conceptualized as a culture of masculinity that exists regarding Western science (e.g. U.K. Archer et al., 2014). This concept seems to be adopted by children at a very early age (U.S.: Baron et al. 2014).

However, gender relations seem to differ between cultural contexts (e.g. Rüschenpöhler & Markic, 2019a). No gender gap seems to be present in certain East Asian countries (e.g. Ng et al., 2012). Further, cultural backgrounds play a role: students of dominant ethnic groups tend to hold stronger science self-concepts than students from non-dominant ethnic groups (Rüschenpöhler & Markic, 2019a).

13.2.3 The Social Context

The role of the teacher for academic self-concept formation has been emphasised (e.g. Raufelder et al., 2015), especially for non-dominant ethnic groups (Wood et al., 2013). Language also plays an important role in chemistry learning and science learning (e.g. Markic & Childs, 2016) although the relation between the perception of linguistic abilities and self-concept seems not to have been investigated.

13.2.4 Chemistry Self-Concept Research

The existing studies on chemistry self-concepts of secondary school students are generally in line with the academic self-concept literature. The study by Nieswandt (2007) suggests a close relationship between chemistry self-concept and achievement in secondary school. Nielsen and Yezierski (2016) found that chemistry and mathematics self-concepts are in most cases positively correlated. Furthermore, a correlation between chemistry and science and technology self-concept has been shown (Sheldrake, 2016). Both findings are in line with the theory that STEM self-concepts are correlated (see Marsh et al., 1988).

In a preliminary study (Rüschenpöhler & Markic, 2019b), we showed that in Germany, the gender gap in chemistry depends on the students' cultural backgrounds. Among students without a migration background, boys tend to have more positive chemistry self-concepts than girls, just as the literature suggests. However, among students with a Turkish background, the girls tended to show stronger chemistry self-concepts than the boys. This study aimed to gain first insights into the chemistry self-concepts of secondary school students and operated with data of a medium-sized sample. The results, therefore, require further support. The preliminary investigation also suggested that the perception of social support and of linguistic abilities in chemistry might affect chemistry self-concept (Rüschenpöhler & Markic, 2019b).

Regarding research instruments, Bauer's Chemistry Self-Concept Inventory (Bauer, 2005) and the Attitude towards the Subject of Chemistry scale (Bauer, 2008; refined: Xu & Lewis, 2011) are available. Both target young adults and would require an adaptation if employed in secondary school. An alternative is to depart from a science-specific self-concept scale which is designed for research with secondary school students and to adapt it to the field of chemistry. Here, several well-tested scales with good measurement properties are available (for a review see Rüschenpöhler & Markic, 2019a) such as the PISA 2006 scale (OECD, 2009b).

13.3 Research Questions

The present study aims to test the findings from the preliminary study (Rüschenpöhler & Markic, 2019b). The following three research questions are addressed concerning secondary school students in Germany, based on the literature reviewed above:

(RQ1) How is chemistry self-concept related to gender and the students' cultural backgrounds?
(RQ2) How is chemistry self-concept related to learning goal orientations?
(RQ3) How is chemistry self-concept related to the students' perception of social support and the perception of their linguistic abilities in chemistry?

The perception of linguistic abilities in chemistry class is defined as the students' feeling of understanding of the scientific language in chemistry. This comprises the feeling of being able to understand texts, the teacher's oral language, and chemical equations (Rüschenpöhler & Markic, 2020).

13.4 Methods

13.4.1 Research Instrument

Data was collected using a questionnaire (full version: Rüschenpöhler & Markic, 2020) with items of the Likert type. It comprised 8 scales, out of which 7 are established in their respective fields and were only slightly adapted to fit the context of chemistry education.

- For (1) the chemistry self-concept scale, we replaced the word 'science' with 'chemistry' in the science self-concept scale of PISA 2006 (OECD, 2009b, Q37).
- We measured the students' perception of social support in chemistry with three indicators.
  - (2) the sense of belonging scale from PISA 2003 (OECD, 2005, Q27, 'My school is a place where' replaced with 'In my chemistry class')
  - (3) the perception of student support scale from the 2013/4 HBSC study (Inchley et al., 2016, MQ61, 'my class' replaced with 'my chemistry class')
  - (4) the perception of teacher support scale from the same study (MQ62, 'teacher' replaced with 'chemistry teacher')
- Learning goal orientations were measured with three indicators as well.
  - (5) six items from Cacioppo and Petty's (1982) need for cognition scale translated to German (Bless, Wänke, Bohner, Fellhauer, & Schwarz, 1994; 'in chemistry' added to the items)
  - (6) Dweck's (2000) incremental theory of intelligence scale ('for chemistry' added to the items)

  - (7) the scale for task persistence from PISA 2012 (OECD, 2014, Q36, 'in chemistry' or 'chemistry' added to the items).

- The scale measuring (8) the students' perceptions of their linguistic abilities in chemistry was designed specifically for the present study because no established instrument was available.

The students' migration background was conceptualized using the definition from the German 2013 census (Statistisches Bundesamt, 2013).

13.4.2 Sample

Participation was based on informed consent by the part of the students, the chemistry teachers, and the parents, following the principles of the declaration of Helsinki (World Medical Association, 2013). In the first step, a pre-test was conducted ($N = 68$). In this pre-test, the comprehensibility of the items was focused on and discussed with the teacher. In the final investigation, data of 585 students from 10 German schools were collected. The students were aged 12–18 ($M = 15$) and enrolled in grades 8–10 in German secondary schools. Females made up for 45.5% (266 students) of the sample, 57.6% (248 students) had a migration background which is typical for the urban regions in which most of the data were collected. The largest group of students with a migration background had a Turkish background (12.3%, 72 students).

13.4.3 Data Analysis

In the first step, the quality of the measurement was investigated, based on oral student feedback during data collection and calculations of Cronbach's α and confirmatory factor analyses. For answering RQ1, the relations of gender and migration background with chemistry self-concept were investigated in 2×2 ANOVAs with type III sums of squares. In this analysis, the focus was laid on a comparison between students with Turkish migration background and students without migration background because only in these groups, sample sizes were sufficient for the ANOVAs. The samples of the other groups were too small for the analysis. For answering RQ2 and RQ3, a multiple linear regression model was tested. In all analyses, negatively worded items were reverse coded, and the values were group mean centred to clean the data from group effects (Enders & Tofighi, 2007).

13.5 Results

13.5.1 Measurement Quality

Cronbach's α (see Table 13.1) ranged between .7 and .8 for most of the scales, just as desired (Kline, 2000), except for the incremental theory of intelligence scale (.65) and the self-concept scale (.91). For the self-concept scale, this was expected based on previous investigations (.88–.94, OECD, 2009b). Regarding the incremental theory scale, difficulties appeared also in the confirmatory factor analysis (Table 13.1). Therefore, the incremental theory scale was excluded from analyses. For more information, see Rüschenpöhler and Markic (2020).

13.5.2 RQ1. The Relation with Gender and Cultural Backgrounds

In the first research question (RQ1), we wanted to know how the students' different migration backgrounds and gender affect their chemistry self-concepts. The concept of migration background, as it is employed in Germany, differentiates between students based on national lines. The number of people falling into the specific categories of migration background is thus smaller compared to the number of people who would identify as Hispanic, Black, or White. For the present analysis, this meant that not all groups of students with their different migration backgrounds could be compared when differentiating between boys and girls because sample sizes were too small. We, therefore, conducted the ANOVA using a subsample of those groups in which sample sizes were acceptable. This was the case for students without migration background and with Turkish migration background: we compared boys ($N = 129$) and girls ($N = 115$) without migration background with boys ($N = 40$) and girls ($N = 32$) with a Turkish migration background. Since Levene's test was significant for gender ($p < .01$) and the interaction of gender and culture

Table 13.1 Results from the confirmatory factor analyses and values for Cronbach's α for all scales, including the number of items, mean values, and standard deviations

	Items	*M*	*SD*	α	*SRMR*	*CFI*
(1) self-concept	6	3.91	1.23	.91	.026	.971
(2) sense of belonging	5	4.85	1.14	.78	.041	.939
(3) perception of student support	3	4.55	1.14	.72	.036	.978
(4) perception of teacher support	3	4.38	1.32	.72	.059	.948
(5) need for cognition	6	3.63	1.43	.76	.039	.951
(6) incremental theory	4	4.24	1.22	.65	.122	.761
(7) task persistence	5	3.81	1.22	.77	.057	.890
(8) perception of linguistic abilities	4	4.30	1.31	.80	.018	.987

($p < .05$), we opted for a robust analysis with bootstrap with 599 repetitions, the modified one-step estimator of location and Malhalanobis distances. Gender and culture alone did not show significant effects ($F_{gender}(1, 312) = 0.04$, $p_{gender} = .843$, $F_{culture}(1, 312) = 2.98$, $p_{culture} = .089$). However, gender relations seemed to differ between the culture groups, as the interaction effect showed: $F(1, 312) = 6.51$, $p < .05$. This was unexpected based on the hypotheses derived from science education literature but confirms the findings from the pilot study.

13.5.3 RQ2 and RQ3. The Relation with Learning Goal Orientations, Perception of Social Support, and Perception of Linguistic Abilities

Besides the effects of gender and culture on chemistry self-concept, we wanted to understand the role of the social environment in chemistry class, including the perception of linguistic abilities in chemistry classroom (RQ3). Further, we wanted to know if chemistry self-concept is related to learning goal orientations in chemistry (RQ2) which could mediate the association of self-concept with achievement. We tested a multiple linear regression model on the whole dataset ($N = 585$) with chemistry self-concept as the outcome variable (see Table 13.2). This analysis showed that chemistry self-concept is strongly related to the two indicators of learning goal orientations that were included in the questionnaire, namely need for cognition and task persistence. Also, the students' perception of their linguistic abilities in chemistry seems to be closely related to their self-concept. The social support the students perceive in chemistry tends to explain less variance. Only the perception of teacher support had a significant effect on chemistry self-concept although with a smaller effect size.

Table 13.2 Results from the multiple linear regression model with chemistry self-concept as the outcome variable

	β	$SE\ \beta$	p
(2) sense of belonging	.052	.038	.172
(3) perception of student support	.044	.034	.191
(4) perception of teacher support	.075	.028	.008**
(5) need for cognition	.197	.035	<.001***
(6) incremental theory	–	–	–
(7) task persistence	.365	.044	<.001***
(8) perception of linguistic abilities	.327	.035	<.001***
Gender	.040	.024	.097
Turkish background	−.091	.035	.009**
No migration background	.062	.048	.193

$R^2 = .629$, ***0.001, **<.01, *<.05

13.6 Discussion and Conclusion

The present study investigated the chemistry self-concepts of secondary school students living in Germany. Based on the evidence on gender relations in other science fields, the goal was to understand gender relations in chemistry self-concept. The assumption was that female students and students with a migration background would show lower chemistry self-concepts than male students and students belonging to the dominant ethnic group. Further, the study sought to understand how chemistry self-concept, being a mental construct, can exert influence on achievement. It assumed that chemistry self-concept is correlated with learning goal orientations which could mediate between chemistry self-concept and achievement.

13.6.1 The Role of Gender and Migration Background

Contrary to self-concept literature on other science domains, gender and migration background did not have significant effects on chemistry self-concept. Instead, an interaction effect appeared in the subsample of students without migration background and with a Turkish migration background. This interaction effect had already been discovered in the pilot study (Rüschenpöhler & Markic, 2019b) and was now confirmed (see Rüschenpöhler & Markic, 2020). Since the effect appeared in two independent samples, it seems to be substantial.

What could explain this finding? We assume that this difference in gender relations might be due to culturally different associations of science with masculinity. In Turkey, there is no gender gap in science-related jobs: women and men are almost equally represented on the job market in science. There are even slightly more women than men working in science (OECD, 2009a). Moreover, female students outperform male students in science in secondary school (Batyra, 2017a, b). While in many Western regions, science is associated with masculinity (Archer et al., 2014) and it is difficult for young women to integrate science aspirations in their identities (Archer et al., 2013), this might be different in Turkey. Our hypothesis is, therefore, that the young people with a Turkish migration background living in Germany are influenced by a different gender conception of science. This 'deviation' from the concept of science as being a male domain could represent an interesting object for studying alternative conceptions of gender and science.

13.6.2 The Role of Learning Goal Orientations and the Perception of Social Support

Further, the study shows that chemistry self-concepts are closely related to learning goal orientations, which aligns well with research in educational psychology (Dishon-Berkovits, 2014). Based on these findings, it would be interesting to design an intervention study supporting students' learning goal orientations in chemistry. The goal would be to establish if supporting learning goal orientations in chemistry could positively influence chemistry self-concept and achievement. The teacher seems to play an important role in the development of positive chemistry self-concepts, a finding which aligns with literature as well (e.g. Raufelder et al., 2015). The proposed intervention could, therefore, unfold its potential if the support of learning goal orientations in chemistry is provided by the teacher, for instance in the form of targeted feedback, and if strategies supporting language learning are included in chemistry classes.

Acknowledgements This research was partly funded by a grant of the internal research funding at Ludwigsburg University of Education.

References

Archer, L., DeWitt, J., Osborne, J., Dillon, J., Willis, B., & Wong, B. (2013). 'Not girly, not sexy, not glamorous': Primary school girls' and parents' constructions of science aspirations. *Pedagogy, Culture & Society, 21*(1), 171–194. https://doi.org/10.1080/14681366.2012.748676

Archer, L., DeWitt, J., & Willis, B. (2014). Adolescent boys' science aspirations: Masculinity, capital, and power. *Journal of Research in Science Teaching, 51*(1), 1–30. https://doi.org/10.1002/tea.21122

Baron, A. S., Schmader, T., Cvencek, D., & Meltzoff, A. N. (2014). The gendered self-concept: How implicit gender stereotypes and attitudes shape self-definition. In P. J. Leman & H. R. Tenenbaum (Eds.), *Gender and development* (pp. 109–132). Psychology Press.

Batyra, A. (2017a). *Gender gaps in student achievement in Turkey: Evidence from the Programme for International Student Assessment (PISA) 2015*. http://en.egitimreformugirisimi.org/wp-content/uploads/2017/03/PISA-%C4%B0NG%C4%B0L%C4%B0ZCE.pdf

Batyra, A. (2017b). *Gender gaps in student achievement in Turkey: Evidence from Trends in International Mathematics and Science Study (TIMSS) 2015*. https://aydindoganvakfi.org.tr/static/media/images/files/TIMSS_ENG.pdf

Bauer, C. F. (2005). Beyond 'student attitudes': Chemistry self-concept inventory for assessment of the affective component of student learning. *Journal of Chemical Education, 82*(12), 1864–1870. https://doi.org/10.1021/ed082p1864

Bauer, C. F. (2008). Attitude toward chemistry: A semantic differential instrument for assessing curriculum impacts. *Journal of Chemical Education, 85*(10), 1440–1445. https://doi.org/10.1021/ed085p1440

Bless, H., Wänke, M., Bohner, G., Fellhauer, R. F., & Schwarz, N. (1994). Need for cognition: Eine Skala zur Erfassung von Engagement und Freude bei Denkaufgaben. *Zeitschrift Für Sozialpsychologie*, 147–154.

Cacioppo, J. T., & Petty, R. E. (1982). The need for cognition. *Journal of Personality and Social Psychology, 42*(1), 116–131. https://doi.org/10.1037/0022-3514.42.1.116

Dishon-Berkovits, M. (2014). A study of motivational influences on academic achievement. *Social Psychology of Education, 17*(2), 327–342. https://doi.org/10.1007/s11218-014-9257-7

Dweck, C. S. (2000). *Self-theories: Their role in motivation, personality, and development.* Psychology Press.

Eccles, J. S., & Wang, M.-T. (2016). What motivates females and males to pursue careers in mathematics and science? *International Journal of Behavioral Development, 40*(2), 100–106. https://doi.org/10.1177/0165025415616201

Enders, C. K., & Tofighi, D. (2007). Centering predictor variables in cross-sectional multilevel models: A new look at an old issue. *Psychological Methods, 12*(2), 121–138. https://doi.org/10.1037/1082-989X.12.2.121

Inchley, J., Currie, D., Young, T., Samdal, O., Torsheim, T., Augustson, L., Mathison, F., et al. (Eds.). (2016). *Growing up unequal: Gender and socioeconomic differences in young people's health and well-being. Health Behaviour in School-Aged Children (HBSC) study. International report from the 2013/2014 survey*. World Health Organization.

Jansen, M., Schroeders, U., & Lüdtke, O. (2014). Academic self-concept in science: Multidimensionality, relations to achievement measures, and gender differences. *Learning and Individual Differences, 30*(11), 11–21. https://doi.org/10.1016/j.lindif.2013.12.003

Jurik, V., Gröschner, A., & Seidel, T. (2013). How student characteristics affect girls' and boys' verbal engagement in physics instruction. *Learning and Instruction, 23*, 33–42. https://doi.org/10.1016/j.learninstruc.2012.09.002

Kline, P. (2000). *The handbook of psychological testing* (2nd ed.). Routledge.

Markic, S., & Childs, P. E. (2016). Language and the teaching and learning of chemistry. *Chemistry Education Research and Practice, 17*(3), 434–438. https://doi.org/10.1039/C6RP90006B

Marsh, H. W., Byrne, B. M., & Shavelson, R. J. (1988). A multifaceted academic self-concept: Its hierarchical structure and its relation to academic achievement. *Journal of Educational Psychology, 80*(3), 366–380.

Marsh, H. W. (1990). Causal ordering of academic self-concept and academic achievement: A multiwave, longitudinal panel analysis. *Journal of Educational Psychology, 82*(4), 646–656. https://doi.org/10.1037/0022-0663.82.4.646

Marsh, H. W., & Craven, R. G. (2006). Reciprocal effects of self-concept and performance from a multidimensional perspective: Beyond seductive pleasure and unidimensional perspectives. *Perspectives on Psychological Science, 1*(2), 133–163. https://doi.org/10.1111/j.1745-6916.2006.00010.x

Ng, K. T., Lay, Y. F., Areepattamannil, S., Treagust, D. F., & Chandrasegaran, A. L. (2012). Relationship between affect and achievement in science and mathematics in Malaysia and Singapore. *Research in Science & Technological Education, 30*(3), 225–237. https://doi.org/10.1080/02635143.2012.708655

Nielsen, S. E., & Yezierski, E. J. (2016). Beyond academic tracking: Using cluster analysis and self-organizing maps to investigate secondary students' chemistry self-concept. *Chemistry Education Research and Practice, 17*(4), 711–722. https://doi.org/10.1039/C6RP00058D

Nieswandt, M. (2007). Student affect and conceptual understanding in learning chemistry. *Journal of Research in Science Teaching, 44*(7), 908–937.

OECD. (2005). *PISA 2003 technical report.* OECD Publishing. https://doi.org/10.1787/9789264010543-en

OECD. (2009a). Chart A4.6 tertiary graduates in science-related fields among 25-34 year-olds in employment, by gender (2009). https://doi.org/10.1787/888932460192

OECD. (2009b). *PISA 2006 technical report*. OECD Publishing.

OECD. (2014). *PISA 2012 technical report*. OECD Publishing.

Raufelder, D., Sahabandu, D., Martínez, G. S., & Escobar, V. (2015). The mediating role of social relationships in the association of adolescents' individual school self-concept and their school

engagement, belonging and helplessness in school. *Educational Psychology, 35*(2), 137–157. https://doi.org/10.1080/01443410.2013.849327
Riegle-Crumb, C., Moore, C., & Ramos-Wada, A. (2011). Who wants to have a career in science or math? Exploring adolescents' future aspirations by gender and race/ethnicity. *Science Education, 95*(3), 458–476. https://doi.org/10.1002/sce.20431
Rüschenpöhler, L., & Markic, S. (2019a). Self-concept research in science and technology education: Theoretical foundation, measurement instruments, and main findings. *Studies in Science Education, 55*(1), 37–68. https://doi.org/10.1080/03057267.2019.1645533
Rüschenpöhler, L., & Markic, S. (2019b). A mixed methods approach to culture-sensitive academic self-concept research. *Education Sciences, 9*(3), 240. https://doi.org/10.3390/educsci9030240
Rüschenpöhler, L., & Markic, S. (2020). Secondary school students' chemistry self-concepts: Gender and culture, and the impact of chemistry self-concept on learning behaviour. *Chemistry Education Research and Practice, 21*(1), 209–219. https://doi.org/10.1039/C9RP00120D
Shavelson, R. J., Hubner, J. J., & Stanton, G. C. (1976). Self-concept: Validation of construct interpretations. *Review of Educational Research, 46*(3), 407–441. https://doi.org/10.3102/00346543046003407
Sheldrake, R. (2016). Students' intentions towards studying science at upper-secondary school: The differential effects of under-confidence and over-confidence. *International Journal of Science Education, 38*(8), 1256–1277. https://doi.org/10.1080/09500693.2016.1186854
Statistisches Bundesamt. (2013). *Bevölkerung und Erwerbstätigkeit Bevölkerung mit Migrationshintergrund: Ergebnisse des Mikrozensus 2013*. Statistisches Bundesamt.
Wood, L. C., Ebenezer, J., & Boone, R. (2013). Effects of an intellectually caring model on urban African American alternative high school students' conceptual change and achievement in chemistry. *Chemistry Education Research and Practice, 14*(4), 390–407. https://doi.org/10.1039/C3RP00021D
World Medical Association. (2013). World Medical Association declaration of Helsinki: Ethical principles for medical research involving human subjects. *Journal of the American Medical Association, 310*(20), 2191. https://doi.org/10.1001/jama.2013.281053
Xu, X., & Lewis, J. E. (2011). Refinement of a chemistry attitude measure for college students. *Journal of Chemical Education, 88*(5), 561–568. https://doi.org/10.1021/ed900071q

Chapter 14
Undergraduate Science Majors' Identity Work in the Context of a Science Outreach Program: Understanding the Role of Science Capital

Alexandre Cavalcante and Allison J. Gonsalves

14.1 Introduction

This paper reports on the initial outcomes of a study that brought together undergraduate science majors (USMs) and youth in a science outreach program. The research reported in this paper will focus on the USMs, and is motivated by findings in the literature that USMs are increasingly leaving the STEM pipeline and require additional retention strategies that include developing science teaching skills to prepare them for the workplace, professional school or graduate school (Rao et al., 2007). Thus, we are interested in learning about USMs' past experiences with science, and how opportunities to engage USMs in teaching experiences may impact their identification with science and may develop skills (e.g., communication skills, confidence and professional preparation) that will serve them later in scientific careers (Nelson et al., 2017). This research sought to understand why USMs choose to engage in science outreach, and how these experiences might impact USMs' thinking about and identification with science.

A. Cavalcante (✉)
University of Toronto, Toronto, Canada
e-mail: alexandre.cavalcante@utoronto.ca

A. J. Gonsalves
McGill University, Montreal, Canada
e-mail: allison.gonsalves@mcgill.ca

O. Levrini et al. (eds.), *Engaging with Contemporary Challenges through Science Education Research*, Contributions from Science Education Research 9,
https://doi.org/10.1007/978-3-030-74490-8_14

14.2 Science Identities and Science Capital

Increasing attention has been paid to the role of students' social backgrounds in shaping their aspirations and relationship with STEM subjects (e.g., Archer et al., 2013), and their experiences in post-secondary STEM education (e.g., Black & Hernandez, 2016). In considering how USMs come to engage in science outreach, the perspectives they take on this endeavour and their aspirations in science, we draw on theoretical constructs that help us to understand the relationship between USMs social backgrounds and their orientation to STEM careers and to teaching science. In this chapter, we are interested in Archer et al.'s (2013) concept of *science capital* which they argue is critical to explaining students' aspirations towards a future in science. According to the researchers, science capital is the "science-related qualifications, understanding, knowledge (about science and 'how it works'), [and] interest and social contacts (e.g. knowing someone who works in a science-related job)" (p. 3) a student may accrue throughout their youth and early years studying science. This is a helpful concept to consider how these three USMs came to science, how they think about science and importantly, how they think about themselves in science. Thus, we also regard participating in a learning community (like science outreach) as a form of identity work. Our conception of identity draws on Holland et al.'s (1998) notion of 'identities in practice', wherein identity is culturally reified in stories one tells about oneself in response to social interactions or in socially mediated practices (e.g., science labs, lessons or outreach activities). Engagement in these practices can result in recognition as a science person, which further reifies the process of identification as an insider to science (Carlone & Johnson, 2007). For example, recognition from a meaningful other (e.g., teacher, parent, peer) can strengthen one's sense of self and belonging in science, prompting further engagement in practices that yield these forms of recognition.

14.3 Methods and Analysis

Teaching Science in the Zone (TSZ) is an afterschool program designed to pair USMs with youth at inner city elementary schools in low-income neighbourhoods of Montreal, Canada. Participants for this study were USMs from different Life Sciences programs (Immunology, Biochemisty and Anatomy). They were selected based on their response to a recruitment notice to the pool of the TSZ volunteers. Their participation in the program meant spending some time preparing and rehearsing science modules to be enacted with the youth throughout the span of three weeks at the end of the TSZ program. They engaged with youth by leading three science modules in an inquiry-based format. The modules were designed by the authors in collaboration with an education undergraduate student and the USMs rehearsed them prior to enacting in the schools. The data reported on here includes interviews (Kvale, 2008) and video diaries (Noyes, 2004) from the three USMs who participated in the TSZ in the first year of the project. Although the three came from similar science programs, their trajectories into science and science outreach are unique and reveal distinct roles played by their experiences in informing their science identity work and orientations to outreach. During the interviews, we asked the USMs about their experiences with science (both in school and out of school) and their perceptions of science and scientists. We also prompted them to discuss their experience in the TSZ program through video diaries recorded after each module. We conducted a thematic analysis of the interview and video diary data to unpack how their choice of careers and orientations towards science and teaching are mediated by their social backgrounds. We transcribed their interviews and video diaries and engaged in multiple rounds of coding to ensure consistency. The first step entailed independent coding of different participants' data across the research team (e.g., Saldaña, 2015). We then compared the codes, refined them, and re-coded a different participant's data. After that, the codes were refined one more time and the research team agreed on the codebook to be used. We refined the following themes: (1) their backgrounds and experiences with science; (2) their ideas about science, science students and scientists; and (3) their orientations to teaching science in outreach. The results are presented in a narrative format following this order.

14.4 Results

DeWitt et al. (2016) describe key dimensions of science capital, many of which we see reflected in our participants' talk of their early experiences with science. In particular, we found that USMs discussed the salience of: knowledge about the transferability of science; access to and participation in out-of-school science learning contexts; family science skills and knowledge; and talking about science in everyday life. In the three cases presented below, we discuss these aspects of science capital in the contexts of the USMs "previous science experiences". We also found that USMs with high levels of science capital narrate a strong *science student identity* which suggests that they can exchange science capital for recognition. Table 14.1 provides an overview of the three USMs discussed in this chapter in relation to the three thematic themes (early science experiences, narratives of identity and orientations towards science outreach). In what follows, we present case descriptions of each of the three participating USMs. In these cases, we discuss the USMs' trajectories into science, and the opportunities each has had to accumulate different forms of science capital. We then discuss how their various forms of science capital have been exchanged for recognition as science people, and how their experiences and identity work might be related to the orientations they each have towards science outreach.

Table 14.1 Summary of participant information

Science trajectories	Lee	Rajesh	Mathew
Previous experiences with science	Abundant opportunities for developing science capital in and out of school. Cultivated a "passion for science", especially immunology	Accumulated a lot of science capital out of school. Pressure to enter engineering. Chose anatomy leading to medicine.	Grew up on a farm in a working-class town. Accumulated science capital in school; saw it as a way for class mobility.
Science identity	Did not see any obstacles in his pursuit of science. Narrated a strong science identity.	Attributed success to hard work, not natural talent. Thinks he still has a lot to learn.	Describes scientists as "brainy and awkward", which is a conflict for his identity work.
Orientations to science outreach	Outreach as "delivering concepts". Saw himself as a role model of passion about science, and the one who has the answers to students' science questions.	Focused on "asking questions and probing for deep explanation". Thus, acted as a facilitator in the outreach context, rather than the expert.	Teaching as a form of science communication that requires "dumbing down" content for students, but also creating opportunities for inquiry.

14.4.1 Lee: The 'Natural' Scientist

Lee is in his second year and studies Microbiology and Immunology. He grew up in an urban middle-class family and his parents, Chinese immigrants to Canada, have followed careers within science. Lee's narrative of his early experiences with science signals many opportunities to accumulate science capital (Archer et al., 2015). He especially discusses the importance of family knowledge in science, and knowledge about transferability of science (vis-à-vis science careers) (e.g., DeWitt et al., 2016). He discusses the early influences that family had on him: "I grew up in China, and my grandparents were science teachers, they exposed me to a lot of science. Students would come over all the time talk about science." In Canada, his parents wanted him to have as many opportunities as possible to engage in different activities/fields, however, they also wanted him to be free to choose to study whatever he felt passionate about.

Lee narrated a background with abundant opportunities for cultivating science capital which included positive experiences with science in elementary and secondary school, participation in science enrichment programs, science Olympiads, and robotics competitions. He was good at school science, and there were many opportunities for him to exchange the science capital he had already accumulated in his out-of-school life for smooth transitions into school science. Lee suggests that learning science was straightforward to him: "there wasn't really much to science, it was just 'oh here's a concept'." As time went by, science classes became "more hands on, more challenging", and he became more interested in life sciences, particularly in topics related to pathogens. He recollects a memorable experience in grade 12 that exemplifies such interest: "It was a biotechnology course. We did a lot of really cool experiments such as we took a Petri plate and transformed some bacteria to make them antibiotic resistant." This experience was formative to his career aspirations. He mentions that "since then, I've been really interested in microbiology and immunology. That was the hardest course I've ever taken, but I think at the end it was the most rewarding one." Lee's important experiences with science extended beyond the classroom space. Encouraged by his parents and other meaningful adults in his life (such as the biotechnology course teacher), he participated in a range of out-of-school-time science activities locally and nationally. Through those experiences, Lee cultivated what he describes as a "passion for science", which ultimately led him to choose to pursue a program in life sciences.

Lee's trajectory into science was uncomplicated, and emerged as a natural choice. He did not see any obstacles in his pursuit of science, and narrated a strong science identity. Wong (2016) described five different 'types' of science participation ranging from science adverse to science prominent. Science prominent students tended to express science-related career aspirations with above average achievement and high levels of interest and capital in science. We see Lee as a science prominent student. He "always knew" he would go to science and explains that doing science always made him "excited about discovering something new." When asked about his perception of himself as a science student, Lee describes himself as a good science student because, beyond grades, he pursues science opportunities on his own time and is always excited about classes. These ideas of what makes him a good science student resonate with his description of *science people*: "if you're talking to someone who really enjoys science, there are those special moments where you can talk about hours on end that you both really enjoy. That shows you're into science, that you really have a passion for what you're doing." For him, anyone can do science, and it is "just the passion that sets two people apart, and what opportunities you decide to take because you're interested in it."

As a science prominent student (Wong, 2016), Lee's unproblematic identity work in relation to science is reflected in his orientation to science outreach. His motivation to partake in the outreach program came from his previous experiences as well as his career aspirations. Previous experiences include not only those in which he participated as a student but also those in which he acted as a mentor. They were inspirational to Lee and made him develop an appreciation for this kind of work. His career aspirations, which include going to medical school and being a science communicator, influence his desire to participate in the outreach program as he describes his desire to develop skills such as conflict resolution, communication and organization – skills he considers important to becoming a successful scientist.

Given his positive trajectory with formal science teaching, it is unsurprising that he viewed science outreach in a similar fashion to traditional forms of science teaching: as "delivering concepts" that students need to know. Hence, he perceives his role as a mentor as someone who delivers the concepts in a way that is simple and drives students to learn the material, close to a science communicator: "it's all about how you can deliver that material in a way that the audience understands." He sees himself as a role model who is passionate about science, and the person in the room who has the answers to students' science questions.

14.4.2 Rajesh: The Hard-Working Scientist

Rajesh is a fourth-year student majoring in Anatomy and Cell Biology. He is also from an immigrant family (Pakistan) and moved to Canada during his childhood. Like Lee, Rajesh has strong family influences in science, exposure to how science skills can translate into careers, and describes many early childhood experiences with science – all influential dimensions of science capital (DeWitt et al., 2016). His father works in the oil industry as a geologist and, according to Rajesh, his profession inspired many "scientific discussions" in their household growing up. It was also due to his father's professional occupation that Rajesh describes having been strongly encouraged by his parents to pursue engineering. "They kept forcing my brother and I to pursue engineering because it's a safe program. My dad had a lot of connections within the engineering field." Despite having applied to some engineering programs, he ended up choosing his current program "because if there's one machine that's made absolutely perfectly it's the human body", something about which his parents "weren't too happy at the beginning", but eventually accepted.

Like Lee, Rajesh narrated a background with strong opportunities for developing science capital both in and out of school. He had a lot of exposure to and interest in science even before starting school. A meaningful friendship "rooted his interest" in science and led to multiple explorations in everyday situations. In school, he did lot of "hands-on science activities" during afterschool programs, and these positive experiences began to develop his orientation towards science as a prospective career option. He experienced a sharp distinction between STEM subjects when compared to humanities subjects in terms of academic performance. A meaningful relationship with a high school teacher inspired him "to question nature and take it further, and learn more and research more and do what you're passionate about." This meaningful relationship also involved recognition of Rajesh as a science person. He narrated that the teacher sent him a postcard after a research project saying "I was very interested in this project you did. I understood the passion that you put. I hope to hear about your research in the future." Rajesh feels strongly about this moment of recognition which inspired him. In fact, he still has that postcard at home and "when I publish my first paper, I'm going to send it to him with that postcard on it."

Rajesh's early experiences have helped him accumulate many forms of science capital, which we argue has helped him to cultivate a strong science identity. We also describe Rajesh as a "science prominent student" (Wong, 2016). He had science-related career aspirations early on, which were connected to his love for science and his early experiences. However, unlike Lee, despite his "natural" interest in science, Rajesh does not describe himself as naturally good. Instead, he attributes success in science to hard work and suggests that he still has a lot of learning to do. He feels challenged in university and refers to the limitations of his knowledge by stating that "I understand science well enough, but as a good scientist always says, they don't know everything. They're always open to know more and they cannot say anything with certainty." This connection between science and work ethic also informs Rajesh's ideas of science people. He describes scientists as hard-working people who are "comfortable with the unknown, with questioning why certain things are a certain way. They don't take things by face value." Taking this further, this work ethic can even reveal a "level of obsession. The really good scientists, to some extent they were obsessed. They don't stop working until they got whatever they wanted to know." He explains other characteristics that he considers important to be successful in science, such as being critical, skeptical and "open to the possibility of changing current dogma. That is a trait for a good scientist."

Rajesh's previous experiences and identity work informs his motivation to participate in science outreach as well as his approach to teaching science in this context. Similar to Lee, he intends to pursue medical school and believes this program can help him develop skills which he considers important to be a good doctor. For him, "you can do as much science as you can but if you can't communicate to someone else as effectively as possible, then you haven't done your job." This finding is supported by research investigating USMs' reasons for engaging in outreach, which suggests that many are motivated by the possibility of personal career advancements and a desire to improve their teaching skills (Anderson et al., 2015). Science undergraduates have also been shown to participate in outreach to improve their communication skills, which are thought to be essential for science careers (Fogg-Rogers et al., 2017). In Rajesh's case, he prioritizes communication, which he believes will improve his practice as a future physician.

These reasons for doing outreach connect with Rajesh's approach to engaging youth in science. His approach is informed by his own perception of science as a "thought process." Instead of emphasizing the learning of certain concepts, he would rather have students understand how science works and which practices are involved in scientific investigation. For him, science is "a method of approach to the world rather than a subject in grade school. So, most of my answers to student questions are tailored towards that understanding of science." This is why his orientation to teaching is focused on "asking questions and probing for deep explanation". Thus, Rajesh acts as a facilitator in the outreach context, rather than the expert.

14.4.3 Matthew: Not a 'True' Science-Person

Matthew is a third-year Biochemistry student in the process of the transferring to the faculty of Education. He narrated a childhood growing up on a farm in a working-class town. His father is a bus driver and his mother a nurse, both of whom he does not identify as "science people". Unlike the other two participants, Matthew did not see any influence from his parents in developing an interest in science. Even though his mother worked in a science-related field, he argued that "yes, it is science, but not really… How much does she know about the structure of the cell?" Although his parents encouraged him to pursue university studies, this did not significantly impact his decision because "they were giving me advice for something that I already wanted to do." Beyond his household, Matthew also did not recognize any meaningful influence toward science in his community; they "didn't care about science."

Matthew's previous experiences with science seem to contribute less to his identity work than those of Lee and Rajesh. Thus, the forms of science capital that Matthew accrued in his early life are much different. We regard Matthew as demonstrating "science intermediate" participation (Wong, 2016). He expresses interest in science, and achieved well in secondary school, but he had limited opportunities to accumulate science capital in his early life, and struggles in his science program when he gets to university. In secondary school, he discusses having taken the "maximum amount of science courses." However, he describes being motivated to take science as a way to forge a career that will help him to avoid blue-collar work. According to him, science was an escape from physical labour for him and his friends, "a small group of nerds that were more academically inclined." Although he lacked in science capital from his out-of-school life, Matthew gained recognition as a science person in secondary school where he was successful in relation to his peers. Matthew suggests that he liked science in secondary school because it was easy, and he did not need to struggle to be successful. He acknowledges that "it was never really something that I struggled with and that's why I liked it at the time." Thus, Matthew was recognized by teachers and peers as a science person in secondary school, which contributed to his trajectory into post-secondary science. However, without significant amounts of science capital to bring to this program, Matthew struggled. Once entering his biochemistry program, Matthew began to question whether he had the "mindset for science". He struggles with the course work complaining that although the "material itself isn't complicated, [...] the exams are notoriously difficult, because they all require to think outside of the box."

When prompted to discuss his identity work as a science student, Matthew explains that he still identifies with science, but no longer considers himself a science person. Although he is in the process of switching programs (to education), he still feels that "even though biochemistry is not where my heart belongs in science, I still think science is part of me." This conflict seems to be rooted in his orientation towards science. He describes scientists as "brainy and socially awkward", which does not align with his own identity. For him, scientists lack personality and "the more I get into it, the more I realise that people become robots." In fact, his ideas about people in science are reflected even in the way he refers to those who are not in science as "normal people". Matthew's experiences of science in the post-secondary context are so contrary to his secondary school expectations, that it raises questions about whether his struggles can be attributed to his relative lack of science-related experiences early on in life. As discussed, Matthew had little to no science capital apart from secondary science courses, and in particular lacked family members or role models in science, and knowledge about science-related careers (e.g., DeWitt et al., 2016).

Matthew's experiences with and ideas of science in his university program have led him to switch to a Secondary Science Bachelor of Education. He describes being still unsure about his future as a science educator but, when thinking about a science career, he decided to engage in something that could connect his interest in science and his personality. He states that "I'm a really social person so I don't want to be trapped in a lab and run experiments all day." Such an orientation seemed to be informative to Matthew's decision to participate in the TSZ program. Not only was it an opportunity to engage with science in a different way, but that was also a chance for him to explore inquiry-based learning. He had read about this teaching approach, and the TSZ seemed "interesting to see." As a future educator, he believes "that's a tool you'll forever have in your pocket."

During the TSZ program, Matthew's previous experiences and identity work in science seems to come together in his approach to outreach. He indicates that what he "liked about inquiry-based learning is that these kids were able to take information they knew and make a guess". At the same time, he reproduces some discourses associated with traditional forms of science teaching when he states that "the appropriate form of instruction is direct, feeding information". For him, the importance of science outreach comes from its use value, i.e., how science can contribute to understanding the world instead of leading to a specific career. At the same time, he argues that teaching and communicating science requires "dumbing down" the content for students. While he strives to create opportunities for inquiry in his practices, he is still "very skeptical about how the inquiry-based model was built, because as a kid I recall being very shy. I would never have participated in this." Matthew's conflicting relationship with stereotypical ideas of science has led him to search for ways to reconcile his long-standing interests in science with the identity conflicts he experienced in his biochemistry program, and the outreach program seemed to act as an opportunity for him to do so.

14.5 Discussion and Conclusions

These three participants have had varying opportunities to accumulate science capital, but we also notice that its exchange value in higher education is dependent on the students' identity work in relation to how they view the culture of the science contexts that they enter. To illustrate, in secondary school, Matthew accumulated science capital in school by positioning himself as expert in relation to others. This form of science capital, however had a low exchange value (Black & Hernandez, 2016) in the post-secondary context given his struggles in exams. He views the culture of his science program as incompatible with his personality: he narrates not having the "personality" of his colleagues, and not identifying with the cultural arbitrary for "brainy" science person (e.g., DeWitt et al., 2013). Thus, his science identity seems to be the most fragile among the three participants. For Lee and Rajesh, opportunities to exchange science capital for recognition came more easily, which led to strong science identities. As they accumulated other forms of science capital, this variety seems to have supported their identity work in times of struggle (e.g. when their program became more challenging), which in turn resulted in recognition of themselves as science people.

In light of these findings, it seems that for these participants, the development of science identity occurred as a result of two processes. As they engaged with science in and out of school growing up, they accumulated science capital in multiple forms. However, such science capital led to strong science identities (in post-secondary contexts) only when it was interpreted as useful to be exchanged for recognition (e.g. their own recognition as science people as a result of academic performance; when they applied for science-related opportunities such as the TSZ program; when they mobilized their previous experiences in conversations with meaningful others such as peers and professors). In these situations, the students interpreted their science capital as compatible with the culture of their post-secondary science contexts. While we recognize that these two processes might occur simultaneously, it seems that accumulation does not lead necessarily to strong science identity by itself. In Matthew's case, we do not argue that he had less science capital compared to the other two USMs, but that the science capital he accumulated was less diverse and therefore less exchangeable. Figure 14.1 shows a visual interpretation of our findings.

For Lee and Rajesh, the TSZ represents an opportunity to continue the process of science capital exchange and accumulation, whereas Matthew experiences this program as an opportunity to engage in different forms of science identity work (i.e., to develop a science teacher identity). We also suggest that these students' early experiences with science, their university experiences and their aspirations for science-related careers shape their orientations to science in ways that influence their science teaching in outreach contexts. For example, Lee (who accumulated science capital easily and refers to scientists as "naturally able") regards science teaching as the delivery of concepts. This bears implications for how science outreach from

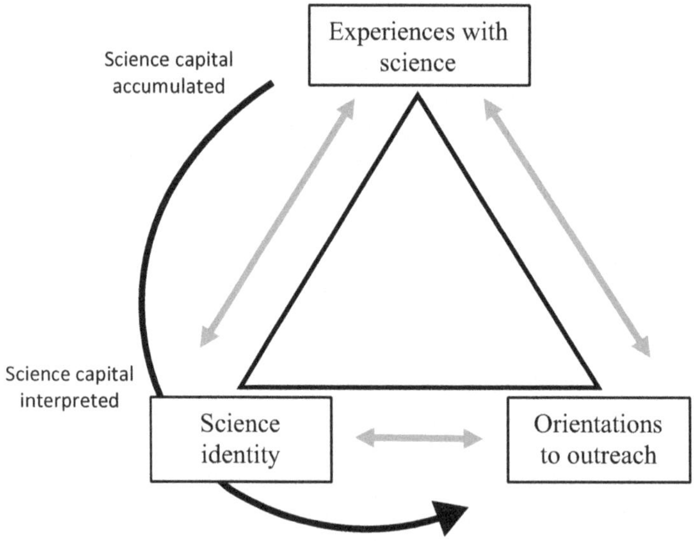

Fig. 14.1 Visual interpretation of the findings

universities is conducted (how students are recruited, what views of science they bring, what teaching practices they engage with, etc.), and the possibilities to shift the traditional "stand and deliver" pedagogical strategies so predominant in outreach contexts.

Acknowledgement This project was funded by the *Fonds de Recherche du Québec – Société et Culture*.

References

Anderson, M. K., Tenenbaum, L. S., Ramadorai, S. B., & Yourick, D. L. (2015). Near-peer mentor model: Synergy within mentoring. *Mentoring & Tutoring: Partnership in Learning, 23*(2), 116–132.

Archer, L., Osbourne, J., DeWitt, J., Dillon, J., Wong, B., & Willis, B. (2013). *ASPIRES: Young people's science and career aspirations: Age 10–14*. Kings College.

Archer, L., Dawson, E., DeWitt, J., Seakins, A., & Wong, B. (2015). 'Science capital': A conceptual, methodological, and empirical argument for extending Bourdieusian notions of capital beyond the arts. *Journal of Research in Science Teaching, 52*, 922–948.

Black, L., & Hernandez-Martinez, P. (2016). Re-thinking science capital: The role of 'capital' and 'identity' in mediating students' engagement with mathematically demanding programmes at university. *Teaching Mathematics and Its Applications: International Journal of the IMA, 35*(3), 131–143.

Carlone, H. B., & Johnson, A. (2007). Understanding the science experiences of women of color: Science identity as an analytic lens. *Journal of Research in Science Teaching, 44*(8), 1187–1218.

DeWitt, J., Osborne, J., Archer, L., Dillon, J., Willis, B., & Wong, B. (2013). Young children's aspirations in science: The unequivocal, the uncertain and the unthinkable. *International Journal of Science Education, 35*(6), 1037–1063.

DeWitt, J., Archer, L., & Mau, A. (2016). Dimensions of science capital: Exploring its potential for understanding students' science participation. *International Journal of Science Education, 38*(16), 2431–2449.

Fogg-Rogers, L., Lewis, F., & Edmonds, J. (2017). Paired peer learning through engineering education outreach. *European Journal of Engineering Education, 42*(1), 75–90.

Holland, D., Lachicotte, W., Skinner, D., & Cain, C. (1998). *Agency and identity in cultural worlds*. Harvard.

Kvale, S. (2008). *Doing interviews*. Sage.

Nelson, K., Sabel, J., Forbes, C., Grandgenett, N., Tapprich, W., & Cutucache, C. (2017). How do undergraduate STEM mentors reflect upon their mentoring experiences in an outreach program engaging K-8 youth? *International Journal of STEM Education, 4*(1), 3.

Noyes, A. (2004). Video diary: A method for exploring learning dispositions. *Cambridge Journal of Education, 34*(2), 193–209.

Rao, S., Shamah, D., & Collay, R. (2007). Meaningful involvement of science undergraduates in K-12 outreach. *Journal of College Science Teaching, 36*(6), 54–58.

Saldaña, J. (2015). *The coding manual for qualitative researchers*. Sage.

Wong, B. (2016). Minority ethnic students and science participation: A qualitative mapping of achievement, aspiration, interest and capital. *Research in Science Education, 46*(1), 113–127.

Chapter 15
Pre-service Teachers' Psychological Distance Towards Environmental and Health Socio-Scientific Issues

Alexander Georg Büssing, Jacqueline Dupont, and Susanne Menzel

15.1 Introduction

Everyday people are confronted with a lot of objects, which are automatically evaluated based on their relevance for survival (Liberman & Trope, 2014). Some of these objects, such as socio-scientific issues (SSI), represent rather abstract phenomena (Zeidler, 2014). People have only few possibilities of experiencing the underlying issues directly and are most often confronted with them in the media (Arbieu et al., 2019). These confrontations nonetheless lead to specific mental representations of the underlying problems as specific or unspecific, which may strongly differ between people (Lee et al., 2015). In combination with personality traits such as attitudes or values, these differences may affect how people experience issues to be problematic at all or which decisions may be right to solve them (Büssing et al., 2019a).

The mental representations towards these issues are also central for teaching and learning. For example, prior studies illustrated how pre-service teachers who felt close towards the biodiversity conservation conflict of returning wolves reported a higher motivation for teaching about the issue (Büssing et al., 2019c). This finding was replicated in another study, in which psychological distance also predicted the enjoyment for teaching about two other issues (Büssing, Dupont, & Menzel, 2020). Similarly to this, another study found systematic differences among the variables

A. G. Büssing (✉)
Institute for Science Education, Leibniz University Hannover, Hannover, Germany
e-mail: buessing@idn.uni-hannover.de

J. Dupont · S. Menzel
Didactics of Biology, Department of Biology/Chemistry, Osnabrück University, Osnabrück, Germany
e-mail: jdupont@uos.de; susanne.menzel@uni-osnabrueck.de

O. Levrini et al. (eds.), *Engaging with Contemporary Challenges through Science Education Research*, Contributions from Science Education Research 9,
https://doi.org/10.1007/978-3-030-74490-8_15

predicting teachers' intention and willingness to teach about cancer depending on whether the teachers had personal experience with disease or not (Heuckmann et al., 2020).

Currently, however, the relation between people and SSIs is only scarcely understood and researchers lack standardized ways of defining and measuring the closeness to specific issues. Therefore, within the current study we (1) applied the construct of psychological distance to three specific SSIs, (2) compared the subjects' experienced psychological distance towards these issues, and (3) investigated possible antecedents within a sample of biology pre-service teachers.

15.2 Theoretical Background

15.2.1 Psychological Distance

Generally, *psychological distance* describes the "subjective experience that something is close or far away from the self, here, and now" (Trope & Liberman, 2010, p. 440). This experience may be elicited either by objects or processes and can be described either as concrete or abstract (Trope & Liberman, 2010). Concrete representations, also called low-level construal, are characterized as highly detailed and fine grained, while abstract representations, or high-level construal, represents rather superordinate mental representations (Trope & Liberman, 2010).

The construal of objects is bound to the four dimensions of psychological distance (Liberman & Trope, 2014). Following this basic principle, people construe their distance towards objects and processes based on their social, spatial, temporal, and hypothetical distance (Liberman & Trope, 2014). This means, people will feel close to an object or process if this process concerns them personally (social), within their close spatial surrounding (spatial), in an immediate moment (temporal), and they appraise the process to be very likely (hypothetical). All of these dimensions may directly be translated to SSIs.

15.2.2 Psychological Distance Towards SSIs

A socio-scientific approach to science education describes a rather progressive instructional paradigm, aiming at evidence-based reasoning and scientific inquiry with the inclusion of moral dimensions within suitable controversial scientific issues (Zeidler, 2014). Such controversial issues are characterized by conflicting key beliefs or values of a substantial number of people about topics, which are not capable of being settled by scientific evidence alone (Levinson, 2006). For the example of biology education, suitable issues often stem from the environmental or health domain, due to their combination of public debates with biological contents (Zeyer & Dillon, 2014; Zeyer & Kyburz-Graber, 2012).

SSI-based instruction has been found to foster student learning and interest, which may be explainable due to the possibility of creating real-world connections of students (Klosterman & Sadler, 2010; Sadler, 2009). While several researchers described these real-world connections of SSIs, the construct of psychological distance may allow to better capture differences in people's experience of closeness towards these issues and thus for understanding peoples' connections with the SSIs. Table 15.1 shows how the idea of psychological distance may be transferred to understand the closeness and distance towards climate change.

15.2.3 Differences Between Issues and Antecedents

Prior studies described climate change as a rather distant phenomenon due to its global and abstract characteristics (e.g. McDonald et al., 2015). However, other SSIs may elicit more concrete mental representations.

Such an example would be the biodiversity conservation conflict of returning wild wolves in Germany (Arbieu et al., 2019). Due to protective regulations, the large carnivore was able to return to several European habitats (Chapron et al., 2014). But especially in rural regions wolves may kill livestock (Enserink & Vogel, 2006), leading to emotional discussions between stakeholders (Ronnenberg et al., 2017). Considering psychological distance may be especially interesting in this topic, because prior studies found the spatial distance towards wolf habitats as critical dimension for the acceptance of the species (Hermann et al., 2013; Karlsson & Sjöström, 2007). It may be possible that pre-service teachers' residency or social environment may be connected to how they construe this issue, which could affect their motivation for teaching about it.

As described above, such a connection was already described for the health topic of cancer, as a study showed how the psychological distance towards the disease may affect teachers' teaching motivation about the issue (Heuckmann et al., 2020).

Table 15.1 The four dimensions of psychological distance with subsequent level of construal exemplified for the socio-scientific issue of climate change

	Level of construal	
Dimension	Low-level construal (concrete)	High-level construal (abstract)
Social distance	Climate change will affect me.	Climate change will affect other people.
Temporal distance	Climate change is happening now.	Climate change will happen in the distant future.
Spatial distance	Climate change will happen in my direct geographical surrounding.	Climate change will happen somewhere else but not here.
Hypothetical distance	Climate change is likely to happen.	Climate change is unlikely to happen.

This may be similar within the health topic of pre-implantation genetic diagnosis (PGD). The issue resolves around the possibility of pre-selecting human embryos before they are implanted and represents a health issue with strong ethical implications based on a scientific issue (Sadler & Zeidler, 2004). Based on the foundation of this issue, only people who were confronted with the decision about applying the technique should have had a direct experience of the issue, other occasions may be media representations or the application of the issue in their own school time. Given the young age of pre-service teachers, they may have had only few experiences of the issue, which may be a good possibility to be compared with more directly concerned issues. Besides this, knowledge may be a general predictor of psychological distance, as more knowledge should entail more concrete mental representations (Liberman & Trope, 2014).

15.2.4 Aim of the Present Study and Research Questions

The aims of the study were to investigate (1) differences between the psychological distance towards the selected issues and (2) antecedents of this psychological distance. Concerning the antecedents, we concentrated on the factors of urbanity, own school time, knowledge, and selected media sources. As described above, it may be reasonable to assume connections with the psychological distance towards the issues. As we intended to draw conclusions for teacher education, we concentrated on pre-service teachers and selected the subject of biology, as all SSI-topics included in the study may be covered in biology lessons. The present study aimed at two specific research questions:

RQ_1: Are there differences between pre-service biology teachers' psychological distance between the selected SSIs?
RQ_2: Which variables predict pre-service biology teachers' psychological distance towards the selected SSIs?

15.3 Methods

15.3.1 Research Design and Sample

As we were interested in differences and connections between variables, we followed a cross-sectional quantitative research design and constructed a paper-pencil-questionnaire. The questionnaire was distributed at lectures for pre-service biology teachers at four different German universities in the North-west, Eastern, and Western part of Germany.

From these universities, overall 189 biology pre-service teachers participated (73.5% female, $M_{age} = 23.45$ years, $SD_{age} = 3.71$). While the higher amount of female pre-service teachers represents a common distribution in pre-service teacher samples in Germany, future studies should try to generalize the results to larger and more representative samples. Nonetheless, for the present study we relied on this sample based on the rather explorative nature.

As study participation was voluntary, the research design guaranteed anonymity, and all participants were aware of the purpose of the research, we did not seek approval by the local ethics committees in line with the regulations of the German Research Foundation (DFG). Standardized informed consent was obtained with an introductory text on the first page of the questionnaire, informing about the purpose of the study, data collection, and saving as well as analysis of the resulting data. Additionally, questions or study withdrawal were allowed at any time.

15.3.2 Questionnaire and Measures

The questionnaire contained a general part asking for demographic data and psychological traits, followed by specific parts for each of the selected SSIs. These specific parts included more measures than could be described in this chapter, the results for these variables have been published elsewhere (Büssing et al., 2020). The order of the specific parts was randomized by distributing the same amount of different versions of the questionnaire, including all possible orders of the three specific parts. Overall, the pre-service teachers needed about 20 min to complete the whole questionnaire.

To measure psychological distance, we selected an already applied measure, which comprises one item per dimension, resulting in four items per SSI (Büssing et al., 2019c). Besides this, we selected urbanity, self-reported knowledge, the application of the topics in the pre-service teachers' own school time, as well as sources of knowledge as possible antecedents of psychological distance. All items were measured on a 6-point scale (1 = do not agree at all, 6 = agree completely), except for urbanity (coded as "on the countryside" (1), "rather on the countryside" (2), "in the city and on the countryside" (3), "rather in the city" (4), "only in the city" (5). As all of these items had to be formulated for the specific issue, we relied on single-item measures for these scales. More information about the applied measures is displayed in Table 15.2.

15.3.3 Statistical Analysis

As a first step, we assessed the measurement probabilities of our instruments by inspecting the Cronbach's alpha as an indicator of scale reliability. Values above .70 were deemed sufficient. Furthermore, we specified a CFA to analyze how well the hypothesized model fitted with the data (Rosseel, 2012). All Cronbach's alpha values were above .70 and the results for the confirmatory factor analysis of the three

Table 15.2 Overview of variables, example items, response format, and descriptive statistics

Variable	Example item	SSI	M	SD	Mdn
Urbanity	"I grew up in the city."	–	2.50	1.26	2.00
Knowledge	"My amount of information about the topic of *the return of wolves* is high."	Wolf	3.26	1.14	3.00
		CC	4.40	1.04	4.00
		PGD	3.83	1.18	4.00
Own school time	"The topic of *climate change* was addressed in my own school time."	Wolf	1.91	1.25	1.00
		CC	4.62	1.43	5.00
		PGD	3.65	1.84	4.00
Psychological distance	"I am personally concerned by *the return of the wolves*."	Wolf	3.68	1.04	3.75
		CC	2.10	.88	2.00
		PGD	4.81	.85	5.00
Sources of knowledge from print/online-media	"Where did you get your information about the topic of *pre-implantation genetic diagnosis*?" "… Print and Online media"	Wolf	4.19	1.41	5.00
		CC	4.85	1.06	5.00
		PGD	4.08	1.46	4.00
Sources of knowledge from books	"… books"	Wolf	2.96	1.35	2.00
		CC	4.03	1.48	4.00
		PGD	3.54	1.60	4.00
Sources of knowledge from school	"… school"	Wolf	2.23	1.25	2.00
		CC	4.57	1.20	5.00
		PGD	3.75	1.63	4.00
Sources of knowledge from the social environment	"… social environment"	Wolf	3.52	1.56	4.00
		CC	4.32	1.28	5.00
		PGD	3.32	1.52	3.00

Note. Italic letters indicate an exemplary topic, which was varied between the scales for the respective items

scales for psychological distance obtained a sufficient fit (χ^2 = 79.239, df = 51, p = .007, CFI = .97, RMSEA = .07, SRMR = .06).

However, we found that the item for the hypothetical dimension of PGD did not load on the respective factor. This result is explainable, as the pre-service teachers could have experienced no sense in indicating the likelihood of a process, which they know is definitely happening. We therefore excluded this item. This increased Cronbach's alpha of the measure for psychological towards PGD from .72 to .87.

Firstly, we investigated the descriptive statistics (Table 15.2) and intercorrelations between all variables with Spearman's rho as a correlation coefficient due to its robustness (Field & Wilcox, 2017). Following this, we calculated subsequent difference tests with robust methods as well as robust regressions. All analyses were computed with RStudio 1.1.456 running R 3.5.1 and several specialized packages such as lavaan (Rosseel, 2012). The supplemental material for the paper including the code, dataset, and intercorrelations between all variables can be found in the Open Science Framework: https://doi.org/10.17605/OSF.IO/QNFTX.

15.4 Results

15.4.1 *Differences in Psychological Distance Between Issues*

First of all, we compared the psychological distance between the issues (see Fig. 15.1a). According to the descriptive differences between these variables (Table 15.2), we subsequently found significant differences between all variables in the robust ANOVA ($F(1.89, 206.19) = 383.951$; $p < .001$).

In the post-hoc tests all variables differed with large effect sizes. While the largest difference was found between the psychological distance towards climate change and the PGD ($t(111) = -26.6315$, $d = 3.13$, $p < .001$), also the psychological distance towards returning wolves and climate change ($t(110) = 15.196$, $d = 1.64$, $p < .001$) as well as between returning wolves and PGD ($t(110) = -13.4258$, $d = 1.19$, $p < .001$) differed with large effect sizes.

As we found correlations between gender and psychological distance towards returning wolves and climate change (see supplemental material), we decided to include gender differences into the analysis. As described in Fig. 15.1b and Fig. 15.1c, we found significant differences between male and female participants concerning their psychological distance towards returning wolves ($t(42.7) = 2.056$, $d = .42$, $p < .05$) as well as climate change ($t(40.38) = 2.2227$, $d = .37$, $p < .05$).

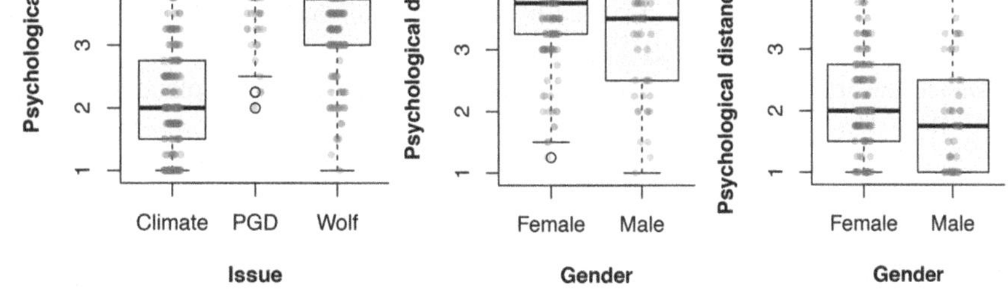

Fig. 15.1 Distribution and standardized results from difference tests (Cohen's *d*) for (**a**) the psychological distance between the socio-scientific issues of climate change (Climate), pre-implantation genetic diagnosis (PGD), and returning wolves (Wolf), (**b**) female and male participants concerning the psychological distance towards wolves, as well as (**c**) female and male participants concerning the psychological distance towards climate change. * = $p < .05$, *** = $p < .001$

Table 15.3 Standardized regression coefficients (β) with standard error (SE) from robust regression analysis for the psychological distance towards returning wolves (Wolf), climate change (CC) and pre-implantation genetic diagnosis (PGD)

	Wolf	CC	PGD
Predictors	β (SE)	β (SE)	β (SE)
Intercept	4.20*** (.43)	4.01*** (.49)	6.14*** (.28)
Gender	−.23 (.18)	−.14 (.15)	−.17 (.14)
Urbanity	.24*** (.06)	.00 (.05)	.02 (.05)
Reported knowledge about issue	−.05 (.08)	−.30*** (.08)	−.02 (.06)
Implementation in own school time	.06 (.09)	−.13 (.10)	−.16** (.05)
Knowledge from print- and online-media	.02 (.06)	−.05 (.07)	−.00 (.04)
Knowledge from books	.00 (.07)	−.00 (.04)	−.08 (.04)
Knowledge from school time	.01 (.10)	.15 (.12)	.05 (.06)
Knowledge from social environment	−.25*** (.06)	−.07 (.07)	−.11* (.04)
R^2 (adjusted R^2)	.29 (.25)	.25 (.21)	.21 (.18)

Note. * = $p < .05$, *** = $p < .001$. Gender was coded as female (1) and male (2). R^2 = Explained variance within the respective dependent variable

15.4.2 Antecedents of Psychological Distance

Table 15.3 shows the results concerning the robust regressions of the predictors for the three SSIs, which largely differed between the contexts. Concerning the return of wild wolves, urbanity ($\beta = .24, p < .001$) and knowledge from the social environment ($\beta = -.25, p < .001$) were significant predictors. The psychological distance towards climate change was only predicted by the reported knowledge about the issue ($\beta = -.30, p < .001$). Finally, psychological distance towards PGD was predicted by the reported implementation of the topic in the own school time ($\beta = -.16, p < .01$) and knowledge from the social environment ($\beta = -.11, p < .05$).

Overall, the model for psychological distance towards returning wolves explained the largest amount of variance in the dependent variable (25%; adj. $R^2 = .25$). While the model for psychological distance towards climate change explained a slightly lower amount of variance (21%; adj. $R^2 = .21$), the model for the psychological distance towards PGD showed with 18% the lowest amount of explained variance (adj. $R^2 = .18$).

15.5 Discussion

We found large differences concerning the psychological distance between the issues. Climate change was found to be the SSI with the lowest psychological distance. Therefore, the surveyed pre-service teachers obviously perceived climate change as more psychologically close than the SSI of returning wolves. These results contradict findings in prior studies, which found climate change to be very abstract and therefore rather distant (McDonald et al., 2015). We believe these results may be explainable by the nature of our sample, as it consisted of biology pre-service teachers. This sample may represent a group with a higher interest in nature-related topics or a more positive attitude towards nature compared to more general samples. We should also consider that our test persons might have perceived the topic as particularly important in educational contexts and may therefore have felt an obligation for teaching about it. Further studies in this regard may be interesting.

As hypothesized before, the issue of PGD was reported to be rather distant, as only people who are confronted with the decision of applying a PGD may be directly concerned by the issue. All other types of confrontation may be limited to media reports, for example when legislators are discussing the regulation of the technique, or the own school time which deals with the topic from a rather technical perspective. This may be further contextualized by the results from our second research question, investigating the possible sources of information about the SSIs.

While the significant factors generally varied between the issues, only very few variables gained predictive power for psychological distance towards the three SSIs. Concerning the topic of returning wolves, only urbanity and knowledge from the social environment gained predictive abilities. This means, people from the countryside and people with more information from their social environment reported a smaller psychological distance towards the issue of returning wolves. These results are in line with prior research that showed how the issue of returning wolves inherits a strong geographical and social focus (Hermann et al., 2013). Particularly, people from wolf habitats and with friends or family with livestock may experience a strong connection to the issue, due to a possible confrontation in everyday life (Enserink & Vogel, 2006).

The regression models for the SSI of climate change showed knowledge to be the only significant predictor for psychological distance towards the issue. We found a negative effect, which is why people with more reported knowledge about climate change reported less psychological distance. While this result makes sense from a psychological perspective, as people with more knowledge about climate change may also inherit more concrete mental representations about the issue (Liberman & Trope, 2014), we are unable to rule out effects of self-report bias. Furthermore, it may be possible that the reported knowledge stronger reflects the self-concept of the person than the actual declarative knowledge, as both are strongly tied to each other (Paulick et al., 2016). Therefore, future studies should further investigate relevant self-concepts in relation to psychological distance. Future studies may also differentiate between knowledge dimensions of teachers (Großschedl et al., 2015) and include better ways of assessing knowledge (Großschedl et al., 2019). However, our study was able to illustrate these connections for the first time, which is why it is nonetheless of merit.

Within this issue of PGD, the own school time and a higher perceived amount of information from the social environment were significant predictors. Similar to the return of wild wolves, indirect experiences from the social environment may therefore affect pre-service teachers' construal of this issue, in combination with the implementation in their own school time. However, both effects were rather small, which was also reflected by the small amount of explained variance. This result is in line with the theoretical suggestions and implies that the pre-service teachers in our study experienced only little personal confrontation with the issue, which may explain the larger psychological distance. But as with the other results, we are unable to rule out effects of self-reporting bias.

15.6 Conclusions and Outlook

Based on the results of our study, biology pre-service teachers may differ for their psychological distance towards SSIs, even when these results are depending on several contextual factors. For example, there may be a difference in perceiving the issue of returning wolves when it is addressed in a wolf habitat, as teachers may be stronger confronted with the issue in their social environment. But as with the other results, the findings of this study should only cautiously be generalized, as the construct of psychological distance is rather new to (science) educational research and our study was of a rather explorative nature.

An important following question may be, if psychological distance can be viewed as a trait or a state (McDonald et al., 2015). Again, this may also be very context-specific, as interventions about climate change, for example, may be generally more successful than about returning wolves, given the global nature of changing climate and an increased level of perceived severity. This underlines the importance of comparative studies to understand how people construe specific processes. Another thought may be a possible connection of psychological distance and emotional perception, as both showed to be strongly tied to each other (Van Boven et al., 2010). As teaching about SSI may be emotional for specific topics (Heuckmann et al., 2018), this could also be interesting in the context of teaching SSI. Other studies may therefore link psychological distance to interest (Ekborg et al., 2013) or decision-making (Parchmann et al., 2006). All results could again be abstracted to other, sometimes also difficult, teaching situations such as teaching in inclusive settings (Büssing et al., 2019b).

Overall, our study was able to illustrate differences and antecedents of biology pre-service teachers' distance towards three selected SSIs. As mentioned above, in the light of many subsequent questions, our study may represent a foundation for further studies. This will hopefully serve to better explain peoples' connections with SSIs, and thus, their learning and teaching behaviors.

Acknowledgements We acknowledge Marina Berling for her helping with the data collection and Dr. Benedikt Heuckmann for valuable comments on prior versions of this paper.

References

Arbieu, U., Mehring, M., Bunnefeld, N., Kaczensky, P., Reinhardt, I., Ansorge, H., & Nowak, C. (2019). Attitudes towards returning wolves (Canis lupus) in Germany: Exposure, information sources and trust matter. *Biological Conservation, 234*, 202–210. https://doi.org/10.1016/j.biocon.2019.03.027

Büssing, A. G., Dupont, J., & Menzel, S. (2020). Topic specificity and antecedents for pre-service biology teachers' anticipated enjoyment for teaching about socio-scientific issues: Investigating universal values and psychological distance. *Frontiers in Psychology, 11*(1536), 11–18. https://doi.org/10.3389/fpsyg.2020.01536

Büssing, A. G., Jannink, N., Scholz, G., & Halbe, J. (2019a). An adapted concept mapping technique to help conservation implementation – Exemplified for wolves returning to Lower Saxony in Germany. *Global Ecology and Conservation, 20*(e00784), 1–12. https://doi.org/10.1016/j.gecco.2019.e00784

Büssing, A. G., Menzel, S., Schnieders, M., Beckmann, V., & Basten, M. (2019b). Values and beliefs as predictors of pre-service teachers' enjoyment of teaching in inclusive settings. *Journal of Research in Special Educational Needs, 19*(S1), 8–23. https://doi.org/10.1111/1471-3802.12474

Büssing, A. G., Schleper, M., & Menzel, S. (2019c). Do pre-service teachers dance with wolves? Subject-specific teacher professional development in a recent biodiversity conservation issue. *Sustainability, 11*(1), 1–24. https://doi.org/10.3390/su11010047

Chapron, G., Kaczensky, P., Linnell, J. D. C., von Arx, M., Huber, D., Andrén, H., … Anders, O. (2014). Recovery of large carnivores in Europe's modern human-dominated landscapes. *Science, 346*(6216), 1517–1519. https://doi.org/10.1126/science.1257553

Ekborg, M., Ottander, C., Silfver, E., & Simon, S. (2013). Teachers' experience of working with socio-scientific issues: A large scale and in depth study. *Research in Science Education, 43*(2), 599–617. https://doi.org/10.1007/s11165-011-9279-5

Enserink, M., & Vogel, G. (2006). Wildlife conservation – The carnivore comeback. *Science, 314*(5800), 746–749. https://doi.org/10.1126/science.314.5800.746

Field, A. P., & Wilcox, R. R. (2017). Robust statistical methods: A primer for clinical psychology and experimental psychopathology researchers. *Behaviour Research and Therapy, 98*, 19–38. https://doi.org/10.1016/j.brat.2017.05.013

Großschedl, J., Harms, U., Kleickmann, T., & Glowinski, I. (2015). Preservice biology teachers' professional knowledge: Structure and learning opportunities. *Journal of Science Teacher Education, 26*(3), 291–318. https://doi.org/10.1007/s10972-015-9423-6

Großschedl, J., Welter, V., & Harms, U. (2019). A new instrument for measuring pre-service biology teachers' pedagogical content knowledge: The PCK-IBI. *Journal of Research in Science Teaching, 56*(4), 402–439. https://doi.org/10.1002/tea.21482

Hermann, N., Voß, C., & Menzel, S. (2013). Wildlife value orientations as predicting factors in support of reintroducing bison and of wolves migrating to Germany. *Journal for Nature Conservation, 21*(3), 125–132. https://doi.org/10.1016/j.jnc.2012.11.008

Heuckmann, B., Hammann, M., & Asshoff, R. (2018). Using the theory of planned behaviour to develop a questionnaire on teachers' beliefs about teaching cancer education. *Teaching and Teacher Education, 75*, 128–140. https://doi.org/10.1016/j.tate.2018.06.006

Heuckmann, B., Hammann, M., & Asshoff, R. (2020). Identifying predictors of teachers' intention and willingness to teach about cancer by using direct and belief-based measures in the context of the theory of planned behaviour. *International Journal of Science Education, 42*(4), 547–575. https://doi.org/10.1080/09500693.2020.1717671

Karlsson, J., & Sjöström, M. (2007). Human attitudes towards wolves, a matter of distance. *Biological Conservation, 137*(4), 610–616. https://doi.org/10.1016/j.biocon.2007.03.023

Klosterman, M. L., & Sadler, T. D. (2010). Multi-level assessment of scientific content knowledge gains associated with socioscientific issues-based instruction. *International Journal of Science Education, 32*(8), 1017–1043. https://doi.org/10.1080/09500690902894512

Lee, T. M., Markowitz, E. M., Howe, P. D., Ko, C. Y., & Leiserowitz, A. A. (2015). Predictors of public climate change awareness and risk perception around the world. *Nature Climate Change, 5*(11), 1014–1020. https://doi.org/10.1038/nclimate2728

Levinson, R. (2006). Towards a theoretical framework for teaching controversial socio-scientific issues. *International Journal of Science Education, 28*(10), 1201–1224. https://doi.org/10.1080/09500690600560753

Liberman, N., & Trope, Y. (2014). Traversing psychological distance. *Trends in Cognitive Sciences, 18*(7), 364–369. https://doi.org/10.1016/j.tics.2014.03.001

McDonald, R. I., Chai, H. Y., & Newell, B. R. (2015). Personal experience and the 'psychological distance'of climate change: An integrative review. *Journal of Environmental Psychology, 44*, 109–118. https://doi.org/10.1016/j.jenvp.2015.10.003

Parchmann, I., Gräsel, C., Baer, A., Nentwig, P., Demuth, R., Ralle, B., & ChiK Project Group. (2006). "Chemie im Kontext": A symbiotic implementation of a context-based teaching and learning approach. *International Journal of Science Education, 28*(9), 1041–1062. https://doi.org/10.1080/09500690600702512

Paulick, I., Großschedl, J., Harms, U., & Möller, J. (2016). Preservice teachers' professional knowledge and its relation to academic self-concept. *Journal of Teacher Education, 67*(3), 173–182. https://doi.org/10.1177/0022487116639263

Ronnenberg, K., Habbe, B., Gräber, R., Strauß, E., & Sibert, U. (2017). Coexistence of wolves and humans in a densely populated region (Lower Saxony, Germany). *Basic and Applied Ecology, 25*, 1–14. https://doi.org/10.1016/j.baae.2017.08.006

Rosseel, Y. (2012). Lavaan: An R package for structural equation modeling. *Journal of Statistical Software, 48*(2), 1–36. https://doi.org/10.18637/jss.v048.i02

Sadler, T. D. (2009). Situated learning in science education: Socio-scientific issues as contexts for practice. *Studies in Science Education, 45*(1), 1–42. https://doi.org/10.1080/03057260802681839

Sadler, T. D., & Zeidler, D. L. (2004). The morality of socioscientific issues: Construal and resolution of genetic engineering dilemmas. *Science Education, 88*(1), 4–27. https://doi.org/10.1002/sce.10101

Trope, Y., & Liberman, N. (2010). Construal-level theory of psychological distance. *Psychological Review, 117*(2), 440–463. https://doi.org/10.1037/a0018963

Van Boven, L., Kane, J., McGraw, A. P., & Dale, J. (2010). Feeling close: Emotional intensity reduces perceived psychological distance. *Journal of Personality and Social Psychology, 98*(6), 872–885. https://doi.org/10.1037/a0019262

Zeidler, D. L. (2014). Socioscientific issues as a curriculum emphasis. In N. G. Lederman & S. K. Abell (Eds.), *Handbook of research on science education* (Vol. 2, pp. 697–726).

Zeyer, A., & Dillon, J. (2014). Science|Environment|Health – Towards a reconceptualization of three critical and inter-linked areas of education. *International Journal of Science Education, 36*(9), 1409–1411. https://doi.org/10.1080/09500693.2014.904993

Zeyer, A., & Kyburz-Graber, R. (Eds.). (2012). *Science|Environment|Health: Towards a renewed pedagogy for science education*. Springer.

Chapter 16
Self-Efficacy of In-Service Secondary School Teachers in Relation to Education for Sustainable Development: Preliminary Findings

Athanasios Mogias, George Malandrakis, Penelope Papadopoulou, and Costas Gavrilakis

16.1 Introduction

The international community is now committed to Sustainable Development (SD), as a vision that incorporates responses to the most pressing modern economic, social, and environmental issues that humanity is facing (UN General Assembly, 2019). United Nations 2030 Agenda for Sustainable Development and its 17 Sustainable Development Goals (SDGs) are certainly the most ambitious and widely accepted political text and roadmap to guide efforts towards this vision (UN General Assembly, 2015). Education and especially Education for Sustainable Development (ESD) is among the most urgent challenges of the twenty-first century and if addressed contains the most promise (Wals, 2012), given that it is both a component of SDGs (target 4.7) and a critical tool to promote the Agenda as a whole. Within this context, educational resources have been developed to support

A. Mogias (✉)
Department of Primary Education, Democritus University of Thrace, Alexandroupolis, Greece
e-mail: amogias@eled.duth.gr

G. Malandrakis
Department of Primary Education, Aristotle University of Thessaloniki, Thessaloniki, Greece
e-mail: gmalandrakis@eled.auth.gr

P. Papadopoulou
Department of Early Childhood Education, University of Western Macedonia, Florina, Greece
e-mail: ppapadopoulou@uowm.gr

C. Gavrilakis
Department of Primary Education, University of Ioannina, Ioannina, Greece
e-mail: cgav@uoi.gr

O. Levrini et al. (eds.), *Engaging with Contemporary Challenges through Science Education Research*, Contributions from Science Education Research 9,
https://doi.org/10.1007/978-3-030-74490-8_16

curriculum developers and teachers in designing relevant strategies, curricula, and courses (Rieckmann, Mindt, Gardiner, Leicht & Heiss, 2017). However, several questions arise, such as whether teachers are prepared to implement ESD and to what extent they feel equipped to fulfill this task. Towards this direction, many teacher education institutions around the world have already integrated or plan to integrate ESD in their curricula, to prepare future teachers of primary and secondary education to address global challenges of sustainability. Various frameworks and guides (e.g., Sleurs, 2008; UNESCO, 2018) have been developed to support such integration – for example the notion of pre-service and in-service teachers' self-efficacy is a critical factor affecting their ability to implement ESD (Moseley, Huss & Utley, 2010).

16.1.1 Teachers' Self-Efficacy for ESD

Based on Bandura's social learning theory, self-efficacy belief indicates one's confidence in her/his ability to organize and execute a course of action, to solve a problem or accomplish a task (Bandura, 1977). Self-efficacy may suggest a belief in the ability to engage in a successful behavioural performance or to achieve a desirable outcome (Moseley & Taylor, 2011). Moreover, self-efficacy may focus either on the performance or on the learning (Ormrod, 2012). Therefore, self-efficacy of teachers, regardless of the subject matter, is closely connected with the teaching-learning process, since it is linked to behavioural patterns that teachers show in the classroom. These patterns can result in marked differences in the type of teaching and the strategies and methodologies used by teachers in their daily practices.

From the early '90s and until the last few years, only two self-efficacy scales had been developed to measure teachers' ability to implement Environmental Education (EE), the precursor of ESD, both deriving directly from different versions of the Science Teaching Efficacy Belief Instrument (STEBI) (Enochs & Riggs, 1990). The first was the Environmental Education Efficacy Belief Instrument (EEEBI) (Sia, 1992), and the second was the Environmental and General Science Teacher Efficacy Assessment (EGSTEA) (Moseley & Taylor, 2011). As science-based, these instruments can hardly cover the wider subject matter of EE/ESD and its holistic and systemic pedagogical approaches. However, despite the plethora of studies worldwide on teachers' self-efficacy (4,742 results in Scopus, January 2020), the main targeted population is that of pre-service teachers (388 studies), while none of the 96 studies focusing on in-service teachers deal with EE/ESD (Wilson & Tan, 2004). In addition, although the self-efficacy of secondary teachers has also been extensively studied (44 results), only nine of them involve in-service teachers, of which none is connected with EE/ESD. Therefore, the literature review that follows is constrained to studies with primary and secondary education teachers (both pre- and in-service) related to self-efficacy in EE/ESD.

In particular, some studies have already used EEEBI to assess teachers' self-efficacy (see Evans, Tomas & Woods, 2016). For instance, Moseley, Reinke and

Bookout (2002) evaluated the effect of a 3-day outdoor EE program on pre-service elementary teachers' self-efficacy. They found that although the program did not change the already high self-efficacy beliefs of the participants, some days after the completion of the program their self-efficacy dropped significantly, probably due to the re-evaluation of their ability to teach, as they learned more about teaching methodologies. Similarly, Moseley et al. (2010) examined the influence on EE teaching self-efficacy of K-12 teachers (both primary and secondary education) after their participation in a two-week intensive summer course about earth systems science, using the GLOBE curriculum. They revealed significant gains in both self-efficacy dimensions of EEEBI - i.e. personal environmental teaching efficacy (PETE) and environmental teaching outcome expectancy (ETOE) - immediately following the workshop. Using the same instrument, Gardner (2009) investigated the self-efficacy of elementary education pre-service teachers in the USA and found that not only do they feel a lack of sufficient knowledge and skills in EE, but also realized that there is an interesting relationship between teachers personal experiences with nature as young children and their current self-efficacy beliefs. Richardson et al. (2014) also used the EEEBI to examine the change of pre-service teachers' EE self-efficacy due to their engagement in a two-year intervention grounded on inquiry-based instruction.

Boon (2011) developed a scale to investigate Australian pre-service (early childhood and primary) teachers' beliefs and their knowledge about ESD. An adjusted version of this questionnaire was also used by Effeney and Davis (2013) to explore relationships between knowledge and efficacy for teaching sustainability in a group of pre-service primary and early childhood education teachers in the same country. They revealed that the participants were confident in their abilities to teach ESD and their self-efficacy was strengthening with increased levels of perceived knowledge. However, perceived knowledge had no relationship with actual knowledge (Effeney & Davis, 2013). It should be mentioned that there has been much discussion in the literature about the relationship between perceived/actual knowledge and self-efficacy (see Mintz et al., 2020). Several studies have revealed correlations between high levels of perceived knowledge and self-efficacy, which, according to Mintz et al. (2020), reminds us that the definition of self-efficacy is one's belief in his/her ability to carry out a task, independently of measures of actual ability (Bandura, 1997). Thus, without disregarding actual knowledge, these findings encourage the use of perceived knowledge as a potential dimension and a handy determinant of teachers' self-efficacy.

Moreover, Dahl's (2019) recent study with pre-service teachers in seven different teacher education programs in Europe, based on an instrument focused on teacher professional competencies, including an item regarding teaching for SD, showed that they do not feel well prepared to educate for sustainability. Another recent study (Tomas et al., 2017), also used only a few items of a wider Likert-style survey to explore pre-service teachers' (early childhood and primary) attitudes toward ESD, and to assess their ESD self-efficacy before and after completing an ESD unit in an Australian university.

Attention should also be paid to a new questionnaire created by Gan and Gal (2018) aiming to evaluate general education pre-service (early childhood and primary) teachers' ability to promote ESD, emphasizing pro-environmental behaviour both inside and outside the classroom (private and public sphere, respectively). The same purpose and the emergent calls for initial teacher-education institutions to integrate ESD competencies into their programs (Sleurs, 2008; Rieckmann, et al., 2017), motivated Malandrakis, Papadopoulou, Gavrilakis, and Mogias (2019) to develop another new teachers' self-efficacy assessment instrument, called Teachers' Self-Efficacy Scale for Education for Sustainable Development (TSESESD). It describes ESD self-efficacy as a belief linked with four domains of competencies, namely (a) values and ethics, (b) systems thinking, (c) emotions and feelings, and (d) actions. This instrument has been inspired by Sleurs' (2008) model which consists of five competence domains: (a) knowledge, that is conceptual, factual and action related, is related to time as well as to space and that is inter-, trans, pluri- or cross-disciplinarily constructed; (b) systems-thinking, meaning the different kinds of systems that are addressed, including interrelationships in time and space; it implies the awareness of being part of the living system, "Earth" in space and time; (c) emotions, since thinking, reflecting, valuing, making decisions, and acting are inseparably tied with emotions; empathy and compassion thereby play a key role; (d) values and ethics, where the main guiding principle of ESD is equity (social, intergenerational, between genders, between communities, between human beings and nature, etc.); and finally (e) action, the process where all the competencies of the other four domains merge to meaningful creations, participation, and networking in SD in all four levels: individual, classroom/school, regional, and global. In TSESESD, knowledge is not included as a domain, since according to the founding description of social cognitive theory (Bandura, 1989), it interacts with environmental and behavioural factors (e.g., self-efficacy) to influence human behaviour. Thus, in Malandrakis et al.'s (2019) study, knowledge was examined as a factor influencing self-efficacy. Specifically, the perceived knowledge was examined as it is more relevant to the self-efficacy founding theory (Bandura, 1997), as it tends to reflect teachers' confidence in what they know and are able to do. In parallel, TSESESD instrument attempts to integrate critical methodological elements of ESD, such as the holistic and interdisciplinary approach of knowledge, critical and systems thinking, emphasis on values clarification and so on, which are needed to deal with socio-economic and political dimensions of environmental and other sustainability issues (Malandrakis, Papadopoulou, Gavrilakis & Mogias, 2016; Malandrakis et al., 2019). TSESESD has already been used with pre-service primary school teachers and checked for face and content validity (Malandrakis et al., 2016), while its construct validity and factor structure has also been examined, revealing good psychometric properties (Malandrakis et al., 2019).

Based on the above and the obvious lack of studies focusing on in-service teachers of secondary education, the goal of the present study is to examine the self-efficacy beliefs of in-service secondary school teachers for ESD teaching, as well as their perceived knowledge of environmental issues.

16.2 Methodology

16.2.1 Research Instrument

The newly launched instrument entitled "Teachers' Self-Efficacy Scale for Education for Sustainable Development" (TSESESD) (Malandrakis et al., 2019) was used in the present study. The conceptual framework for the development of the instrument was based on the relevant literature from the fields of Science Education and Environmental Education / Education for Sustainable Development (EE/ESD) (e.g., Nolet, 2009; Sia, 1992; Sleurs, 2008). The instrument encompasses the above mentioned four domains of competencies, the magnitude of which portray teachers' belief in their ability to implement them in ESD. Moreover, following OECD's (2002) recommendations for a broader framework in terms of competencies, incorporating not only social and behavioural components, but also knowledge, cognitive, and practical skills, this study also investigated secondary school teachers' perceived Content Knowledge (CK) about specific ESD concepts, like the greenhouse effect, climate change, ozone layer depletion, ecological footprint, and biodiversity loss, among others. Moreover, their perceived Pedagogical Content Knowledge (PCK) was also studied taking into consideration the relevant literature (e.g., Sleurs, 2008), with the addition of three more dimensions, those of interdisciplinarity, ESD curricula, and assessment.

As a result, the whole instrument was composed of 24 items in the self-efficacy scale, not equally distributed among the four domains, and 14 and 17 item subscales focusing on CK and PCK, respectively (Tables 16.1 and 16.2). Furthermore, participants' characteristics regarding gender, scientific specialization, residency, years of service, age, high school level of teaching, previous experience in ESD, and

Table 16.1 In-service secondary teachers' self-efficacy scale for ESD (TSESESD), perceived knowledge scale, and their sub-domains indices

Domain	Items	Range	Mean	SD	Cronbach α	Skewness	Kurtosis
Values and ethics	6	1–7	4.77	1.49	0.944	−0.508	−0.246
Systems thinking	5	1–7	4.15	1.54	0.936	−0.159	−0.614
Emotions and feelings	3	1–7	4.61	1.33	0.885	−0.417	−0.023
Actions	10	1–7	4.31	1.47	0.962	−0.314	−0.319
Total	**24**	**1–7**	**4.43**	**1.47**	**0.975**	**−0.343**	**−0.325**

Table 16.2 In-service teachers' knowledge scale and its sub-domain indices

Domain	Items	Range	Mean	SD	Cronbach α	Skewness	Kurtosis
Perceived knowledge							
Content knowledge	14	1–7	4.33	1.73	0.949	−0.156	−0.088
Pedagogical content knowledge	17	1–7	3.84	1.65	0.966	−0.749	−0.720
Total	**31**	**1–7**	**4.06**	**1.68**	**0.971**	**−0.022**	**−0.733**

the frequency of information sources used about general environmental and/or sustainable development issues were also included.

16.2.2 Participants

Two hundred sixty-seven Greek in-service secondary education teachers from eight mainland and island cities participated in the present study. Fifty-six percent of the participants were females, while the majority (62.5%) had less than 20 years of service, although almost half of the participants were over 50 years of age. The 48.3% came from science disciplines (mainly physics teachers), 64.3% were serving in junior high schools (grades 7–9) at the time of the study, and 35.7% in upper high schools (grades 10–12). Their previous experience with ESD, either by attending relevant training seminars or implementing ESD projects in their schools, was limited (39.2% and 25.4%, respectively). Finally, participants reported the internet as the main source of their environmental information with a mean value of 4.14 (±1.12) in a 5-point Likert scale, while TV documentaries, specialized journals, and books followed (3.13 ± 1.25, 2.35 ± 1.40, and 2.33 ± 1.34, respectively).

16.2.3 Data Analysis

Teachers' answers were assigned to numbers from 1 (*"not at all"/"not sure at all"*) to 7 (*"very good"/"absolutely sure"*), with lower scores indicating lower teachers' self-efficacy and perceived knowledge levels, and vice versa. Data analysis involved (a) descriptive statistics applied to portray mean values (± standard deviation) of the 24 self-efficacy and 31 perceived knowledge items (CK and PCK), (b) regression coefficients to determine probable perceived knowledge effects on self-efficacy scores, and (c) implementation of inferential statistics to further investigate the potential effect, in terms of significant differences, of background factors on teachers' self-efficacy and knowledge scores. For all statistical tests, the significance level was predetermined at a probability value of 0.05 or less.

16.3 Results

Greek in-service secondary school teachers were found to report rather moderate self-efficacy scores in the TSESESD domains (4.43 ± 1.47), while they presented slightly lower scores on the knowledge scale (4.06 ± 1.68), indicating also moderate perceived content and pedagogical content knowledge of certain environmental and/or sustainability issues (Tables 16.1 and 16.2). More specifically, for the Self-efficacy scale, "Values and Ethics" presented the highest mean score among the domains (4.77 ± 1.49), while "Systems thinking" the lowest (4.15 ± 1.54) (Table 16.1). Regarding the perceived knowledge scale, CK presented the highest mean score (4.33 ± 1.73), while PCK exhibited an evident low score (3.84 ± 1.65) (Table 16.2). Moreover, all domains showed excellent internal consistency values, revealing an overall Cronbach α value of 0.975 for Self-efficacy, and 0.971 for perceived knowledge, respectively (Tables 16.1 and 16.2). Inter-correlations both within scale domains and between self-efficacy and perceived knowledge scales were also calculated (Table 16.3).

In particular, inter-correlations of domains in the self-efficacy scale ranged between 0.678 and 0.789, in the perceived knowledge was 0.706, while a strong correlation was also revealed between the two scales (0.775). All correlations were statistically significant at the 0.01 level (Table 16.3). Additionally, regression analysis was performed to determine the effect of perceived knowledge in self-efficacy scores. The analysis revealed that 60% of the observed variance ($R^2 = 0.600$) in teachers' self-efficacy scores can be explained through perceived knowledge, portraying a good association between the two variables.

Independent *t*-tests and One-Way Analysis of Variance were further performed to investigate probable significant differences in terms of the participants' gender, specialization, seminar training, previous experience in ESD, and years of service. Although male teachers appeared to be slightly more knowledgeable, their female counterparts showed rather higher self-efficacy values, but in both cases, the differences were not statistically significant (Fig. 16.1a). Science teachers, although exhibited significantly higher scores in perceived knowledge, they were slightly surpassed by their colleagues from humanities studies in self-efficacy domains (Fig. 16.1b). Seminar training and previous experience in ESD appeared to positively influence both teachers' perceived knowledge and self-efficacy ($p \leq 0.001$) (Fig. 16.1c and d). Finally, an interesting finding, which acquires a more in-depth investigation, is that secondary education teachers with limited teaching experience (less than 6 years of service) as well as those with more than 30 years of class experience, were found to report higher mean values than their peers in both perceived knowledge and self-efficacy (Fig. 16.1e).

Table 16.3 Correlation indices among the domains under study

	1	1a	1b	1c	1d	2	2a	2b
1. TSESESD		0.887**	0.896**	0.840**	0.946**	0.775**	0.574**	0.833**
1a. Values & Ethics			0.771**	0.686**	0.736**	0.732**	0.561**	0.772**
1b. Systems thinking				0.678**	0.777**	0.758**	0.577**	0.803**
1c. Emotions & Feelings					0.789**	0.559**	0.391**	0.620**
1d. Actions						0.697**	0.503**	0.761**
2. Perceived knowledge							0.909**	0.937**
2a. Content knowledge								0.706**
2b. Pedagogical content knowledge								

** Correlation is significant at the 0.01 level

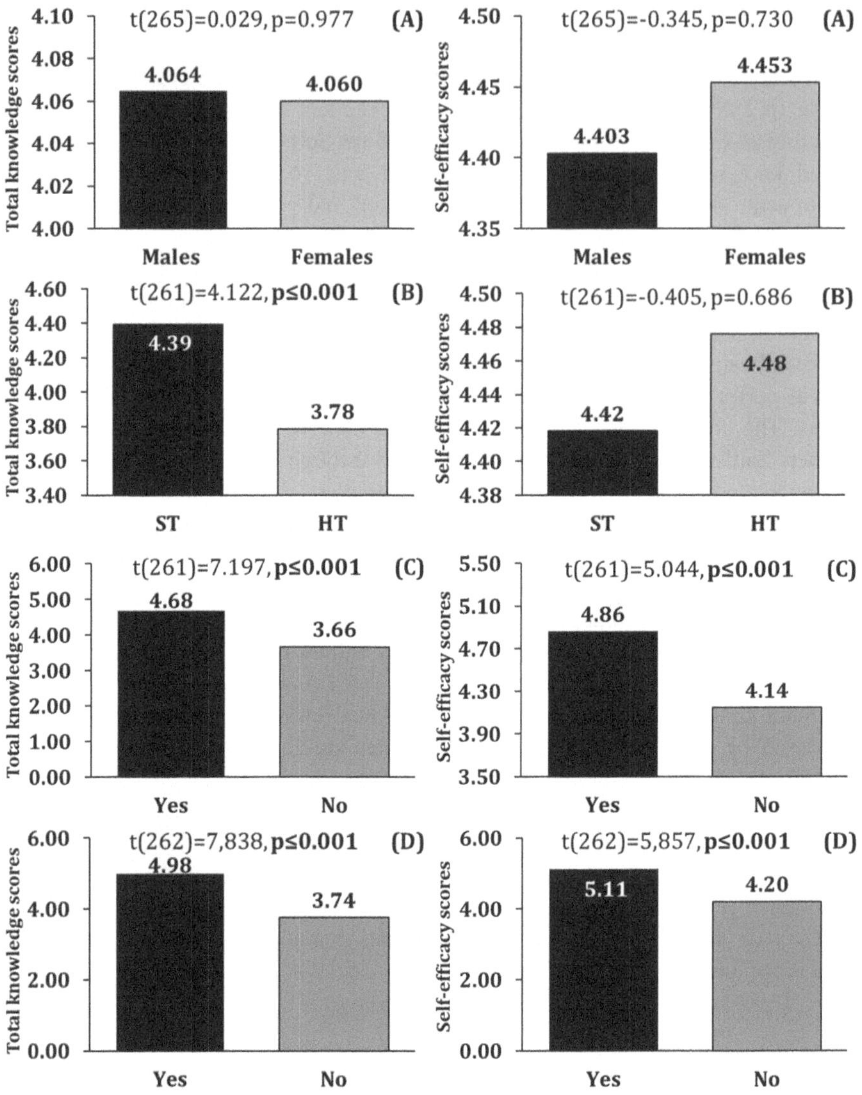

Fig. 16.1 Mean values of Knowledge and Self-efficacy scores concerning participants' gender (**a**), general specialization field (ST: Science Teachers; HT: Humanities Teachers) (**b**), in-service seminar training (**c**), previous experience in ESD (**d**), and years of service (**e**)

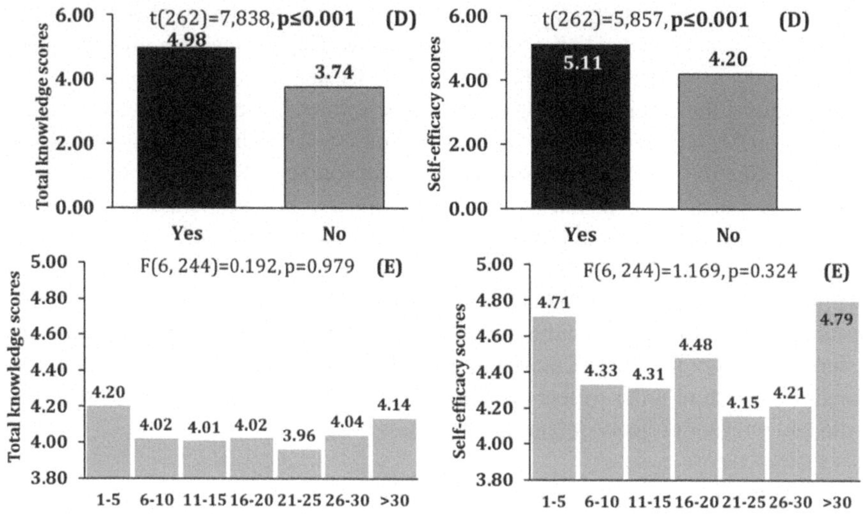

Fig. 16.1 (continued)

16.4 Discussion and Conclusions

Analysis indicates that TSESESD for secondary in-service teachers has good psychometric properties, having excellent internal consistency scores, along with strong and significant correlations among all domains.

The implementation of TSESESD with Greek secondary in-service teachers revealed that they exhibit moderate scores in both their self-efficacy beliefs to teach ESD issues and in their perceived content and pedagogical content knowledge of certain environmental and/or sustainability issues. Their self-efficacy scores are relatively lower than those reported for both Greek pre-service and in-service elementary teachers using the same scale (Malandrakis et al., 2019). For perceived knowledge the situation is mixed, as the secondary in-service educators of the present study gain about the same scores with their in-service colleagues of elementary schools, but higher than the scores of the pre-service elementary teachers (Malandrakis et al., 2019). However, the direct comparison of our findings with those in the literature is difficult, since the previous studies not only use tools other than TSESESD (mainly EEEBI) that are based on very different conceptual frameworks, but also due to the focus of these studies on pre-service teachers.

Another important finding is that, within secondary teachers, the science-oriented group seems to possess significantly higher perceived knowledge scores than their humanities colleagues, but the latter exhibit higher self-efficacy scores in teaching ESD. Also, the seminars and the previous experience in ESD, along with the years of teaching experience seem to have a critical role in teachers' self-efficacy and perceived knowledge, with those having either too few years of service or too much to exhibit the higher scores in both scales. However, despite the differences, these secondary in-service teachers' scores are considered as moderate. This finding can partially be explained due to the lack of relevant training during their undergraduate studies and a respective shortage in their in-service support on these topics.

In light of the significant progress that has been occurring in ESD lately at the international level, many teacher training institutions have already integrated, or are in the process of integrating corresponding novel courses. These programs are often content-oriented (e.g., physics, biology, chemistry, history, language) to some extent, following traditional pedagogic approaches and, therefore, leave little or no space for learning and training about effective teaching methods and techniques within the framework of ESD. Therefore, the proposed instrument is expected to establish a baseline for providing detailed information to university programs and course designers in terms of incorporating all the necessary competencies teachers need to acquire in order to feel capable of planning and implementing ESD curricula and relevant projects or learning activities worldwide.

Acknowledgments The authors would like to thank all high school teachers for their substantial assistance in the participation of the research.

References

Bandura, A. (1977). Self-efficacy: Toward a unifying theory of behavioral change. *Psychological Review, 84*(2), 191–215.

Boon, H. (2011). Beliefs and education for sustainability in rural and regional Australia. *Education in Rural Australia, 21*(2), 37–54.

Dahl, T. (2019). Prepared to teach for sustainable development? Student teachers' beliefs in their ability to teach for sustainable development. *Sustainability, 11*, 10p.

Effeney, G., & Davis, J. (2013). Education for sustainability: A case study of pre-service primary teachers' knowledge and efficacy. *Australian Journal of Teacher Education, 38*(5), 32–46.

Enochs, L., & Riggs, I. (1990). Further development of an elementary science teaching efficacy belief instrument: A pre-service elementary scale. *School Science and Mathematics, 90*(8), 694–706.

Evans, N., Tomas, L., & Woods, C. (2016). Impact of sustainability pedagogies on pre-service teachers self-efficacy. *Journal of Education for Sustainable Development, 10*(2), 243–261.

Gan, D., & Gal, A. (2018). Self-efficacy for promoting EfS among pre-service teachers in Israel. *Environmental Education Research, 24*(7), 1062–1075.

Gardner, C. C. (2009). *Self-efficacy in Environmental Education: Experiences of elementary education preservice teachers*. College of Education, University of South Carolina.

Malandrakis, G., Papadopoulou, P., Gavrilakis, C., & Mogias, A. (2016). Designing and testing an education for sustainable development self-efficacy scale for pre-service teachers: Preliminary findings. In J. Lavonen, K. Juuti, J. Lampiselkä, A. Uitto, & K. Hahl (Eds.), *Electronic proceedings of the ESERA 2015 conference, science education research: Engaging learners for a sustainable future, part 9* (pp. 1280–1292). University of Helsinki.

Malandrakis, G., Papadopoulou, P., Gavrilakis, C., & Mogias, A. (2019). An education for sustainable development self-efficacy scale for primary pre-service teachers: Construction and validation. *The Journal of Environmental Education, 50*(1), 23–36.

Mintz, J., Hick, P., Solomon, Y., Matziari, A., Ó'Murchú, F., Hall, K., Cahill, K., Curtin, C., Anders, J., & Margariti, D. (2020). The reality of reality shock for inclusion: How does teacher attitude, perceived knowledge and self-efficacy in relation to effective inclusion in the classroom change from the pre-service to novice teacher year? *Teaching and Teacher Education*. https://doi.org/10.1016/j.tate.2020.103042

Moseley, C., & Taylor, B. (2011). Analysis of environmental and general science teaching efficacy among instructors with contrasting class ethnicity distributions: A four-dimensional assessment. *School Science and Mathematics, 111*(5), 199–208.

Moseley, C., Reinke, K., & Bookout, V. (2002). The effect of teaching outdoor environmental education on preservice teachers' attitudes toward self-efficacy and outcome expectancy. *The Journal of Environmental Education, 34*(1), 9–15.

Moseley, C., Huss, J., & Utley, J. (2010). Assessing K–12 teachers' personal environmental education teaching efficacy and outcome expectancy. *Applied Environmental Education & Communication, 9*(1), 5–17.

Nolet, V. (2009). Preparing sustainability-literate teachers. *Teachers College Record, 111*(2), 409–442.

OECD (2002). Definition and Selection of Competences (DeSeCo): Theoretical and conceptual foundations. Strategy Paper. DEELSA/ED/CERI/CD (2001).

Ormrod, J. E. (2012). *Human learning* (6th ed.). Pearson.

Richardson, G. M., Liang, L. L., & Wake, D. G. (2014). Examining the durability of environmental education self-efficacy beliefs in preservice teaching. *Applied Environmental Education and Communication, 13*(1), 38–47.

Rieckmann, R., Mindt, L., Gardiner, S., Leicht, A., & Heiss, J. (2017). *Education for sustainable development goals – Learning objectives*. UNESCO. https://developmenteducation.ie/resource/education-sustainable-development-goals-learning-objectives. Accessed 15 Oct 2018

Sia, A. P. (1992). *Preservice elementary teachers' perceived efficacy in teaching environmental education: A preliminary study*. National Center for Research on Teacher Learning.

Sleurs, W. (ed.) (2008). *Competencies for ESD (Education for Sustainable Development) teachers. A framework to integrate ESD in the curriculum of teacher training institutes*. Comenius 2.1 project 118277-CP-1-2004-BE-Comenius-C2.1. ENSI-CSCT.

Tomas, L., Girgenti, S., & Jackson, C. (2017). Pre-service teachers' attitudes toward education for sustainability and its relevance to their learning: Implications for pedagogical practice. *Environmental Education Research, 23*(3), 324–347.

UN General Assembly (2015). *Transforming our world: the 2030 Agenda for Sustainable Development.* Resolution adopted by the General Assembly on 25 September 2015. A/RES/70/1. https://www.un.org/ga/search/view_doc.asp?symbol=A/RES/70/1&Lang=E. Accessed 20 Oct 2018.

UN General Assembly (2019). *Political declaration of the high-level political forum on sustainable development convened under the auspices of the General Assembly*. Resolution adopted by the General Assembly on 15 October 2019. A/RES/74/4. https://undocs.org/en/A/RES/74/4. Accessed 12 Sept 2018.

UNESCO. (2018). *Integrating education for sustainable development (ESD) in teacher education in South-East Asia: A guide for teacher educators*. UNESCO.

Wals, A. E. J. (2012). *Shaping the education of tomorrow: 2012 full-length report on the UN Decade of Education for Sustainable Development*. UNESCO, DESD Monitoring & Evaluation.

Wilson, P., & Tan, G.-C. I. (2004). Singapore teachers' personal and general efficacy for teaching primary social studies. *International Research in Geographical and Environmental Education, 13*(3), 209–222.

Part III
Developing Innovative Theoretical Perspectives and Methodologies

Chapter 17
Where Are We? Syntheses and Synergies in Science Education Research and Practice

Bruce Sherin

17.1 Introduction

I begin with some personal history. I graduated from my undergraduate institution with a B.A. in physics, and the full intention of obtaining a Ph.D. in that same discipline. However, after about 2 years of graduate school I determined that I was in the wrong place; it turned out that I greatly enjoyed *learning* physics, but that I was less interested in spending many years of my life working in one tiny corner of the field. It was thus that I found my way to educational research.

It was an exciting time both for me and for educational research in general. The cognitive revolution (Gardner, 1987) was gaining traction, and computers seemed to have the potential to be transformational tools for education. In my new graduate program in educational research I was beginning to read material that was quite different than what I had read as a physics student. There was Piaget, the basics of cognitive science and artificial intelligence, and the new applications of this work to the study of learning.

But I was struggling to make it all fit together. For example, I did not fully understand what Piaget meant by "logical structures" or "operations" (e.g., Inhelder & Piaget, 1958). Nor did I understand how these ideas were similar or different than, for example, what the most current cognitive scientists were writing. I expected learning theory to be like physics – maybe not exactly, but at least close. But I gradually came to understand that this expectation was not one that would be fulfilled. There weren't principles or even definitions that researchers agreed upon. Researchers used terms such as "concept," "metacognition," and "strategy," that I thought were intended to be technical terms, and hence well-defined. But even

B. Sherin (✉)
Northwestern University, Evanston, IL, USA
e-mail: bsherin@northwestern.edu

O. Levrini et al. (eds.), *Engaging with Contemporary Challenges through Science Education Research*, Contributions from Science Education Research 9,
https://doi.org/10.1007/978-3-030-74490-8_17

within individual journal articles, these terms were neither explicitly nor consistently defined. This was hard news for the junior physicist in me 30 years ago.

The even worse news is that the situation has not substantially improved. Thirty years ago, it seemed possible that the cognitive revolution would provide a basis for a real, shared science of science learning. But that has not happened. We have learned—we have made some progress. But there is certainly not a shared theory and well-defined terminology. The purpose of this paper is to do some work in that direction, or at least to name some of the bigger problems.

The talk on which this paper is based was itself based on work done for a book, *Converging perspectives on conceptual change*, which was guided and assembled by Olivia Levrini and Tamer Amin (Amin & Levrini, 2018; Sherin, 2018). Tamer Amin has himself been an author of some of the best attempts to synthesize the range of existing work on conceptual change in science, and his work has shaped the view presented here (Amin, 2009; Amin et al., 2014).

For the earlier book, I was asked to put together a section on the nature of concepts and conceptual change. Multiple authors contributed, and my job was to synthesize those contributions. However, unlike the earlier book, here I focus on an individual, cognitive level of analysis, since I understand that to be my charge.

17.2 What Is "Conceptual Change"?

Conceptual change has no precise and shared definition. It can be taken to apply to any learning events in which there are changing "concepts." When used in that way, conceptual change is essentially a synonym for learning. But, more frequently, we use the phrase *conceptual change* to apply to a subset of learning—learning that is dramatic in scope or, in some manner, difficult to attain. So, for example, learning to understand Darwin's theory of evolution might well require conceptual change; learning that mammals give birth to live young probably does not.

In the field of science education, we often mean something still more specific when we speak of conceptual change. The idea that we begin with is that students have knowledge of the natural world that they gain prior to formal instruction, both from direct interaction with the world, as well as communication with other humans. The story of conceptual change is often taken to be the story of how this pre-existing knowledge changes; it is the process through which a learner makes difficult or extensive change to existing knowledge about the workings of the world.

Understood in this more narrow sense, research on conceptual change in science is concerned with three questions:

1. What is the nature of students' pre-instruction knowledge of the natural world? (i.e., what is the "stuff" that is changing?)
2. How does this knowledge change as learners develop and scientific expertise is acquired? (i.e., What are the processes of change?)
3. What is *difficult* about this change?

What I'll attempt to show in the rest of the paper is that there is no consensus about the answers to these questions. This is particularly striking given how long we have been seeking answers. In fact, I first encountered them in a graduate course on conceptual change, taught by Andrea diSessa and Ann Brown, sometime around 1990.

The remainder of this paper will have three parts. In the first part I'm going to give something like an illustrated tour of different ways the three questions have been answered. Then, in the second part I'm going to talk about impediments to progress, that is, why we have not achieved consensus on answers to the questions. Then, in the last section I will briefly discuss how we might make further progress. I feel I must warn the reader before beginning. Identifying the problems is a relatively easy endeavor. But offering solutions remains difficult, and my suggestions, at best, offer only small steps forward.

17.3 An Illustrated Tour

17.3.1 Conceptual Change in Physics

Our tour begins in physics. The domain of physics has been one of the main foci of research in conceptual change. This is partly historical accident, but also partly because of features of the domain. We have rich everyday experiences of the physical world and learning formal physics does seem to require a dramatically new way of understanding the world. One of my favorite examples to illustrate this point is that, from the point of view of Newtonian physics, an automobile moves forward because the road pushes it forward. Even diehard physicists must admit that there is something intuitively bizarre about this notion.

Much of the early research on physics learning concerned how students understand the motion of projectiles. I'm going to illustrate some issues with a bit of data. It's drawn from project that that I was part of with Andrea diSessa and David Hammer, while I was graduate student, and it has become classic data—it has been discussed in many prior journal articles and chapters (e.g., diSessa & Sherin, 1998). The episode begins with the interviewer (diSessa) asking the university student, who he calls "J", to talk about the forces acting on a ball that is tossed straight up and falls back down. J immediately answers as follows:

> Not including your hand, like if you just let it go up and come down, the only force on that is gravity. And so it starts off with the most speed when it leaves your hand, and the higher it goes, it slows down to the point where it stops. And then comes back down. And so, but the whole time, the only force on that is the force of gravity, except the force of your hand when you catch it.

If you know your physics, then you will have noticed that J's answer here seems unproblematic. Most notably, she says that the only force acting on the ball is the force of gravity. But the interview immediately took a dramatic turn. The interviewer asked J to focus on what happens at the peak of the toss, and J gave a somewhat meandering response:

> …when you throw it, you're giving it a force upward, but the force can only last so long against air and against gravity ... But so you give this initial force, and it's going up just fine, slower and slower because gravity is pulling on it and pulling on it. And it gets to the point to the top, and then it's not getting any more energy to go up. You're not giving any more forces, so the only force it has on it is gravity and it comes right back down. ...

I want to first draw your attention to the last part of J's statements here. The key point is that, in her initial explanation, J said there was only one force acting on the ball, the force of gravity. But, by the end of the second passage, she is saying that there are *two* forces: gravity and a force that is imparted by the hand.

This latter passage provides a nice entry into one way in which conceptual change has been talked about in physics. The idea, according to some researchers, is students such as J can be said to have a "naïve theory" of physics; in particular, J's answer might be seen as consistent with what has been called the "impetus theory" (e.g., McCloskey, 1984). Like any good theory, the impetus theory is seen to have its own principles:

- Motion requires a causal explanation.
- Forces can be stored up in objects. These stored forces are the "impetus."
- The motion of an object depends, in some manner, on a combination of the internal and external forces.
- Sometimes the stored force just dies away on its own. Sometimes an external force can cause the stored force to die away.

Hopefully, you can see how J's response might be seen to align with these principles. She says that, when a ball is tossed upward, the ball is given an upward force, which is then diminished by gravity.

This way of understanding what is going on in the episode provides us with our first set of answers to the three questions about conceptual change:

What is changing in conceptual change? The idea is that students, such as J, begin with a theory—the impetus theory. This is the "stuff" that must change.

What are the processes of change? The process of change, at least at a high level, is one of *theory replacement*; the impetus theory must be replaced by Newtonian theory.

Why is this change hard? It seems reasonable to expect that replacing one theory with another would be difficult. In addition, we are told that the impetus theory is particularly stubborn and resistant to change.

Now I want to present an alternative perspective on what is going on in this episode with J, and it is one that is offered by diSessa himself (diSessa, 1996). diSessa's view is rooted in the idea that, in episodes such as this one, student explanations are often constructed, in the moment, built out of smaller ideas. In particular, he draws attention to a number of small schemas that he calls phenomenological primitives or, more affectionately, p-prims (diSessa, 1993). These p-prims are cued — brought to mind — for reasons that can depend sensitively on what is going on in the moment. They include, for examples, notions having to do with balancing and equilibrium, force and agency, and constraint phenomena.

Now let me play out diSessa's analysis of J's explanation of the tossed ball. The central idea is that asking J to focus on what is happening at the peak cues her to think about *balancing*. The top of the toss, for many learners, is strongly suggestive of balancing. Talking about the peak also probably cues a typical *overcoming* scenario. There are two competing influences, one of which ultimately prevails, overcoming the other. But this leaves J with a problem to solve. She only has one force, the force of gravity; thus, she needs a second force. Over the course of her response, she considers a few alternatives, but ultimately settles on the introduction of an impetus-like force.

diSessa's perspective yields another set of answers to our three questions:

What is changing in conceptual change? There is a large system of elements that includes, many p-prims, as well as other unnamed elements.

What are the processes of change? According to diSessa, this large system of elements gets tuned up so that we use the right p-prims at the right time. Some become less important, some more important. In addition, some p-prims are adapted to play special roles in formal physics.

Why is this change hard? The answer to this question is notably different than the one given by theory-theorists. Change, according to diSessa, is not difficult because there is a single large structure that is difficult to displace. Rather, it's difficult because many small coordinated changes must be made to a system consisting of numerous elements.

17.3.2 Conceptual Change in Astronomy

Intuitive astronomy is another domain that has played a central role in thinking about conceptual change. Again, I believe the reasons for its importance are partly historical; but there are also properties of the domain that have made it an interesting focus for study. As in the domain of physics, we have some everyday access to astronomical phenomena. We can see the sun and stars. However, our access seems somewhat more remote than our experience, for example, with projectiles. We cannot directly observe the shape of orbits, and there are profound issues of scale involved.

More importantly, social inputs at least *seem* to be more important in astronomy. Even when children have not yet learned about the stars and planets in school, they have likely heard about them as part of their everyday lives, in children's books and stories, and beyond.

I am going to again illustrate this part of my tour with a brief bit of transcript. Here I draw on some prior work done with colleagues at Northwestern University, where we asked U.S. middle school students to explain the causes of the Earth's seasons (Sherin et al., 2012). This interview topic has had a long history in science education research. Part of this question's power is that it *seems* mundane, and most people seem to feel that they *should* be able to answer accurately. But the answer is actually quite subtle. Here, I began by asking a student, Angela, "why it's warmer in the summer and colder in the winter." She answered:

> That's because the like sun is in the center and the Earth moves around the sun and the Earth is at one point like in the winter, it's on it's like farther away from the sun and towards the summer it's closer it's near, towards the sun.

In some ways, Angela's answer is unsurprising, and it is not too difficult to speculate about the origins of her answer. Angela knows that the sun is a source of warmth, and she certainly knows that if you're closer to a heat source, then you feel the heat of it more strongly. On top of that, she has probably heard that the Earth orbits the sun in an ellipse, which implies that it is sometimes closer to the sun and sometimes farther away. Thus, Angela's explanation is a very sensible construction, one that in fact includes many elements of the accepted explanation.

In my research group's interviews, Angela's answer was one among a variety of explanation that we observed, which in turn were among the larger set reported in the literature. We called explanations like Angela's *closer-farther* explanation. We also saw *side-based* explanations, in which seasons are explained by the fact that the Earth rotates, and the side facing the sun experiences summer, and tilt-based explanation, in which the hemisphere tilted toward the sun experiences summer.

An important point here is that many of these explanations seem like an amalgam of informally gained knowledge and school knowledge. For example, Angela's explanation combines everyday knowledge about heat sources with knowledge that the Earth's orbit is elliptical.

Now I want to turn to one prominent way in which conceptual change in astronomy has been understood. The work of Stella Vosniadou and colleagues provides an important reference point (Vosniadou & Brewer, 1992; Vosniadou & Skopeliti, 2014), and she was one of the contributors to the book from which this paper is descended. Here I am going to draw on some of her own words from the edited book (Vosniadou, 2018).

At the core of Vosniadou's account is what she calls a *framework theory.* Like McCloskey and others, she uses the term "theory." However, she means something quite different than these prior authors. She emphasizes that framework theories are not "well formed."

> A 'framework theory' is not a well formed, explicit and socially shared scientific theory. Rather, it is a skeletal conceptual structure that grounds our deepest ontological commitments and causal devices in terms of which we understand a domain.

Another of Vosniadou's central assumptions is that, in the moment of an interview, a framework theory can guide the construction of mental models. Thus, the notion of framework theory is consistent with the idea that, during an interview, a learner might do significant work to construct a response. This is something that I believe is often missed in critiques of Vosniadou's work.

Another hallmark of Vosniadou's perspective is the notion that students develop what she calls *synthetic* or *hybrid* concepts. These are the "amalgams" that I mentioned earlier, where prior knowledge is combined with new inputs, for example, when students encounter ideas in formal schooling. Finally, a last important characteristic of Vosniadou's perspective is the belief that students are sensitive to issues of systematicity and coherence. She believes that learners are driven to forge coherence in their ideas.

We can now sum up by looking at how these ideas from the study of intuitive astronomy suggest we should answer our three questions about conceptual change:

What is changing in conceptual change? The short answer is framework theories.

What are the processes of change? Here I'll again quote Vosniadou from the edited volume:

> Learning science requires many conceptual changes in the framework theory, such as changes in categorization, in representation, and in epistemology and the creation of new concepts and new reasoning processes.

Thus, a framework theory must change, as a whole, but there is apparently some sub-structure to framework theories that can be adjusted. In addition, along the way, we should expect to encounter synthetic conceptions.

Why is this change hard? Here the answer is somewhat different than the answers offered by diSessa or McCloskey; Vosniadou might be seen to occupy a midpoint between them. She says:

> The conceptualization of initial understandings as a framework theory, rather than as singular and isolated units, explains why conceptual change is not a sudden replacement of intuitive conceptions with scientific ones.

Thus, like diSessa, Vosniadou argues that conceptual change is difficult because many coordinated changes are required. But, for Vosniadou, some of the difficulty also arises from the fact that initial knowledge is organized into a structure with its own intrinsic coherence and which is resistant to change.

17.3.3 Intuitive Biology

Next I take up the domain of biology. The important touchstone here is the work of Susan Carey and I'm going to use as a point of reference her 1985 book, *Conceptual change in childhood* as well as a shorter summary of that work (Carey, 1985, 1988). In that book, Carey examines the acquisition of biological knowledge between the ages of 4 and 10. She argues that there is a profound restructuring of biological knowledge over these years, and in support of her argument she musters a wide variety of psychological evidence. Here I will give you only the tiniest taste of the larger argument she builds.

Some of Carey's strongest evidence comes from what she calls "patterns of attribution." For example, children of differing ages were asked what things in the world can breathe, sleep, eat, etc. The data showed that 4-year-old children attribute all of these behaviors to people, but that the attribution of these properties to other entities depended on their similarity to people. This was something that she argued across of range of data, that young children answer what we would think of as biological questions by reference to humans and the domain of human experience.

In contrast, 10-year-old children answer these questions essentially as we would expect adults to answer. They know, for example, that all animals eat, breath, and have babies, but they restrict bones and hearts to vertebrates. So, unlike the younger children, 10-year-olds do not answer biological questions by reference to the domain of human experience. They reason from what they know about biological categories, and the functions that all organisms must perform. Carey sums up the big point this way (Carey, 1988):

> If I am correct, there is no domain of phenomena that are strictly biological for the 4-year-old. Phenomena such as eating, breathing, and sleeping are part of the domain of human activities. They are phenomena of the same sort as playing bathing, talking. (p. 23)

> The whys and wherefores of these matters, as the child understands them, include individual motivation (hunger, tiredness, avoiding pain, seeking pleasure) and social conventions. Asked why people eat, 4-year-olds answer "because they are hungry," or "because it is dinner time." (p. 23)

Now let's lay out answers to our three questions for Carey, and the domain of biology:

What is changing in conceptual change? Carey argues that children possess "only a few theory-like cognitive structures, in which their notions of causality are embedded and in terms of which their deep ontological commitments are explicated" (p. 25).

What are the processes of change? The change process is a kind of theory emergence. The idea is that biology understanding emerges out of other pre-existing theory-like structures having to do with human activities and psychology.

Why is this change hard? She emphasizes that this type of conceptual change is hard, and takes time, because it requires the acquisition of specific knowledge about such things as internal organs and biological functions. It's the weight of accumulation of this garden-variety knowledge that leads intuitive biology to bud off from intuitive psychology.

17.3.4 The Ontological Perspective

The final perspective I consider is drawn primarily from the work of Michelene Chi and colleagues (Chi, 1992, 2005). Chi's work is animated by a question that is perhaps slightly different than those addressed by the researchers I have so far discussed. She asks *Why are some misconceptions robust?* Her answer, in a nutshell, is that misconceptions are robust when a concept tends to be treated as belonging to the wrong ontological category.

An ontological category is, essentially, a category of kind of entity in the world. Two categories are ontologically distinct when the predicates of one category cannot be sensibly applied to entities in the other. For example, we can ask about the duration of an *event*, but not of a *physical object*. Thus, events and physical objects are ontologically distinct categories.

Chi and her colleagues pin some of the most important problems in science learning on one particular ontological confusion, between what they call *sequential* and *emergent* processes. In a sequential process, there are agents of distinct types, playing distinct roles. An example is the human circulatory system. In contrast, in an emergent process there are many equivalent agents, and simultaneous local interactions among these agents that result in a macro-scale pattern. An example of an emergent process is diffusion.

The ontological perspective suggests that we should answer our three questions as follows:

What is changing in conceptual change? Ontological categories and the assignment of entities to these categories.

What are the processes of change? Chi and colleagues argue that, in some cases new ontologies will need to be created. In the case of emergence and sequential ontologies, for example, the problem is not only that student mis-categorize emergent processes as sequential, they may actually lack the emergent category.

Why is this change hard? The lack of an ontological category can be a strong barrier to conceptual change.

17.4 Why Are We Stuck: Impediments to Consensus

I now move to the second main part of this paper, in which I try to lay out some of the reasons we have been stuck, and what the impediments are to achieving consensus. At this point, the reader might be wondering if we really are all that far from a consensus view. The perspectives presented above might seem to employ similar ideas and with some significant areas of overlap. Nonetheless, these perspectives are generally taken to be at odds, and I myself believe that many of the apparent similarities do not go too much deeper than the terminology employed.

So what are the problems? I think that there are a number of impediments keeping us form moving quickly to a consensus. I'll lay them out in the rest of this section.

17.4.1 Different Target Phenomena

The easiest impediment to explain is the one I call the problem of *different targets*. The issue is a simple but important one; namely, we are not all working on the same question, and we are not always aware that this is the case.

For example, one difference is that we are often concerned with processes at different timescales. For example, most of the researchers discussed above are concerned with conceptual change that occurs during classroom science instruction. These changes occur over an instructional timescale that can span days, weeks, or occasionally months. Typically, researchers who study this flavor of conceptual change focus somewhere in the middle of this range of timescales, on changes that occur over weeks. The focus here is on changes driven by formal instruction.

However, not all conceptual change researchers are concerned with the instructional timescale. In particular, researchers such as Carey are concerned with changes that occur over developmental time —the months or years during which a child matures. These changes may be driven by a wide range of impetuses, including maturation, everyday experience in the world, cultural transmission, and formal instruction.

17.4.2 Theoretical Incommensurabilities

Some more subtle impediments pertain to what I'll call *theoretical incommensurabilities*. Our theoretical accounts do align in some places, and often sound similar. But in many cases this surface similarity masks deep differences. I will start with an easier point, and then work up to some more difficult ones.

To begin, it is clear that we are looking at analyses that are, in a sense, at different levels. Some researchers are focused on constructs that attempt to capture change at the level of small, constituent levels of knowledge. Others are focused on change at the level of larger systems–ensembles–of knowledge. Indeed, quite a few other researchers have made a similar point (Amin et al., 2014). This might not be too difficult a problem to fix; in fact, it might not even be a true incommensurability, just a difference in focus.

Another important contrast is between theoretical terms that capture entities constructed in-the-moment versus terms that capture established knowledge structures. In the interview with J that I described, we saw that her explanation evolved over the course of a few minutes, as she pieced together an explanation that made sense. In cases such as that, in which a student assembles an explanation or model in the moment, we can choose to focus either on the model that is constructed, or on the existing knowledge out of which that model was built. This seems to me to be a relatively straightforward point. But I believe that there is quite a lot of confusion that could be eliminated if we are simply careful in this way.

But there are some theoretical issues that are more thorny and subtle. One phrase I use to capture some of these issues is *ontological slippage*. I believe that the nature of our own theoretical constructs slides around in profound ways. I'm going to do my best to explain what I'm talking about here. But I'm going to have to meander a bit and hope that it comes together for the reader by the end.

To start, I'll back up some. The terms in our theories get their meaning–and thus their ontological kind–from the larger theories in which they participate. For example, in Newtonian physics, *force* can be understood to be defined by the roles that it plays within that theory, notably its role in the eq. F = ma.

Now, in one class of accounts of conceptual change, our theories are cognitive models, and the entities in our theories are mental representations. So, suppose that I were to talk, in one of my papers, about the "concept of mammal." What might I mean by that? The point is that terms such as *concept* get their meaning from the theoretical framework within which they participate. In a cognitive framework, mental entities (representations) get their meaning from how they participate in models of cognitive processes. Thus, a mental entity corresponding to the concept of mammal would be defined by its causal role in our models of cognition.

Another possibility might be something that we might call "socially-shared" ideas or "ideas in the air." When I talk about the concept of mammal, I might be talking about a shared idea that exists in a field of study and is captured in textbooks. Using the term "concept" in this way is quite reasonable. But it means something very different than a cognitive-concept-of-mammal, which takes its meaning from the role it plays in a cognitive model. If you are an author writing about science "concepts," and I am reading your paper, then I expect to be able to tell which of these meanings of "concept" (or some other meaning) you have in mind. If you slip between multiple meanings, then you are guilty of ontological slippage.

Now, consider the case in which we say that a student "has the impetus theory" or "has Newtonian theory." Note that such a statement generally does not say anything explicit about mental representations–elements of a cognitive model. Rather, I think it is best understood as saying that a learner behaves in a manner that is consistent with a particular set of ideas-in-the-air. It is thus a different kind of account than a cognitive account that seeks to identify elements that are defined by their roles in cognitive models. The entities are ontologically different. This isn't necessary a bad thing. But it is very bad if we are not aware of these differences.

Here is a last point that might be controversial: I believe that Chi's ontological distinction between emergent and sequential processes is primarily an analysis given in terms of ideas-in-the-air. The distinction, as I understand it, is intended to give us a way of opening up a science textbook, looking at the subject matter, and determining which topics will be difficult.

17.4.3 Empirical Differences

The last set of impediments are empirical differences. The interesting point is that it is actually quite difficult to find many true empirical disagreements in the literature, places where researchers have obtained contradictory experimental results. Where these contradictory results do exist, most of these relate to the infamous coherence debate (diSessa, 2013), which often seems to boil down to the question of whether learners hold consistent models and give consistent answers.

Even in these cases, however, I believe the differences often end up being theoretical rather than empirical. In many cases, for example, different experiments produce contradictory results because they use different methods; thus, they ultimately reduce down to questions about the right way to ask questions and interpret results.

In the end, empirical tests are not the right way to go about settling the debates here. This debate, and others, will be settled by looking at the differences in explanatory power of theoretical perspectives across a range of experimental results.

17.5 Toward a Synthesis and Further Progress

Is it possible for us to make progress toward a shared synthesis? Based on our past history, I must admit to not being particularly optimistic, but I do believe some small steps forward are possible. First, I believe we should recognize that the big problems are theoretical and not empirical. Our problem is conceptual clarity.

The "different targets" problem points to some easy steps we can take to begin to achieve this conceptual clarity. Forging a synthesis will mean, in some cases, recognizing that we are sometimes working on relatively distinct sub-parts of a larger endeavor. This does not necessarily mean that these sub-disciplines are unrelated, however. For example, work on developmental phenomena can constrain our cognitive models of the sort of conceptual change that occurs on a shorter timescale.

Theoretically speaking, there is something of an emerging consensus that we should acknowledge; we should recognize that some claims in the literature constitute theorizing about elements, whereas others constitute theorizing about systems of elements. (I like to use the terms "element" and "ensemble.") We should also recognize, and be very clear, that sometimes we are talking about entities constructed, by learner, in the moment. (I like the term "dynamic mental construct" or "DMC" as a name for these in-the-moment constructed mental entities.) It is not clear how far all of this very general language will get us. But we should settle whatever issues we can.

What can we say about why conceptual change is difficult? We saw that different researchers have given quite different answers to this question. As an initial step I think we should assume that the answer to this question will differ across domain and context. In some cases, conceptual change will be hard because it requires many small but coordinated changes. In other case, conceptual change might be had because there is some inherent coherence and resilience of existing knowledge. Yet another answer is one proposed by Carey, that conceptual change is hard because a significant quantity of new knowledge must accumulate. A last possibility is the one of which I'm most skeptical, that learners might be lacking a general and fundamental conceptual resource, such as an ontological category.

Thus, in sum, we can make progress toward more consensus in research on conceptual change if we:

- Recognize that our main problems are theoretical, not empirical.
- Recognize that we are, in some cases, working on quite different problems.
- Adopt some minimal consensus language (e.g., elements and ensembles, dynamic mental constructs)
- Assume that conceptual change will be difficult for a variety of reasons

References

Amin, T. G. (2009). Conceptual metaphor meets conceptual change. *Human Development, 52*(3), 165–197.

Amin, T. G., & Levrini, O. (2018). *Converging perspectives on conceptual change*. Routledge.

Amin, T. G., Smith, C. L., & Wiser, M. (2014). Student conceptions and conceptual change: Three overlapping phases of research. *Handbook of Research in Science Education, 2*, 57–81.

Carey, S. (1985). *Conceptual change in childhood*. MIT Press.

Carey, S. (1988). Reorganization of knowledge in the course of acquisition. In S. Strauss (Ed.), *Ontogeny, phylogeny, and historical development* (pp. 1–27). Ablex.

Chi, M. T. H. (1992). Conceptual change within and across ontological categories: Examples from learning and discovery in science. In R. Giere (Ed.), *Cognitive models of science: Minnesota studies in the philosophy of science* (pp. 129–186). University of Minnesota Press.

Chi, M. T. H. (2005). Commonsense conceptions of emergent processes: Why some misconceptions are robust. *Journal of the Learning Sciences, 14*, 161–199. https://doi.org/10.1207/s15327809jls1402

diSessa, A. A. (1993). Toward an epistemology of physics. *Cognition and Instruction, 10*(2 & 3), 105–225.

diSessa, A. A. (1996). What do "just plain folk" know about physics? In D. R. Olson & N. Torrance (Eds.), *The handbook of education and human development: New models of learning, teaching, and schooling* (pp. 709–730). Blackwell.

diSessa, A. A. (2013). A bird's-eye view of the" pieces" vs" coherence" controversy (from the" pieces" side of the fence"). *International Handbook of research on conceptual change*, 31–48. https://dialnet.unirioja.es/servlet/articulo?codigo=5205177

diSessa, A., & Sherin, B. L. (1998). What changes in conceptual change. *International Journal of Science Education*. http://www.citeulike.org/group/9538/article/4212333

Gardner, H. (1987). *The Mind's new science: A history of the cognitive revolution*. Basic Books.

Inhelder, B., & Piaget, J. (1958). *The growth of logical thinking from childhood to adolescence: An essay on the construction of formal operational structures*. Psychology Press.

McCloskey, M. (1984). Naive Theories of Motion. *Eric.Ed.Gov*. http://eric.ed.gov/ERICWebPortal/recordDetail?accno=ED223417

Sherin, B. (2018). Elements, ensembles, and dynamic constructions. In T. Amin & O. Levrini (Eds.), *Converging perspectives on conceptual change* (pp. 61–78). Routledge.

Sherin, B. L., Krakowski, M., & Lee, V. R. (2012). Some assembly required: How scientific explanations are constructed during clinical interviews. *Journal of Research in Science Teaching, 49*(2), 166–198.

Vosniadou, S. (2018). Initial and scientific understandings and the problem of conceptual chang. In T. Amin & O. Levrini (Eds.), *Converging perspectives on conceptual change*. Routledge.

Vosniadou, S., & Brewer, W. (1992). Mental models of the earth: A study of conceptual change in childhood. *Cognitive Psychology*. http://homepage.psy.utexas.edu/homepage/class/Psy394N/Woolley/9%20Mar%2021%20Reasoning%20&%20Problem%20Solving/Vosniadou%20Brewer%201992.pdf

Vosniadou, S., & Skopeliti, I. (2014). Conceptual change from the framework theory side of the fence. *Science & Education, 23*(7), 1427–1445.

Chapter 18
Processes of Building Theories of Learning: Three Contrasting Cases

Andrea A. diSessa and Mariana Levin

18.1 Introduction

For at least three decades, the relatively poor state of theory in STEM education and the learning sciences—and what we might do about it—has been a topic of concern (diSessa, 1991a). While the field of science education broadly understands the need for theoretical frameworks in order to have a principled basis for designing instruction and testing hypotheses about the efficacy of instructional approaches, we are still far from having a deep and refined understanding of knowledge and learning processes (and thus, how to best support learning via instructional design). We contend that what is needed in science education are theories of learning that are accountable to moment-by-moment details of processes of reasoning and learning. This is necessary because this is the level at which instructional decisions are made and at which students learn or don't learn. We believe that one difficulty towards building such theories of learning is that methods of developing theoretical constructs and theories are not well-understood. Our aim in this essay is to contribute to an effort to understand processes of building theories of learning.

Early on it was recognized that there are really substantially different kinds of theories that inform and might arise from educational design. For example, diSessa and Cobb (2004) enumerate the following kinds of theories of diverse scope and character: (1) Grand theories (e.g., Piagetian stages of intellectual development), (2) Orienting frameworks (e.g., constructivism, semiotics, sociocultural theory), (3)

A. A. diSessa (✉)
Graduate School of Education, University of California, Berkeley, CA, USA
e-mail: disessa@berkeley.edu

M. Levin
Department of Mathematics, Western Michigan University, Kalamazoo, MI, USA
e-mail: mariana.levin@wmich.edu

O. Levrini et al. (eds.), *Engaging with Contemporary Challenges through Science Education Research*, Contributions from Science Education Research 9,
https://doi.org/10.1007/978-3-030-74490-8_18

Frameworks for action (e.g., Realistic Mathematics Education) and (4) Domain-specific theories of instruction (e.g., how students learn about sampling or fractions).

Theories in the learning sciences—in particular those that are produced and deployed in the context of educational designs—have distinctive characteristics, unlike generic psychological or social theories. They may be tied to specific (technical, topical, social) learning environments and local (e.g., site and culturally specific) goals. Such apparent specificity may put them on the more "humble" end of the theoretical spectrum. However, "humble" educational theory building also has important strengths. It tends to be far more applicable than, say, grand theories. And, building theory out of real-world learning and performance data can generate novel theoretical ideas, often of wide scope. Such theory building has regularly resulted in *ontological innovations*—"theoretical constructs that empower us to see order, pattern, and regularity in the complex settings in which we conduct design experiments" (diSessa & Cobb, 2004, p. 84).

In terms of the above taxonomy of types of theories in educational work, we will invoke an "orienting framework." In particular, we present three cases of theory building that all stem from a specific orienting framework concerning the nature of knowledge and learning—Knowledge in Pieces (diSessa, 1993, 2018). Knowledge in Pieces (KiP) is a name for a broad class of theoretical models of knowledge—models in which knowledge is seen as consisting of a *complex system of elements of diverse types*. Ontological innovations may occur at the system level or at the type level, or both. More extensive lists of principles and counter-principles for studying knowledge in a way that is consistent with Knowledge in Pieces are given in diSessa, Sherin & Levin (2016).

The three cases we present in this paper show notably distinct trajectories and patterns of relations to instructional design and real-world learning. Minimally, this diversity cautions us not to restrict theoretical investigation to stereotypical paths. Maximally, their development provides models of theory development and examples of general theoretical moves to add to learning sciences' repertoire. We will (1) gloss the theories, (2) describe their origins and development, and (3) mark salient specific strategies and moves related to theory building. Besides enriching expectations about theory building, our schematization of these different types aims to open up practical moves that science educators can take with regard to theory development and use. We have written the cases, below, in the first person to make some particulars of each situation of theory building easier to express.

18.2 P-Prims (diSessa)

P-prims are a kind of knowledge underlying intuitive and often unschooled ideas, but which play important roles in learning, both counterproductive and productive. P-prims constitute simple abstractions of common phenomena—say, *pushing things harder makes them go faster or farther*—toward which people's attitude is that such happenings are natural and simply the way the world works. Hence, p-prims are "primitive" in that, when they apply, nothing much more needs to be said. "That's just how things are." There are hundreds or thousands of p-prims and they show limited connections with each other.

18.2.1 Context of Early Development

Work on p-prims started in reaction to recognized and important observations related to the world of practice. Students were not "blank slates" onto which we need to write the knowledge of science, but begin with persistent ideas about how the world works. In those very early days, almost everyone believed student ideas were systematic, and may even constitute coherent "intuitive theories." See diSessa (in press) for a review of the history. Aside from the attractiveness of assimilating student learning to theory change—which is "obviously" difficult even for scientists—early work on conceptual change was strongly influenced by the philosophy of science. In particular, Thomas Kuhn's ideas on the structure of scientific revolutions were cited about the stark incompatibility of pre- and post-revolutionary scientific theories.

Thus, widespread instructional experience really set the original problematic. Most researchers quickly accepted the presence of student ideas, and listing and characterizing them was common. However, the mainstream followed a "garden path" of assuming that student learning followed patterns shown in professional science. They also assumed that intuitive ideas were categorically false, and universally constituted impediments to learning.

As a teacher of physics, I had noticed student ideas, quite distinct from what I was teaching. However, from the beginning, I treated at least some of these ideas as productive, adapting my instruction to engage them. With that orientation from my own practice, I was skeptical about the then common framing of intuitive ideas as both theoretical in nature and universally unhelpful. My experience was that there were many more ideas than could be fit into a naïve theory. I also felt confident that (1) students were often very different one from another (hence, a "common theory" could, at best, make limited sense of individual students), and (2) those ideas were often transient and hard to get hold of, unlike what one expects of a theory. Finally, I was convinced that terms like "theory" hide more than they reveal, most prominently because of their ambiguity and empirical intractability.

18.2.2 First Stages

I undertook a multi-year project to investigate intuitive physics, built primarily on a large database of interviews of students learning physics. The database was essential for building the theory that eventuated. I needed many of experiences with student thinking to characterize it because (1) the nature of elements was an important part of the theory-building (unlike assuming or just asserting that "students' ideas are theories"), and (2) the evolving theory demanded empirical description of many elements. Pushing student ideas into the "box" of a naïve theory was tempting, but they kept not fitting.

The first full presentation of the theory occurred about a decade after my initial set of interviews (diSessa, 1993). In the meantime, I developed an elaborated analytical framework to organize and solidify the theory. The framework specified a range of perspectives that I felt were all needed and provided interlocking constraints on the developing theory. What is the nature of individual knowledge elements? What are systemic relations among elements, if any? How do they work in episodes of students thinking? What are the nature and underlying mechanisms of development? The first technical paper on the theory dealt only speculatively with instructional issues. But, I think, one may need to be modest in demanding practical results from our theories right away, even if, in the end, improving real-world instruction is non-negotiable.

18.2.3 Later Stages

The two most notable later developments of the theory both related to instruction, but in different ways. The first was taking the model seriously enough to track moment-by-moment student learning. The importance of this possibility is that what happens in moments of student learning is both highly differentiating of theories (hence important to our development of good theories), and also highly germane to instruction. Some recent work does both of these things. diSessa (2017) uses moment-by-moment analysis of learning to argue against competing theories of conceptual change, and also to show new instructional pathways, which appear to have felicitous properties. Kapon and diSessa (2012) explained exactly why different students may learn in radically different ways from the same instruction. It thus engages the practical problem of teaching in a way that respects diversity. On a theoretical plane, this work also showed how technical aspects of p-prim theory can be used in detailed and empirically accountable explanations of specific learning trajectories.

The second late-stage development was a deeper deployment of the theory in the design of instruction. In a recent project, my group designed instruction globally and systematically—including the very choice of topics to be taught—out of an understanding of student ideas (Swanson, 2019; see also the instructional sections and those relating to Ohm's p-prim and agency in diSessa, 2014).

18.3 Coordination Classes (diSessa)

This model of a type of knowledge—a model of a particular kind of well-formed concepts, including scientific quantities—had a radically different development than p-prims. Also in contrast to p-prims, coordination classes (CC) highlight knowledge at the system level, rather than elements, including important issues of large-scale organization.

18.3.1 Context of Early Development

The motivation and origins of the idea of coordination classes couldn't be farther from the pattern set by p-prims. Instead of an extended and deeply data-driven first stage, a sketch of coordination class theory emerged in a single paper as a "thought experiment" that was used to critique and establish new directions concerning some very basic issues of cognition. In particular, I developed a critique of one particular part of a well-known enveloping hypothesis about cognition, Newell (1980) and Simon's Physical Symbol System (PSS) Hypothesis. Roughly, a PSS is "what it takes to do symbolic computation," for example, what's inherent in every symbolic computer. The *PSS Hypothesis* is precisely that human intelligence is built (with caveats) on the same architecture.

Information processing psychology, to which Newell and Simon were early and important contributors, was the first to model moment-by-moment thinking in a precise enough way to actually accomplish thinking (problem solving). My negative reaction to their work was that I did not think their models were realistic and insightful of specifically *human* thinking. The point of contention concerned, in their terms, "the relation between symbols and the world," which they describe only vaguely. What I thought necessary to add was a recognition that the world and our sensory access to it is highly diverse and contextual. Thus, for example, to identify a person as Joe (to connect the symbol "Joe" to a person in the world) might require different strategies in different circumstances: an audio mode (hearing him from the next room) or a visual one (recognizing his face). But we also need "alignment," that all such strategies must determine the same thing (i.e., be able to recognize Joe accurately). Scientific concepts have exactly the same requirements as symbols. Since we need to be able to see them in a huge range of contexts, a lot of context-particular learning is demanded. But, also, all these determinations need to be aligned. Learners will encounter many learning difficulties in making sure their context-specific ways of seeing align, that is, determine exactly the same thing.

18.3.2 First Stages

The CC model as described in early work (diSessa, 1991b) is high-level, but asks a necessary question: How *do* people get information from the world? The question also appears empirically tractable—examining specifically what people attend to and infer across different contexts. However, the model initially lacked data of any sort. There was no basis to assert that it applied to any element of mind, much less specifically to scientific concepts. That said, I soon began to realize that scientific quantities, in particular, matched the basic assumptions of the model. Such concepts require the learner to look at a wide range of contexts, probably using a wide range of specialized strategies, to determine the same information about the world. To "see" a force, one sometimes senses it proprioceptively, sometime visually (contact of bodies or changing motion), and sometimes by stipulation (all objects attract one another gravitationally).

18.3.3 Later Stages

The work of checking the CC model in the context of real-world data, showing both its relevance and also its insightfulness, remained. It turned out to be surprisingly successful. Gradually, researchers provided empirically well-supported and detailed analyses of learning where the core theoretical difficulties of creating a CC matched well with what was empirically seen (see Barth-Cohen & Wittmann, 2017; Jacobson & Izsák, 2014; Levin & diSessa, 2016; Levrini & diSessa, 2008; Levrini, Levin & Fantini, 2018, 2020; Lewis, 2012; Parnafes, 2007; Wagner, 2006). In testing against real-world data, the elements of the theory came to be better articulated and were extended to encompass initially unanticipated situations. For example, it became clear that, sometimes, a group of CCs-in-the-making enter into a symbiotic relationship, characterized by mutual bootstrapping (Parnafes et al., 2006).

18.4 Strategy Systems (Levin)

The following case illustrates how a heuristic epistemological framework such as KiP can shape theory building processes not only by sharing high-level principles (such as the idea of articulating complex systems of knowledge or the productivity of intuitive knowledge), but also by offering a set of reference models that can be intentionally extended and elaborated in new contexts and for new purposes. The reference models that are extended in this case are p-prims (as an example of knowledge elements) and coordination classes (as an example of knowledge systems), and thus, this case builds directly upon the previous two sections, demonstrating a KiP-distinctive way that heuristic epistemological frameworks can be generative.

18.4.1 Context of Early Development

The empirical focus of this case is the emergence of a novel mathematical strategy and is based on an in-depth microgenetic learning analysis (Parnafes & diSessa, 2013) of a pre-algebra student, Liam, who largely independently constructed a deterministic and essentially algebraic algorithm for solving algebra word problems (Levin, 2018). Liam's strategy gradually emerged over the course of his work on several problems in the context of semi-structured sessions with a tutor/researcher. While his earlier strategy was based simply on "guessing and checking" trial values and converging to a solution to problems, his later strategy can be recognized as the method of double false position (a problem solving approach known in antiquity, rediscovered across cultures and centuries, and described in Berlinghoff & Gouvêa, 2004).

The strategy change literature (see Siegler, 2006 for a review) has long used microgenetic methods to track observable changes in strategy usage through labeling the problem solving actions of individuals and coding the work of students over a large number of items. While such accounts are well-tuned for describing patterns in strategy use and choice, these methods alone could not capitalize on details of the data relevant to how or why any particular strategy emerged. In order to understand Liam's case and to make a model of the underlying process by which new knowledge is constructed through activities such as problem solving, I needed to have theoretical and methodological tools that would allow me to identify and characterize the content, form, and dynamics of the knowledge that individuals draw upon as they solve problems, not just the appearance of specified strategies.

18.4.2 Early Stages – Initial Contact Between Epistemological Framework and Data

Though Liam's beginning and ending strategies were quite different, Liam's reasoning throughout the sessions showed significant threads of continuity. One of the things that immediately stood out to me was that in written form, Liam's initial and final strategies would be indistinguishable (they both ostensibly involved recording a sequence of trial values in chart form and testing each trial). In the earliest version of his strategy, Liam attended only to whether a specific trial value was "too high" and "too low." However, in the later strategy, he extracted and leveraged a much more complex set of determinations and inferences, notably including quantifications of how changes in an input variable related to changes in the output variable. I was aware of Orit Parnafes' work using Knowledge in Pieces and Coordination Class Theory to give a moment-by-moment account of how individuals attend to and extract information from computational representations, and the implications of this for their developing understanding. Both the accountability to moment-by-moment details of understanding and the focus of KiP, and CC Theory in particular, on the role of what individuals were attending to during their reasoning processes, convinced me that the KiP perspective could be a helpful starting place for me in understanding strategy development processes like Liam's. However, partly because I was modeling a learning process in mathematics as opposed to physics, where most prior work in KiP and CC Theory had been done, and partly because I was modeling a different kind of learning process (strategy refinement as opposed to concept learning), it was not immediately obvious how to "apply" the perspective to my data.

In initial attempts to negotiate between my empirical data and KiP, I started by conjecturing that Liam's problem solving strategies might be construed as compositions of smaller, sub-strategic and sub-conceptual knowledge elements. Further, I hypothesized that the smaller pieces might potentially "belong" to many concepts and could be appropriately or inappropriately activated according to context, and that the development across the sessions could be traced in terms of these sub-strategic and sub-conceptual pieces becoming increasingly coordinated.

18.4.3 Later Stages – Reformulation: Strategies as Complex Systems

Eventually, I came to consider a strategy to be a particular kind of complex knowledge system in which the function of the system is to coordinate many, diverse kinds of knowledge for the purpose of solving problems. While coordination classes are concerned with what individuals attend to and infer as they determine a concept-specific class of information, in all but the simplest of cases what an individual needs to attend to in order to determine the solution to a problem is potentially much broader (including possibly the coordination of multiple coordination classes). Since this focus on perception and inference appeared to be productive, I retained and elaborated these aspects in modeling problem solving processes and how they change. However, I also recognized in my data two other important aspects missing from the classical coordination class model: strategic attention to what information one should assemble in order to solve a problem and knowing what actions one can take as a result of inferences made during problem solving. The strategy system model thus eventually contained a focus on four components: strategic frame, categories of attention, inferential relations, and allowable actions.

Let me briefly illustrate these key ideas. **Strategic frame**: Liam was aware that there were definite ways of approaching a recognizable class of problems. By the end of the arc of learning I observed, he had a good sense of a new and fool-proof set of steps that he could implement. **Categories of attention**: Liam parsed the problem-solving situation into things like: (1) an "input," number that is adjustable with a determinable effects; (2) an "output," the result of that determination in a particular case; (3) a "goal number" that will be stipulated in the problem statement. **Inferential relations**: These turned out to be gradually developing and core to

Liam's learning. In particular, he developed the category of "change in output for a unit change in input," which then made deterministic calculation (a kind of inferential relation) of how to change the input to achieve the goal straightforward. **Allowable actions**: These are exemplified by the steps he took to determine the "unit change in output," or the steps he took to determine "net required change in input to meet the goal output."

In sum, the KiP epistemological framework guided my analysis both top-down by suggesting the reformulation of my analysis in terms of complex knowledge systems and bottom-up by looking at what Liam attended to and inferred in activity to make schematizations of candidate knowledge elements and systems. In presenting an explicit characterization of a new knowledge ontology—a strategy system—the research was not only guided by the Knowledge in Pieces framework, but elaborated it.

18.5 Discussion and Recommendations

We sketched the theoretical genesis of three quite different theoretical constructs: p-prims, coordination classes, and strategy systems. Each path of genesis illustrated a very different pattern of theory building. P-prims show how extensive empirical work can lead to a new theoretical construct that offers powerful and alternate explanations for deeply problematic learning phenomena such as student "misconceptions." Coordination classes offer an example of how a construct can be hypothesized to address issues that are foundational to any perspective on learning—how individuals get information from the world. In this sense, the focus on the theory is given *a priori* and the empirical development and testing of the theory came later. The strategy system case illustrates how to use previous theoretical constructs synthetically—in a generative, and not prescriptive way—in order to model new phenomena, not covered by the original constructs.

It is worth reflecting on a surprising fact. The three theories—with (1) different foci (often-transient intuitions, stable and coherent concepts, strategic capability), (2) different structure, and (3) distinctive paths of development—emerged from the same heuristic framework, Knowledge in Pieces. The fact of the matter is that KiP has stimulated a long history of refinement, new foci, and further ontological innovations. In Lakatos's terms, it has constituted a long-lived "progressive scientific programme." Our own reflection on KiP's success centered on the property of *generativity*. It seems that the framework accepted, even stimulated, change, while at the same time providing resources that could help innovation along. Many of our recommendations, below, connect directly to generativity. And they seem to us plainly to extend beyond the KiP case.

Herewith is a set of general recommendations, drawn from our three case studies, that we feel can enhance theory building generally in science education and the learning sciences more generally. Key elements are in bold type.

The three cases of theory building all respond to the **grave need to develop the epistemological toolkit** available to researchers. Commonsense terms related to knowledge and learning are easily available: "idea," "concept," "reason," "know." But these often suppress or replace precise and operational technical terms of well-developed theory. As a preamble to introducing the idea of coordination class, diSessa and Sherin (1998) argued that "concept" is not a technical term, but instead is used in research in very many underspecified and contradictory ways. More generally, diSessa, Sherin & Levin (2016) identified "skepticism of commonsense knowledge terms" as a core principle of Knowledge Analysis, and, in particular, of KiP. One needs to guard against the "subset model of expertise," that novices have the same *kind* of knowledge as experts, but just less of it. Instead, ontological innovation is necessary to understand more precisely and explicitly how exactly processes of perception and construal work in both learners and experts.

A common benchmark for the power of theoretical ideas concerns whether new constructs allow us to understand phenomena in unexpected and deeper ways than those associated with earlier constructs and ideas. **Theoretical innovations that are subversive in this way may be particularly powerful.** Of course, being intentionally subversive is not the point. But ideas that do not follow conventional wisdom might be especially powerful and warrant correspondingly greater effort to develop.

One of the ways that researchers can set themselves up to make new observations and adequately refine and test theory is to **ensure access to a rich and potentially distinctive empirical channel**. Although perhaps difficult to implement, the heuristic is not essentially more than looking at more and different data. For example, work guided by Knowledge in Pieces almost always focuses on moment-by-moment processes of thinking and learning—which is rare in conceptual change research and also in science and mathematics education more broadly. KiP therefore "consumes" massive amounts of video data in the development and refining of theory, and in its testing and deployment. Data reduction can be important, but it is not the natural first step in theory building.

One of the ways that building theory in education and the learning sciences may differ from prototypical theory building in the hard sciences is an **orientation towards diversity**. Knowledge in Pieces began with two knowledge ontologies: p-prims and coordination classes. However, in addition to "diversifying" these models in adapting them to contexts beyond physics, the genesis of new kinds, such as strategy systems and symbolic forms (Sherin, 2001), illustrates the assumption that understanding knowledge and learning will involve multiple *forms* of knowledge with diverse *functions*. This perspective is starkly different to theoretical perspectives that assume homogeneity of knowledge form and function, for example, modeling all knowledge in terms of "schemes" (Piaget) or "productions" (Newell & Simon). An orientation towards diversity provides an impetus for new theoretical development, and a home for complex and extensive empirical development.

Finally, we recommend developing theory within **frameworks that admit of more than one level of detail**. KiP has a broad, general, and heuristic level, common to the individual theories (postulating the generativity and usefulness of naïve knowledge in learning, pursuing details of complex knowledge systems, accepting potentially high degrees of contextuality of knowledge). But each theory is highly specific in the phenomena to which it applies. Our third example, strategy systems, built on both levels of previous KiP work; it used the general framework, but also elements from the specific models, p-prims and coordination classes. We are skeptical of isolated, highly specialized theories at a time when we clearly need a lot more theory development. And, we are also skeptical of frameworks that are only general—even if inspirational—but simply do not get down to the brass tacks of learning and instruction in particular cases.

References

Barth-Cohen, L. A., & Wittmann, M. C. (2017). Aligning coordination class theory with a new context: Applying a theory of individual learning to group learning. *Science Education, 101*(2), 333–363.

Berlinghoff, W. P., & Gouvêa, F. Q. (2004). *Math through the ages: A gentle history for teachers and others*. MAA Press.

diSessa, A. A. (1991a). If we want to get ahead, we should get some theories. In *Proceedings of the 13th annual meeting of the North American Chapter of the International Group for the Psychology of Mathematics Education* (Vol. 1, pp. 220–239).

diSessa, A. A. (1991b). Epistemological micromodels: The case of coordination and quantities. In J. Montangero & A. Tryphon (Eds.), *Psychologie génétique et sciences cognitives* (pp. 169–194). Archives Jean Piaget.

diSessa, A. A. (1993). Toward an epistemology of physics. *Cognition and Instruction, 10*(2–3), 105–225.

diSessa, A. A. (2014). The construction of causal schemes: Learning mechanisms at the knowledge level. *Cognitive Science, 38*(5), 795–850.

diSessa, A. A. (2017). Conceptual change in a microcosm: Comparative analysis of a learning event. *Human Development, 60*(1), 1–37.

diSessa, A. A. (2018). A friendly introduction to "knowledge in pieces": Modeling types of knowledge and their roles in learning. In *Invited lectures from the 13th international congress on mathematical education* (pp. 65–84). Springer.

diSessa, A. A. (in press). A history of conceptual change research: Threads and fault lines. In K. Sawyer (Ed.), *Cambridge handbook of the learning sciences* (3rd ed). : Cambridge University Press.

diSessa, A. A. & Sherin, B. (1998). What changes in conceptual change? *International Journal of Science Education, 20*(10), 1155–1191.

diSessa, A. A., & Cobb, P. (2004). Ontological innovation and the role of theory in design experiments. *The Journal of the Learning Sciences, 13*(1), 77–103.

diSessa, A. A., Sherin, B., & Levin, M. (2016). Knowledge analysis: An introduction. In A. diSessa, M. Levin, & N. Brown (Eds.), *Knowledge and interaction: A synthetic agenda for the learning sciences* (pp. 30–71). Routledge.

Jacobson, E., & Izsák, A. (2014). Using coordination classes to analyze preservice middle-grades teachers' difficulties in determining direct proportion relationships. In *Research trends in mathematics teacher education* (pp. 47–65). Springer.

Kapon, S., & diSessa, A. A. (2012). Reasoning through instructional analogies. *Cognition and Instruction, 30*(3), 261–310.

Levin, M. (2018). Conceptual and procedural knowledge during strategy construction: A complex knowledge systems perspective. *Cognition and Instruction, 36*(3), 247–278.

Levin, M., & diSessa, A. A. (2016). "Seeing" as complex, coordinated performance: A coordination class theory lens on disciplined perception. In A. A. diSessa, M. Levin, & N. J. S. Brown (Eds.), *Knowledge and interaction: A synthetic agenda for the learning sciences* (pp. 191–211). Routledge.
Levrini, O., & diSessa, A. A. (2008). How students learn from multiple contexts and definitions: Proper time as a coordination class. *Physical Review Special Topics-Physics Education Research, 4*(1), 010107.
Levrini, O., Levin, M., & Fantini, P. (2018). Personal, deeply affective, and aesthetic engagement with science content. In T. Amin & O. Levrini (Eds.), *Converging perspectives on conceptual change: Mapping an emerging paradigm in the learning sciences* (pp. 305–312). Routledge.
Levrini, O., Levin, M., & Fantini, P. (2020). Fostering appropriation through Designing for multiple access points to a multidimensional understanding of physics. In *Physical Review Special Topics – Physics Education Research, 16*, 020154.
Lewis, C. M. (2012). *Applications of out-of-domain knowledge in students' reasoning about computer program state* (Doctoral dissertation, UC Berkeley).
Newell, A. (1980). Physical symbol systems. *Cognitive Science, 4*(2), 135–183.
Parnafes, O. (2007). What does "fast" mean? Understanding the physical world through computational representations. *The Journal of the Learning Sciences, 16*(3), 415–450.
Parnafes, O., & diSessa, A. A. (2013). Microgenetic learning analysis: A methodology for studying knowledge in transition. *Human Development, 56*(1), 5–37.
Parnafes, O., diSessa, A., Wagner, J., Mestre, J., Thaden-Koch, T., & Sherin, B. (2006, June). Theory in pieces: the communal development of a theory. In *Proceedings of the 7th international conference on Learning sciences* (pp. 1078–1083). International Society of the Learning Sciences.
Sherin, B. L. (2001). How students understand physics equations. *Cognition and Instruction, 19*(4), 479–541.
Siegler, R. S. (2006). Microgenetic analyses of learning. In W. Damon & R. M. Lerner (Series Eds.) & D. Kuhn & R. S. Siegler (Vol. Eds.), *Handbook of child psychology: Volume 2: Cognition, perception, and language* (6th ed., pp. 464–510). Hoboken: Wiley.
Swanson, H. (2019) Refining student thinking through scientific theory building. In E. Manalo (Ed.) *Deeper learning, dialogic learning, and critical thinking: Research-based strategies for the classroom*. Abingdon-on-Thames: Routledge.
Wagner, J. F. (2006). Transfer in pieces. *Cognition and Instruction, 24*(1), 1–71.

Chapter 19
Understanding the Role of Image Schemas in Science Concept Learning: Can Educational Neuroscience Help?

Tamer G. Amin

19.1 Introduction

In the last two decades, a body of research has been examining the neural basis of the representation of conceptual knowledge and processes of conceptual change in science (see Mareschal, 2016; Tolmie & Dundar-Coeke, 2020 for reviews). One line of research (Dunbar, Fugelsang, & Stein, 2007; Houde, Zago, Mellet, Moutier, Pineau, Mazoyer & Tzourio-Mazoyer, 2000; Mareschal, 2016; Masson, Potvin, Riopel & Brault-Foisy, 2014) has focused on the early core insight of research on science concept learning that initial, intuitive conceptions and beliefs need to be overcome and replaced (Driver & Easley, 1978; Strike & Posner, 1985). This work on the neural basis of conceptual change in science learning has embraced the two systems model of cognition where a fast, heuristic, parallel, implicit and evolutionarily old processing system coexists with a slow, rule-based, declarative system that supports abstract logical and hypothetical thinking (Evans, 2011; Kahneman, 2011). Consistent with this model, it is claimed that one aspect of science concept learning is the inhibition of prior intuitive, implicit knowledge. Moreover, it is suggested that the brain regions that perform this inhibitory function have been documented. One theoretical conclusion that has been drawn from this work is that science concept learning involves the suppression, rather than the *replacement*, of intuitive conceptions (Mareshal, 2016).

These attempts to examine the neural underpinnings of intuitive conceptions and conceptual change are to be encouraged. However, a focus on the inhibition of intuitive knowledge within a two-system model cannot easily account for more recent developments in developmental psychology, learning sciences and science

T. G. Amin (✉)
Department of Education, American University of Beirut, Beirut, Lebanon
e-mail: tamer.amin@aub.edu.lb

O. Levrini et al. (eds.), *Engaging with Contemporary Challenges through Science Education Research*, Contributions from Science Education Research 9,
https://doi.org/10.1007/978-3-030-74490-8_19

education research on science concept learning. Research on science concept learning across these fields over the last five decades has seen the emergence of numerous theoretical perspectives on the nature of learners' pre-instruction conceptual understanding and the processes of conceptual change (Amin & Levrini, 2018; Amin, Smith & Wiser, 2014; Vosniadou, 2013). As the field has matured, a less dichotomous picture has emerged of the differences between how novice learners and scientists understand the world. In Amin et al. (2014), we argued that, despite subtle differences among theoretical perspectives, most researchers have come to view science concept learning in terms of a heterogenous network of knowledge elements. We now understand that science concept learning implicates reorganization in various types of interacting knowledge elements including perceptual schemas, concepts, mental models, domain specific beliefs, ontological assumptions, and metacognitive and epistemological knowledge and beliefs (Amin et al., 2014). These knowledge elements vary in representational format, including iconic representations (e.g., perceptual schemas and mental models) and propositional representations (e.g., natural language and mathematical symbol systems). From this perspective, pre-instruction concepts, sub-conceptual elements, mental models and beliefs are not simply replaced, or suppressed, but are revised, refined and often put to new use in scientific thought.

This means that a two-system model of cognition in which knowledge within the fast, parallel, heuristic system is inhibited in favor of the slow, declarative system is inadequate. Ultimately, our model of the neurocognitive underpinnings of science concept learning will have to acknowledge a greater degree of continuity between the learner and the expert; it will also have to incorporate an account of how multiple knowledge elements in different representational formats are coordinated and how that coordination changes with the acquisition of greater expertise. Tolmie and Dundar-Coeke (2020) have argued that this knowledge network model of science concept learning is consistent with some current work in cognitive neuroscience indicating that concepts and category representations involve multiple distinct modalities and that parallel associative networks integrate these multimodal representations, with language playing a particularly important organizing role (Reber, Stark & Squire, 1998; Rips, Smith & Medin, 2012; Thomas, Purser & Mareschal, 2012).

A great deal of theoretical and empirical work will be needed to elaborate a neurocognitive model consistent with, and integrated into, our current knowledge network picture of science concept learning (see Tolmie & Dundar-Coeke, 2020 for a recent review and discussion). In this chapter, I focus narrowly on just one, but particularly important, aspect of this larger endeavor: understanding the role of sensorimotor experience in science concept learning. I present an argument suggesting that an interdisciplinary line of research in educational neuroscience integrating theories and methods from embodied cognition, the learning sciences and neuroscience could help us understand this role. I suggest that this research is likely to contribute to productive theory building and to advances in educational design. More specifically, I will point out that a number of lines of research converge in suggesting that *image-schemas*, which are abstracted from our sensorimotor experieces, continue to play an important role in language comprehension, understanding of abstract (including scientific) concepts and reasoning. But while the converging evidence is compelling, this is best seen as a working hypothesis that an interdisciplinary program in educational neuroscience could develop further and support with additional, more direct empirical evidence. Thus, the construct of image schema can serve as a theoretical linchpin linking research in the learning sciences and cognitive neuroscience.

19.2 Sensorimotor Experience and Science Concept Learning: Contributions from Embodied Cognition and the Learning Sciences

Continuity between the learner and the expert scientist has been the focus of research on science concept learning in a number of lines of research in the learning sciences (Amin, Jeppsson & Haglund, 2015/2017; diSessa, 1993; Lindgren & Johnson-Glenberg, 2013; Smith, diSessa & Roschelle, 1992). Some of this work has been inspired by the theoretical perspective of embodied cognition, in which higher level cognition is understood as grounded in sensorimotor knowledge, to design mixed reality environments that support science learning (Lindgren & Johnson-Glenberg, 2013). For example, learners can enact the motion of physical objects (e.g., a meteor orbiting a planet) while interacting with a digital simulation of a physical system. Whole-body learning experiences such as these have been shown to be more effective in supporting the learning of scientific concepts (e.g., Newtonian force and motion) compared to "less embodied" experiences (e.g., using a regular desktop simulation) (Lindgren, Tscholl, Wang, & Johnson, 2016).

This work demonstrates *that* the sensorimotor system indeed does seem to be implicated in higher-level cognition, including developing an understanding of abstract scientific concepts. It is less clear, however, *how* engaging the sensorimotor system supports science concept learning. As a result, there is limited guidance from this work regarding exactly how pedagogical design features translate into science concept learning.[1] Other work in the learning sciences has described various specific roles that intuitions that emerge from sensorimotor experiences might play in science concept learning. These roles include understanding scientific concepts, the construction of explanatory models, grounding comprehension of the abstract language of science, and supporting the meaningful interpretation of equations. My goal in this section will be to point out that researchers examining these various

[1] It is true that the idea of "action-concept congruence" has been proposed as a design tenet for embodied learning to be successful (Holbert & Wilensky, 2012; Lindgren & Johnson-Glenberg, 2013). That is, the idea is that what is enacted by the learner needs to have some congruence to an aspect of an expert's knowledge representation in the conceptual domain in question. This is a very interesting idea, but requires further development. Understanding the roles of image schemata as discussed here can contribute to this.

roles that intuition plays in science learning seem to be converging on the idea that abstractions over sensorimotor experience – i.e. image schemas - can contribute to developing scientific understanding and reasoning in various ways. Given its importance, the construct of image schema must be scrutinized and precise hypotheses about it must be formulated and tested. Therefore, to help with this, I will turn next to research that has already begun on the neural underpinnings of image schemas.

19.2.1 "P-Prims," "Core Intuitions" and Explanatory Scientific Models

A good place to start this survey is the notion of a *phenomenological primitive* (*p-prim*), introduced into the learning sciences literature by diSessa (1983, 1993, 2000) in the context of his Knowledge-in-Pieces account of science concept learning. P-prims are understood as intuitive building blocks of conceptual knowledge. They are *phenomenological* because they are said to emerge from our experiences interacting with the world; and they are *primitive*, because they constitute a fundamental level of understanding the world. Consider, as an example, *Ohm's p-prim* – a schematic knowledge structure incorporating "an agent that is the locus of an impetus that acts against a resistance to produce some sort of result" (diSessa, 1993, p. 126). This schema is understood as emerging from our experiences as agents trying to move, lift, or detach objects that resist our efforts. This schema grounds our intuitive inferences that greater resistance will demand more effort on our part, achieving a more substantial result will demand more effort for the same resistance etc.

diSessa (1993) catalogues dozens of such p-prims (e.g., *continuous force*, *force as mover*, *dynamic balance*). He argues that our intuitive "sense of mechanism" comes from the activation of p-prims. These schemas emerge from our experience – often our bodily experiences interacting with the physical world - but once formed will be activated and guide our interactions with, and inferences about, new similar experiences. The inferences we draw seem obvious once we activate a p-prim; it is in this sense that p-prims support "intuitions" about the physical world. Learning science concepts, on this account, involves, in part, learning to activate sets of p-prims appropriately in particular contexts so as to formulate explanations and make predictions that are consistent with canonical scientific understanding.

Brown (2014, 2018) has drawn on a construct very similar to p-prims – what he calls "core intuitions" - in his account of learners' developing explanatory models in science. Brown includes core intuitions within a broader taxonomy of conscious and implicit conceptual resources. Conscious resources include verbal-symbolic knowledge, conscious images of specific situations, and conscious explanatory models; implicit resources include implicit mental models and implicit core intuitions. For example, a learner trying to make sense of an electric circuit might draw on verbal-symbolic knowledge encountered in a textbook or science class – e.g.

V = IR; she might invoke an image of the internal structure of a bulb; she might explicitly construct an explanatory model of electric current in terms of unobservable electrons flowing in a wire under the influence of a potential difference; she might just implicitly model the flow of electricity as analogous to the flow of liquid in a pipe; and she might implicitly invoke an intuitive sense of the agency of a battery, needed to account for the lighting of a bulb. Brown uses this framework of five knowledge types to account for a learner's thinking with different degrees of coherence and varying in how closely it resembles canonical science. He characterizes greater scientific expertise in terms of the coherent integration of these different conceptual resources to form canonical explanatory models (Cheng & Brown, 2010).

Core intuitions are the causal heart of an explanatory model in Brown's framework. Learners might implicitly attribute various types of agency to the entities being thought about. Interpreting the various entities in terms of core intuitions can lead to the construction of an explicit explanatory model; this will be scientifically accurate to the extent that it is appropriately constrained by conscious verbal-symbolic knowledge in the domain. Brown (2018) lists a number of core intuitions varying in their attributions of agency – e.g. an *initiating agent*, that has its own causal power; a *reactive agent*, which reacts with a degree of agency to an external agency acting on it; or an *instrumental responder*, which functions as a kind of medium to transmit agency. Brown illustrates how these core intuitions, and others, can be activated by learners in the context of reasoning about an electric circuit, but these intuitions transcend this particular domain and could be activated in a wide range of different contexts that call for some kind of causal explanation.

Brown considers core intuitions to be "implicit" and "imagistic" and he contrasts them to conscious verbal-symbolic knowledge. Moreover, his recent account of this framework builds on his earlier work that clearly assumes core intuitions to be closely linked to action. That is, Brown and Clement's (1989) seminal account of bridging analogies was based on the premise that effective analogies in science learning are those that help learners recruit intuitions that emerge from physical experiences. For example, it is a feature of our action-based intuition that when you push down on a spring with your hand, the spring in turn "pushes back", demonstrating its own agency. In the instructional strategy of bridging analogies, an intuitive sense of agency can be recruited as an "anchoring intuition" to guide learners to conceptualize agency in a situation in which they wouldn't typically attribute agency (e.g. a table exerting a normal force upward on a book placed on it). An intermediate situation sharing features of both the anchor and the target (e.g. a book placed on a springy, thin piece of wood) can serve as a bridge between the two. Ultimately, the instructional goal would be the construction of a canonical explanatory model of the target phenomenon that incorporates the intuitive sense of agency (e.g. the wood of which a table is made comes to be conceptualized in terms of microscopic springy particles pushing up against a book placed on it).

Brown's (2014, 2018) core intuitions such as *initiating agent*, *reactive agent*, *instrumental responder* greatly resemble diSessa's (1983, 1993) p-prims such as *force as mover*, *continuous force* and *Ohm's p-prim*; that is, implicit core intuitions and p-prims seem to be very similar knowledge types. Indeed, Brown and Clement's

(1989) account of bridging analogies has been very naturally elaborated in terms of a Knowledge-in-Pieces account by Kapon and diSessa (2012). Moreover, the experiential origin of core intuitions has become particularly clear in more recent work by Mathayas and Brown (2018) who have used currently available technological innovations to expand the ways in which core intuitions can be incorporated into explanatory models. Using motion detectors, learners can now use gestures to activate a computer-based simulation. For example, they can use their hands to activate the moving wall of a container containing a gas. In this way, a learner can project his sense of agency onto the physical system being simulated, enriching his mechanistic explanatory model of the system's behavior. This is reminiscent of White's (1993) earlier work in which learners used a joystick to impart "impulses" to a dot moving in a Newtonian microworld, recruiting an action-based intuition to ground scientific understanding of force and motion.

Common to the work surveyed so far in this section is the idea that intuitions emerging from physical interactions with the world can contribute to scientific understanding. In particular, we saw that a key goal is to help learners incorporate action-based intuitions into explanatory causal models. Various instructional strategies have been designed and technological tools assembled with this learning goal in mind. But science is a symbolically rich technical domain with its models, theories and laws formulated in language and in mathematical representations. I have said little so far about how linguistic and mathematical representations might be understood. Brown's interpretive framework does acknowledge, of course, that verbal-symbolic resources are a key element in scientific understanding. But how are the linguistic and mathematical representations that express scientific knowledge understood? And does intuition emerging from sensorimotor experience, of the kind described so far, play any role in this understanding? These are big questions, but research has begun to address them.

19.2.2 "Image Schemas," "Conceptual Metaphors" and "Symbolic Forms": Understanding Language and Mathematical Representations

Research in the field of cognitive linguistics, embracing the perspective of embodied cognition, has, for a number of decades now, been suggesting that intuitions grounded in sensorimotor experience play an important role in our understanding of language (Dancygier, 2017). Specifically, the theory of conceptual metaphor has made two central claims (Lakoff & Johnson, 1980, 1999). The first is that analysis of natural language reveals vast systems of implicit metaphorical mappings between conceptual domains. For example, time can be construed as a location in space (e.g. "We've *arrived at* a critical moment"; "I can't wait to *get to* the weekend"); emotional and other abstract states can be construed as containers (e.g. "He fell *deep into* a depression"; "The country is *in* a recession"); and abstract causes can be understood as physical forces (e.g. "The limited supply *pushed up* the prices"). The second claim is that understanding abstract ideas is grounded in knowledge that emerges from sensorimotor experience. In the examples just listed, the abstract ideas of time, states and abstract causes are understood in terms of *location in space*, *containers* and *physical forces* – notions that emerge from our concrete physical experiences interacting with the world. It is argued that ultimately abstract concepts are understood in terms of "image schemas," which are generalizations over patterns of sensorimotor experience. Image schemas are knowledge gestalts that support intuition. When image schemas are mapped metaphorically onto abstract conceptual domains, image schema based intuitive inferences are projected to those abstract domains. For example, if depression is construed as a container, the *deeper* the depression, *the harder it is to get out of it*.

There is an extensive literature on the systems of conceptual metaphor implicit in language and the image schemas that ground understanding of abstract conceptual domains and project intuitive inferences on those abstract domains (Kövecses, 2010) This perspective has now been applied to the language of science and to understanding how scientific concepts are learned (Amin, 2009, 2015; Beger & Smith, 2020). Researchers have documented how abstract scientific concepts like energy (Amin, 2009, 2020), entropy (Amin, Jeppsson, Haglund & Stromdhal, 2012; Jeppsson, Haglund, Amin & Stromdhal, 2013) and potential well (Brookes & Etkina, 2007) are construed in terms of image schemas. For example, forms of energy can be construed as locations (e.g. 'the energy was transferred *from* kinetic *to* potential energy') and spontaneous processes are construed as agentive/sentient movement (e.g. 'the second law of thermodynamics … determines the *preferred direction* of the system'). Learning science concepts from this perspective is understood as appropriating the metaphorical mappings already implicit in the language of science and cued by linguistic forms. The selection and design of instructional analogies, representations and simulations has been inspired by identifying, through language analysis, the image schemas that are projected conventionally onto the abstract scientific concepts that are the target of instruction (see contributions to Amin, Jeppsson & Haglund, 2015/2017).

Experience-based intuition can also contribute to the understanding of mathematical representations. Sherin (2001, 2006, 2018) has argued that the acquisition of expertise in physics involves coming to interpret elements of physics equations in terms of intuitive conceptual resources and to refine the use of p-prims through problem solving. Sherin (2001) introduced the construct "symbolic form" to describe how an expert gives meaning to a physics equation and connects this mathematical representation to a qualitative understanding of the physical world. Each symbolic form is composed of a symbol template (e.g. □ - □; □ = □) and a conceptual schema (e.g. *opposing forces*, *balancing*). Interpreting the terms of an equation in terms of symbolic forms gives meaning to it components which can then be mapped onto the regularities in physical phenomena that are captured by the equation. Sherin (2006, 2018) has also shown how using equations in the context of problem solving can refine the activation of p-prims – reducing the probability of activating a less useful p-prim and increasing the activation of another more likely to contribute to canonical scientific reasoning.

We can now add "image schemas" (implicit in language use) and "conceptual schemata" (which are components of "symbolic forms") to the other constructs – "p-prims" and "core intuitions" - that have been proposed to capture intuitive sense-making in science. Many image schemas and conceptual schemata resemble p-prims and core intuitions. Together these constructs and the examples their proponents offer us, suggest ways that sensorimotor experiences and the intuitions that emerge from them can contribute to understanding scientific concepts, constructing meaningful explanatory models and understanding scientific language and mathematical representations. These might suggest directions to deepen our understanding of how sensorimotor experiences contribute to our understanding of scientific concepts, often through mixed reality simulations (Lindgren & Johnson-Glenberg, 2013; Lindgren, Tscholl, Wang, & Johnson, 2016). But the resemblance between these constructs (indeed, they might be referring to the same kind of knowledge element!) suggests that some more careful theoretical and empirical work might also be required. Recent developments in cognitive neuroscience might make a useful contribution here.

19.3 Image Schemas in the Brain

Implicit in the survey of research on science concept learning offered in the last section was a hypothesis I would like to now make explicit. Specifically, the claim is that intuitive knowledge structures that emerge as abstractions over sensorimotor experience contribute to a number of aspects of scientific understanding: understanding abstract scientific concepts, construction of explanatory models, grounding comprehension of the abstract language of science, and supporting the meaningful interpretation of equations. That is, the claim is that many "p-prims," "core intuitions," the source domains of many conceptual metaphors in the language of science and some of the conceptual schemas of symbolic forms are the same kind of knowledge structure – namely, *image schemas*.

I have pointed briefly to similarities between the various constructs already in the course of the survey above. There are also indications that p-prims and core intuitions have been recognized explicitly by their proponents as resembling images schemas in significant ways: diSessa (1993, 2000) noted the similarity in format between p-prims and image schemas as described by Johnson (1987) and Talmy (1988), but distinguished them functionally; and Brown (2018) has identified his core intuitions as image schemas, viewing "imagistic construals," the interplay between conscious images and implicit core intuitions, as analogous to "linguistic construals," involving the interplay between conscious linguistic expressions and implicit image schemas. More extended textual evidence from the publications of key authors could be offered to make a more compelling case that the same construct is being appealed to. This short chapter is not the place to pursue this. Instead, my goal here is to explore how one might begin to provide empirical support for such a hypothesis. More specifically, I suggest in this section that research in cognitive neuroscience can point us in the right direction. In the concluding section, I will then go on to argue that engaging with cognitive neuroscience in this way is not just of theoretical interest but also can have valuable practical pedagogical implications that can be recognized through multidisciplinary research in educational neuroscience.

The research in cognitive neuroscience that I will briefly review here is that which has provided support for a view of cognition as embodied – that is, has supported the idea that aspects of cognition that we have typically considered "higher-level" such as conceptual understanding, language comprehension and reasoning are grounded in sensorimotor experience. The construct image schema, that I have already introduced, has been key in establishing the connection between these higher-level cognitive functions and our sensorimotor system (Johnson, 1987; Lakoff, 1990). In this work, an image schema is understood as the schematization of recurrent patterns of sensorimotor experience. For example, the CONTAINER image schema consists of an *exterior*, an *interior* and a *boundary* (which is a schematization of many recurrent experiences of different kinds: putting things in and taking them out of boxes or other containers, walking into or out of rooms, receiving food into our mouths etc.); the SOURCE-PATH-GOAL image schema consists of *a mover*, *source* (starting location), *goal* (goal location) and *path* (a series of locations between source and goal); the GRASP image scheme consists of *reaching out in some direction*, followed by *manual manipulation* of a *desired object*. These descriptions of image schemas illustrate a number of their key characteristics: as schematizations of recurrent experience, they are gestalts that consist of a fairly small number of parts or components; they are multimodal – i.e. the experiences are combinations of visual, tactile, kinesthetic and other sensorimotor modes, not limited to one modality; and as structured gestalts, they support inferences.

Cognitive neuroscience research has explored the neural basis for image schematic structures. Various neural clusters have been identified in premotor and parietal brain areas that serve secondary functions (as opposed to primary sensory and motor functions), integrating representations of actions, the locations to which those actions are directed and the objects acted on (see Gallese & Lakoff, 2005 and Rohrer, 2005 for review and discussion). These clusters are multimodal and embedded within the sensorimotor system. So while serving integrative functions, they are not a separate amodal module external to the sensorimotor system receiving input from it. The firing of some specific neurons and neural clusters in these areas correspond to different *schematic* components not *specific* sensory experiences or actions (e.g. the general purpose of an action, its manner of execution or the phase of a temporal sequence of phases of an action).

Two features of the functioning of these secondary neural clusters are particularly important for considering the role that they might play in cognition beyond perceiving and acting in the world. While some premotor neurons will only fire when actions are actually performed, others (referred to as "canonical neurons") also fire when an object is seen that *could be* acted on in some way, even if the action is not performed (Fogassi, Gallese, Buccino, Craighero, Fadiga & Rizzolatti, 2001). In other words, some part of the neural activity associated with an action has been identified even when the action is not carried out: seeing an object can involve simulating possible actions on it. Moreover, so-called "mirror neurons" fire not just when an action is performed but also when the action of another is *observed* (Rizzolatti & Sinigaglia, 2016). Mirror neurons differ in the congruence of their pattern of firing – some will fire when the action observed is exactly like the one performed (e.g. a precision grip), but most will fire when observing a wide range of similar actions. These findings show that embedded within sensorimotor brain regions are neural clusters that simulate motor activity (in the absence of the action itself) and can do so by generalizing across a *class* of similar actions. Overall, this has led to the simulation hypothesis: that understanding objects and the actions of others involves the simulation of neural activation associated with one's own actions (Gallese & Sinigaglia, 2011).

The simulation hypothesis has been extended further to language comprehension. Neuroimaging studies have provided evidence that understanding sentences including verbs referring to different body parts (e.g. *hold*, *kick*) activates parietal-premotor circuitry responsible for the actual actions of those body parts (Tettemanti et al., 2005). The action-sentence compatibility experimental paradigm has also been used to link action and sentence comprehension. Performing actions prior to a sentence comprehension task facilitated comprehension when the actions were compatible with actions referred to in sentences (Glenberg & Kaschak, 2002). Moreover, an active area of investigation is also examining the activation of the sensorimotor system to ground understanding of non-literal language (Yang & Shu, 2016). While this research has produced mixed results, it is revealing that subtle variations in features of linguistic stimuli, including the relationship between abstract and literal concrete meanings, impact whether or not the sensorimotor system is activated.

Gallese and Lakoff (2005, p. 456) extract a broad conclusion from such findings: "that a key aspect of human cognition is neural exploitation – the adaptation of sensory-motor brain mechanisms to serve new roles in reason and language, while retaining the original functions as well."

19.4 Image Schemas, Science Concept Learning and Educational Neuroscience

This quick review of the neuroscience research on image schematic representations and the theory of cognition as embodied simulation in the context of the earlier discussion of the roles of experience-based intuition in science concept learning raises a number of issues. The converging nature of the findings lends greater plausibility to similar claims across both literatures: claiming that intuitions grounded in sensorimotor experience support understanding of scientific concepts, and claiming that sensorimotor brain regions are implicated in observing actions of others, as well as understanding literal and non-literal sentences, are mutually reinforcing. In addition, when seen together these literatures suggest interesting new directions for further research.

As mentioned above, research in the learning sciences has identified a number of knowledge elements that seem to have a great deal in common: p-prims, core intuitions, image schematic source domains of conceptual metaphors in the language of science and some of the conceptual schemata of symbolic forms. One central question is: are these all image schemas in the sense discussed above? Is the same kind of knowledge structure being activated to enrich mechanistic explanations, to ground understanding of abstract scientific language and to give meaning to equations? If this is the case, we should expect the activation of similar secondary integrative neural clusters underlying the tasks learning scientists have suggested implicate these various types of intuitive knowledge structures. The methods of cognitive neuroscience can be used to test this hypothesis.

If marrying the theoretical questions of learning scientists investigating science concept learning and the methods of cognitive neuroscientists investigating the embodied nature of cognition shows promise, this will have valuable pedagogical implications. In particular, this can contribute to identifying the important design features of instructional interventions inspired by the broad principles of embodied cognition and the assumption that embodied experiences can contribute to scientific understanding. For example, if experiences with a whole-body mixed reality simulation leads to improved performance on items in the *Force Concept Inventory*, we should expect the activation of secondary integrative neural clusters while learners respond to these items. Moreover, tweaking design features with the goal of activating particularly powerful image schematic structures should lead to better performance on these items and should be reflected in corresponding changes in neural activation. These hypotheses can be tested with valuable theoretical and pedagogical outcomes. The key here is that specific instructional design features can be linked to outcomes on an important diagnostic assessment and to underlying neural activation.

Other lines of investigation following this same pattern can be pursued: Do certain problem solving experiences help learners make sense of equations by activating symbolic forms? Does some carefully selected analogy enrich learners' mechanistic explanations of a phenomenon by activating a core intuition? And is a passage in a textbook carefully crafted to highlight a particularly important conceptual metaphor for this domain more easily understood than others, not so carefully written? In all these cases, the underlying instructional design assumption – that the intervention encourages activation of a particularly useful image schema – can be investigated using the methods of cognitive neuroscience.

The discussion in this short chapter suggests that research investigations in the educational neuroscience of science concept learning along the lines just described is likely to be very productive. There are already some examples of research of this kind in the educational neuroscience of mathematics education (Tsang, Rosenberg-Lee, Blair, Schwartz, & Menon, 2010). Tsang et al. have provided fMRI evidence that finding the mid-point between two integers recruits occipital brain regions otherwise activated for the perception of spatial symmetry. They have gone on to design hands-on activities and simulations building on this finding. Schwartz, Blair and Tsang (2012) have used this example to illustrate what they call Culture B educational neuroscience, where learning scientists and cognitive neuroscientists collaborate to address basic theoretical questions with important pedagogical implications. This is the kind of educational neuroscience of science concept learning I am advocating in this chapter. Specifically, I am suggesting that examining the role of image schemas in science concept learning can be a productive entry point for this kind of collaboration. This would capitalize on converging developments in the learning sciences and the cognitive neuroscience of embodied cognition finding that knowledge structures emerging from sensorimotor experiences can be put to a wide range of uses in higher level cognition, including scientific understanding and reasoning. This would allow our exploration of the cognitive neural underpinnings of science concept learning to move beyond an emphasis on the inhibition of intuitive knowledge within a two-system model of cognition and embrace the continuity between the learner and expert scientist that has been recognized in the learning sciences.

Acknowledgements I would like to thank Mariana Levin, Olivia Levrini and Andy Tolmie for very helpful feedback on an earlier version of this chapter.

References

Amin, T. G. (2009). Conceptual metaphor meets conceptual change. *Human Development, 52*(3), 165–197.

Amin, T. G. (2015). Conceptual metaphor and the study of conceptual change: Research synthesis and future directions. *International Journal of Science Education, 37*(5–6), 966–991.

Amin, T. G. (2020). Coordinating metaphor in science, learning and instruction: The case of energy. In A. Beger & T. H. Smith (Eds.), *How metaphors guide, teach and popularize science. Figurative Thought and Language Series* (pp. 73–110). John Benjamins.

Amin, T. G. & Levrini, O. (Eds.) (2018), *Converging perspectives on conceptual change: Mapping an emerging paradigm in the learning sciences.* Routledge, UK.

Amin, T. G., Jeppsson, F., Haglund, J., & Strömdahl, H. (2012). Arrow of time: Metaphorical construals of entropy and the second law of thermodynamics. *Science Education, 95*(5), 818–848.

Amin, T. G., Smith, C., & Wiser, M. (2014). Student conceptions and conceptual change: Three overlapping phases of research. In N. Lederman & S. Abell (Eds.), *Handbook of research in science education, Vol II* (pp. 57–81). Taylor and Francis.

Amin, T. G., Jeppsson, F, and Haglund, J. (Eds.) (2015/2017). *Conceptual metaphor and embodied cognition in science learning.* (Routledge, UK) [First published as special issues (5–6, 2015) in the *International Journal of Science Education.*]

Beger, A., & Smith, T. H. (Eds.). (2020). *How metaphors guide, teach and popularize science. Figurative thought and language series.* John Benjamins.

Brookes, D. T., & Etkina, E. (2007). Using conceptual metaphor and functional grammar to explore how language used in physics affects student learning. *Physical Review Special Topics – Physics Education Research, 3*, 1–16.

Brown, D. E. (2014). Students' conceptions as dynamically emergent structures. *Science & Education, 23*(7), 1463–1483.

Brown, D. E. (2018). Implicit conceptual dynamics and students' explanatory model development in science. In T. G. Amin & O. Levrini (Eds.), *Converging perspectives on conceptual change: Mapping an emerging paradigm in the learning sciences* (pp. 105–112). Routledge.

Brown, D., & Clement, J. (1989). Overcoming misconceptions via analogical reasoning: Abstract transfer versus explanatory model construction. *Instructional Science, 18*(4), 237–261.

Cheng, M. F., & Brown, D. E. (2010). Conceptual resources in self-developed explanatory models: The importance of integrating conscious and intuitive knowledge. *International Journal of Science Education, 32*(17), 2367–2392.

Dancygier, B. (2017). *The Cambridge handbook of cognitive linguistics*. Cambridge University Press.

diSessa, A. A. (1983). Phenomenology and the evolution of intuition. In D. Gentner & A. L. Stevens (Eds.), *Mental models* (pp. 15–33). Erlbaum.

diSessa, A. A. (1993). Toward an epistemology of physics. *Cognition and Instruction, 10*(2&3), 105–225.

diSessa, A. A. (2000). Does the mind know the difference between the physical and social worlds? In L. P. Nucci, G. B. Saxe, & E. Turiel (Eds.), *Culture, thought, and development* (pp. 141–166). Erlbaum.

Driver, R., & Easley, J. (1978). Pupils and paradigms: A review of the literature related to concept development in adolescent science students. *Studies in Science Education, 5*, 61–84.

Dunbar, K., Fugelsang, J., & Stein, C. (2007). Do naïve theories ever go away? Using the brain and behavior to understand changes in concepts. In M. Lovett & P. Shah (Eds.), *Thinking with data* (pp. 193–205). Lawrence Erlbaum Associates.

Evans, J. S. B. T. (2011). Dual-process theories of reasoning: Contemporary issues and developmental applications. *Developmental Review, 31*, 8–102.

Fogassi, L., Gallese, V., Buccino, G., Craighero, L., Fadiga, L., & Rizzolatti, G. (2001). Cortical mechanism for the visual guidance of hand grasping movements in the monkey: A reversible inactivation study. *Brain, 124*, 571–586.

Gallese, V., & Lakoff, G. (2005). The brain's concepts: The role of the sensory-motor system in conceptual knowledge. *Cognitive Neuropsychology, 22*(3–4), 455–479.

Gallese, V., & Sinigaglia, C. (2011). What is so special about embodied simulation? *Trends in Cognitive Science, 15*(11), 512–519.

Glenberg, A. M., & Kaschak, M. P. (2002). Grounding language in action. *Psychonomic Bulletin and Review, 9*, 558–565.

Holbert, N. & Wilensky, U. (2012). Representational congruence: Connecting video game experiences to the design and use of formal representations. *Proceedings of Constructionism*, Athens, Greece, 2012.

Houde, O., Zago, L., Mellet, E., Moutier, S., Pineau, A., Mazoyer, B., & Tzourio-Mazoyer, N. (2000). Shifting from the perceptual brain to the logical brain: The neural impact of cognitive inhibition training. *Journal of Cognitive Neuroscience, 12*(5), 721–728.

Jeppsson, F., Haglund, J., Amin, T. G., & Strömdahl, H. (2013). Exploring the use of conceptual metaphors in solving problems on entropy. *Journal of the Learning Sciences, 22*(1), 70–120.

Johnson, M. (1987). *The body in mind: The bodily basis of meaning, imagination, and reason*. University of Chicago Press.

Kahneman, D. (2011). *Thinking, fast and slow*. Farrar, Straus & Giroux.

Kövecses, Z. (2010). *Metaphor: A practical introduction* (2nd ed.). Oxford University Press.

Lakoff, G. (1990). The invariance hypothesis: Is abstract reason based on image-schemas? *Cognitive Linguistics, 1*, 39–74.

Lakoff, G., & Johnson, M. (1980). *Metaphors we live by*. University of Chicago Press.

Lakoff, G., & Johnson, M. (1999). *Philosophy in the flesh*. Basic Books.

Lindgren, R., & Johnson-Glenberg, M. (2013). Emboldened by embodiment: Six precepts for research on embodied learning and mixed reality. *Educational Researcher, 42*(8), 445–452.

Lindgren, R., Tscholl, M., Wang, S., & Johnson, E. (2016). Enhancing learning and engagement through embodied interaction within a mixed reality simulation. *Computers and Education, 95*, 174–187.

Mareschal, D. (2016). The neuroscience of conceptual learning in science and mathematics. *Current Opinion in Behavioral Sciences, 10*, 114–118.

Masson, S., Potvin, P., Riopel, M., & Brault-Foisy, L. M. (2014). Differences in brain activation between novices and experts in science during a task involving a common misconception in electricity. *Mind, Brain and Education, 8*, 44–55.

Mathayas, N. and Brown, D. E. (2018). Exploring multimodal scaffolds supporting middle-school students' construction of causal-mechanistic scientific explanations. *Proceedings of the Conference of the International Society of the Learning Sciences*, 2018.

Reber, P. J., Stark, C. E. L., & Squire, L. R. (1998). Contrasting cortical activity associated with category memory and recognition memory. *Learning and Memory, 5*, 420–428.

Rips, L. J., Smith, E. E., & Medin, D. L. (2012). Concepts and categories: Memory, meaning and metaphysics. In K. J. Holyoak & R. G. Morrison (Eds.), *The Oxford handbook of thinking and reasoning* (pp. 177–209). Oxford University Press.

Rizzolatti, G., & Sinigaglia, C. (2016). The mirror mechanism: A basic principle of brain function. *Nature Reviews Neuroscience, 17*, 757–765.

Rohrer, T. (2005). Image schemata and the brain. In B. Hampe & J. Grady (Eds.), *From perception to meaning: Image schemas in cognitive linguistics* (pp. 165–196). Mouton de Gruyter.

Schwartz, D. L., Blair, K. P., & Tsang, J. M. (2012). How to build educational neuroscience: Two approaches with concrete instances. *British Journal of Educational Psychology Monograph Series II, 8*, 9–27.

Sherin, B. (2001). How students understand physics equations. *Cognition and Instruction, 19*(4), 479–541.

Sherin, B. (2006). Common sense clarified: The role of intuition knowledge in physics problem solving. *Journal of Research in Science Teaching, 43*(6), 535–555.

Sherin, B. (2018). Conceptual change and symbol use in physics. In T. G. Amin & O. Levrini (Eds.), *Converging perspectives on conceptual change: Mapping an emerging paradigm in the learning sciences* (pp. 81–88). Routledge.

Strike, K. A., & Posner, G. J. (1985). A conceptual change view of learning and understanding. In L. H. West & A. L. Pines (Eds.), *Conceptual structure and conceptual change* (pp. 189–210). Academic Press.

Talmy, L. (1988). Force dynamics in language and cognition. *Cognitive Science, 12*, 49–100.

Tettamanti, M., Buccino, G., Saccuman, M. C., Gallese, V., Danna, M., Scifo, P., Fazio, F., Rizzolatti, G., Stefano, F., Cappa, F., & Perani, D. (2005). Listening to action-related sentences activates fronto-parietal motor circuits. *Journal of Cognitive Neuroscience, 17*(2), 273–281.

Thomas, M. S. C., Purser, H. R. M., & Mareschal, D. (2012). Is the mystery of thought demystified by context-dependent categorization? Towards a new relation between language and thought. *Mind and Language, 27*(5), 595–618.

Tolmie, A., & Dundar-Coeke, S. (2020). Lifespan conceptual development in science: Brain and behavior. In M. S. C. Thomas, D. Mareschal, & I. Dumontheil (Eds.), *Educational neuroscience: Development across the life span* (pp. 195–222). Routledge.

Tsang, J. M., Rosenberg-Lee, M., Blair, K. P., Schwartz, D. L., & Menon, V. (2010, June). Near symmetry in a number bisection task yields faster responses and greater occipital activity. Poster presented at the 16th annual meeting of the Organization for Human Brain Mapping. Barcelona, Spain.

Vosniadou, S. (2013). *International handbook of research on conceptual change* (2nd ed.). Routledge.

White, B. (1993). ThinkerTools: Causal models, conceptual change, and science education. *Cognition and Instruction, 10*(1), 1–100.

Yang, J., & Shu, H. (2016). Involvement of the motor system in comprehension of non-literal action language: A meta-analysis study. *Brain Topography, 29*(1), 94–107.

Chapter 20
Emotional Engagement in the Application of Experimental Activities with Young Children

Kellys Saucedo and Maurício Pietrocola

20.1 Introduction

Science is, above all, a social activity, promoting interactions, which are integrated and transformed, resulting from social negotiations. Developing hypotheses, working in groups, respecting ideas, building explanations, and paying attention to norms and rules are part of scientific activity. Learning science involves being in a social practice strongly infused with emotions (Bellocchi, 2017; Bellocchi, Quigley, & Otrel-Cass 2017). The world of young children, which is filled with new situations, experiments, curiosities, and questions, is also characterized by an interest in natural phenomena. Therefore, at the beginning of schooling, children find themselves exploring an unfamiliar space and start building new social and emotional bonds there. This set of nuances that make up children's experiences can support their engagement in scientific activity. In the last decades, the preparation to do science has become an issue for international educational reforms. Among them we can mention *Science for All* of the American Association for the Advancement of Science (AAAS, 1989); *La main à la pâte*, in France (Charpak, 1996) and *Wissenschaft im Dialog,* in Germany (Stifterverband Für Die Deutsche Wissenschaft, 1999).

In countries, such as Brazil, however, what we have observed on the part of educational policies for the first stages of schooling, especially in the so-called Literacy

K. Saucedo (✉)
School of Education, University of São Paulo, São Paulo, Brazil
e-mail: kellys@alumni.usp.br

M. Pietrocola
School of Education and Institute for Advanced Studies, University of São Paulo, São Paulo, Brazil
e-mail: mpietro@usp.br

O. Levrini et al. (eds.), *Engaging with Contemporary Challenges through Science Education Research*, Contributions from Science Education Research 9,
https://doi.org/10.1007/978-3-030-74490-8_20

Cycle (Brazil, 2019), is a systematic exclusion of actions aimed at training in science and technology. In the National Common Curricular Baṣe (Brazil, 2018), there are very shy mentions about science teaching in Primary School. The profile that is being drawn is very close to technical training models and not to what science education researchers defend, especially on engagement emotional. In scientific research what we have observed, with this age group, in general, is mainly related to the internal aspects of personality, cognition and behavior of the individual. The cognitive aspect is associated with the student's attempts to understand complex concepts and master difficult skills (Herrenkohl & Guerra, 1998). The aspect behavioral refers to the actions of the student in the classroom, as for example in measuring "time on tasks" (Tobin & Capie, 1982).

Even works in Social Psychology, in what Piaget characterizes as argument among peers (Piaget, 1965), they have their emphasis on the social situations for the functioning of essential psychological and anthropological processes, such as: perception, interest and cognition (Goodwin, 2005; Hidi, 1990). These works delimit their analysis in the formation of the personality, in terms of feelings, emotions and moods. Sutton and Wheatley (2003) present an extensive literature review on the topic. We are, however, interested in expanding our analysis of emotions for the social universe, considering structure and social interactions and how they mutually limit, inhibit, and stimulate emotions, and how they may transform culture.

Concerned with the scenario described, we started to consider the world of young children as a space for science education research that has a lot to offer. We are interested in aspects related to emotional engagement within groups conducting science activities. This is an important area of attention as Hampden-Thompson and Bennett (2013) conclude, "[…] greater levels of student motivation, enjoyment, and future orientation towards science were found in classrooms where students reported that various measures of interaction, hands-on activities, and applications in science took place frequently" (p. 1340). This means that one way to attract students to learning science and technology is to develop and keep their interest and engagement in science activities in the early years.

In science education, emotions should be given the same level of importance as cognition in research on learning (Zembylas, 2005). The affective and cognitive domains can be viewed as connected and mutually dependent (Dearden et al., 1972). Siry and Brendel (2016) explore "[…] the inseparable role of emotions in the teaching and the learning of science at the primary school level" (p. 803) and, at the same time that they develop theoretical foundations, they elaborate personal experiences, strengthening the notion of inseparability between emotions and science learning.

The emotions are woven into the fabric of classroom life (Aultman et al., 2009). The role of emotions may be evaluated when we attend to the fact that the "[…] incentive for acquiring new information and skills does not emerge only from intrinsic interest in the content, or from a system of rewards and punishments such as that offered by grades and testing, but also from the desire to contribute as valued members of a community" (Olitsky, 2007, p. 34). The author suggests that to better understand learning in a social context more research is needed that looks into the structures and conditions that influence group members. Hargreaves (2005) argues that stronger emotional bonds between teachers and students could be a basis for high-quality learning. The emotional engagement refers to students' reactions to and interest in colleagues, teachers, curricular content, school, and extra-curricular activities (Alsop & Watts, 2003; Fredricks et al., 2004).

In the 1990s, the studies on emotions gained new vigor with the advances made possible, above all, by neuroscience (Franks, 2006) and the studies in the microsociology of emotions (Summer-Effer, 2006). Despite a long tradition of research in Psychology and Sociology, the methods and approaches for investigating emotions in natural environments are very recent. Studies on emotions require special methods. Tobin and colleagues are researchers who have brought new methods to the investigation of emotions in collective settings. They have the emotions as a primary focus (Bellocchi, 2017; Olitsky & Milne, 2012; Ritchie et al., 2011; Ritchie & Tobin, 2018). The methods used in research on emotions, combine facial expression analysis (Ekman & Friesen, 2003), verbal and nonverbal conduct (Bellocchi, 2015; Harrigan, 2008), the prosodic analysis (Juslin & Scherer, 2008) and variables such as pulse rate (Tobin & Ritchie, 2012). For instance, online training tools (http://www.ipsp.ucl.ac.be/recherche/projets/FaceTales/en/Home.htm) and automated facial coding software (Facial Action Coding System – FACS) allow researchers to develop skills of identifying emotions and subsequently analyze them considering the socio-cultural context in which they occur. Another method for identifying and characterizing emotions includes the combined analysis of body, eyes and proxemic movements (interpersonal and environmental space) (Bellocchi & Ritchie, 2015).

Our interest in examining the emotional engagement for understanding the ways in which groups sustain solidarity and social ties has led us to the *Interaction Ritual Chains Theory – IRC* (Collins, 1987, 2004). Wilmes and Siry (2018) used Interaction Ritual Theory to examine a pluri-lingual student's participation in inquiry-based science. They conducted an analysis of the interaction ritual of students during small-group science investigations. One of their conclusions is that their inquiry-based science pedagogy created space for students to form successful interaction rituals that, in turn, supported the focal student's science engagement and language development. The *Mutual Focus and Emotional Entrainment Model* developed by Collins (2004) also allow us to examine face-to-face interactions and consider the role of context and emotions to understand emotional engagement, in an approach that, until recently, was not used in the scientific field. There is not much research that deals with learning science as a collective achievement, supported by emotions, in which the individual participants "do" science in interaction with others. Milne and Otieno (2007) are among the first disseminators of the possibilities and limits of analyzing emotional engagement among students in chemistry classes, with scientific demonstrations as places of interaction.

The volume of research aimed at creating conditions to stimulate the interest of young children in scientific knowledge (Schreiner & Sjøberg, 2007; Siry, 2012 and Siry et al., 2012) is much less expressive. In Early Childhood Education, Hargreaves (2005) emphasizes the importance of emotions in children's adaptation to the new social spaces in which they are inserted, as well as to the development of interest in acquiring knowledge. It is necessary for children to act in ways that result in strengthened bonds of emotional engagement, including teacher-student, student-student and student-knowledge engagement. Indeed this emotional engagement is integral to the development of the child in the subsequent stages of Basic Education. When aiming to understand the role of science in early childhood, it is important to not overestimate the importance of learning concepts. Even if it is important to pay attention to what children are learning at this age, it is more important to encourage them to participate in the activities that are designed for them. That is, children play to learn and learn to play. And both the play and the learning take place in group activities, where emotions are the glue that sustain the collective.

We are thus interested in studying the role of emotional engagement in group science activities. Therefore, this study adopts a sociological perspective, understanding that sociological structures consider social practices essential to any learning process. The act of learning takes place in the relationship between people that are active in the world. Consequently, "[…] learning science is a collective achievement, as individual participants "do" science in interaction with others" (Siry et al., 2012, p. 313). This view excludes the perspectives that conceive learning as a static and internalized process.

We investigate the emotional engagement of children between 2 and 5 years of age in an experimental science activity and the fluency of their non-verbal student-student, student-teacher interactions. We are guided by the following question: How do emotions produce solidarity and corroborate children's emotional engagement as they conduct scientific activities? We adopt the methodological design used by Collins (1987, 2004) where it is possible to understand how the mutual focus of attention and the emotions shared by children in social interaction produce positive emotions. More precisely, we aim to investigate how emotional language, expressed in nonverbal interactions, operates for the emotional engagement of children in hands-on science activities.

20.2 Theoretical Framework

Emotional engagement is an important notion in rituals theories (RT), as it frames all social dynamics as focused rituals. In these theories, emotion is understood as the "glue" that unites people in society, being co-produced by people engaged in social interactions. Durkheim (1912/2001) was one of the first to propose a theory of rituals and emotions, the fundamentals of which were based on ethnographic studies of the behavior rituals of aborigines in Central Australia. In Durkheim's theory of rituals, emotions are part of a process of collective arousal which he referred to as *collective effervescence.* Collective effervescence is experienced as a heightened awareness of group membership as well as a feeling that an outside powerful force has sacred significance. The sentiment experienced by the group is transferred to symbols at the center of the ritual.

More recently, other theories of rituals were developed to deal with modern societies. Goffman (1967) revives the idea of Interaction Ritual to interpret the face-to-face encounters that take place in everyday life. Randall Collins (1981, 1990, 2004) was an intellectual descendent of Durkheim and Goffman, who conceptualized his own theory of rituals and applied it to the informal and secular life of everyday face-to-face encounters. For him, any social encounter could be seen as an IR of some kind where a common focus of attention (mutual focus) determines a strong emotional attunement (emotional entrainment) among the participants. In these cases, IRs produce a process of internal feedback that generates strong emotional experiences. These moments are loaded with cultural significance and at the same time produce motivational poles that attract or repel member participation. It is the emotional experiences that create, reproduce, reinforce, or transform a culture, creating or reinforcing symbols.

Four ingredients are essential conditions for the beginning of an IR (Collins, 2004): (i) **Group assembly:** the physical co-presence of at least two people or more, so that they are affected by each other); (ii) **Barrier from outsiders:** a barrier (physical or not) that separates the participants from those who do not participate in the collective meeting; (iii) **Mutual focus of attention:** all share a focus of mutual attention to an object or activity, where participants gain mutual awareness of the focus of attention of one another; (iv) **Shared mood:** all participants experience the same emotional state.

All the components of the ritual are variable in their intensity, just as their variation also produces changes in the effects of the ritual of interaction. The processes involved in the ritual evolve over time, generating flows of microevents that can last for an instant, minutes, hours or days. When the four elements are present the emotional state is shared among participants and the production of a collective effervescence occurs, what Collins as calls "Emotional Energy" (EE). (Collins, 2004, p. 61).

Tobin (2010) states that synchronization and fun – which include laughter, applause, supportive signals, vibration – establish emotional microclimates, and when this happens in science classes students can create positive expectations and increase their interest in content. That aspect of IR in Collins' perspective is called the feedback process in time where one ritual loads participants with EE and EE becomes available to be recuperated in future IRs. The feedback process in time implies the combination of ingredients of ritual interaction (i) copresence, mutual awareness, common mood, shared focus of attention, (ii) entrainment (rhythmic activity); and effects of interaction: (i) EE – solidarity and meaningful symbols –, and (ii) build "on symbols for EE and for finding future EE generating interaction rituals" (Summer-Effer, 2006, p. 139). Regarding this process, Collins (2004) emphasizes that EE, originated in previous meetings, creates, and maintains social interactions.

The IRs are not isolated in time but constitute "chains", that they feed the social life of people. For that reason, Collins' theory includes the notion of Interaction Ritual Chains (IRC). He describes ritual as "[...] a mechanism of mutually focused emotion and attention, producing a momentarily shared reality, which thereby generates solidarity and symbols of group membership" (Collins, 2004, p. 7). The intensity of the EE produced determines the bonds of group solidarity and generates symbols. According to Collins (2004), even the most common things may become symbols. We believe that the experimental science activity has the potential and the necessary resources to foster interaction rituals capable of producing emotions and focused attention, which allow us to understand the role of emotions in young children's emotional engagement.

There has been some recent interest in the sociology of emotions highlighting emotional engagement centered on social situations, unrelated to the individual's cognitive development, and forms of language-interaction (Goodwin, 2005). However, there has been little research on science practices in Early Childhood Education from the perspective of emotional engagement. Young children, in particular, use more non-verbal language to express themselves than adults (e.g. gesticulation, facial expressions, body and eye movements). In this view, the main contribution of IRC is the possibility of identifying ingredients and effects that promote successful interactions which would easily go unnoticed in those studies focused on verbal language. To address this gap, we analyzed the emotional engagement of children between 2 and 5 years of age in an experimental science activity and the fluency of non-verbal student-student, student-teacher interactions.

20.3 Research Method and Findings

Field research took place in a holiday camp in July 2017, where ten children aging from 2 to 5 years old participated in hands-on science activities with a mentor/teacher. Most of the children knew the teacher/mentor from regular school classes. An event was selected of a hands-on activity called "dye drop explosion", was selected to illustrate the IRC methodology and the insights it can lead to. This activity involves dye bubbles, which burst when they pass through an oil layer, turning into spinning tapes. The activity was videotaped and then the event was analyzed to identify situations that showed a transformation in the emotional engagement and interaction. For the analysis, we used interpretative methods, based on hermeneutic phenomenology, and constructs from the Theory of Interaction Ritual Chains.

Following Tobin (2010), we started by watching the recorded video and selecting "events" that break the flow of interaction. They are short, between 2 and 3 min. Events are central records for us to understand something that transforms individuals, collectives, and institutions. "Events are defined in terms of contradictions that arise as culture is enacted" (Tobin & Ritchie, 2012, p. 118). The event selection is analogous to using a zoom lens. Two researchers were engaged in that task, one of them being the mentor/teacher. For identifying salient events they analyzed in videos outward expressions of children' emotions and "[…] other significant interactions, such as facial expressions (e.g., a puzzled look), emotional vocalizations (e.g., a shrill tone of voice or laughter), gestures (e.g., placing one's head in their hands), or actions (e.g., withdrawing from a group)" (Tomas et al., 2016, p. 242). After that, the event was presented to a group of researchers (5 persons) to be validated, who then confirmed the event which was explained based on the Model of Mutual Focus and Emotional Entrainment (Collins, 2004).

Hands-on activities are generally a challenge for teachers, as their success requires engagement on the part of students. On many occasions, activities of this type end up focusing only on the teacher's action with minimal attention to student engagement (Milne & Otieno, 2007). The Model of Mutual Focus and Emotional Entrainment contributes to identifying moments of shared experience capable of generating bonds of solidarity and synchrony, constituting the evidence of engagement in practical scientific activity.

The sequence of pictures below illustrates how we verified a group of facial expressions and gestures showing the strong emotional impact experienced by A9 and the emotional involvement of the other participants (Fig. 20.1). They share the same emotional state and mutual focus of attention. At the beginning of the activity, the children were more introspective, focused on the materials and the mentor's speech. Nevertheless, at this point we identified an intense agitation/vibration of A9: its expressions mix basic emotions of assertion-happiness (of medium/high intensity) with basic emotions of aversion-fear (of high intensity) (Turner, 2007). The emotions expressed on A9's face and hand gestures combine rejoicing and high anxiety and alter the culture enacted in the interaction. The arrows in Fig. 20.1b indicate the exact moment of rupture identified by the researchers.

The practical scientific activity begins with the teacher presenting the materials. Children pay attention to the teacher assembled in a semi-circle, producing a barrier separating them from "outsiders." The strategies and interactions are conducted at a specific site, e.g., demonstration on the table. As soon as the teacher started adding liquids in the glass container, we noticed that the children's attention was attracted by the oil that emerges in the water. A2 rises from the chair and A8 tilts her body forward with the armrest, followed by A7 and A10, who also come closer to the glass containers. At this point, all the children are staring at and facing the demonstration. The mutual focus of attention, the barriers to outside involvement, the synchronized body and eye movements indicate the beginning of a chain of interactions (Fig. 20.2).

With all the liquids added, one of the children notices that the oil does not submerge. She asks: "Doesn't it go down?" (A2) and the teacher confirms: "– It does not go down", and then asks the group why, with inaudible collective responses. A9 evaluates: "– You put, put the oil on top". A8 is bending over the table, as his eyes

Fig. 20.1 Series of images that indicates a salient event in the video during hands-on science activities

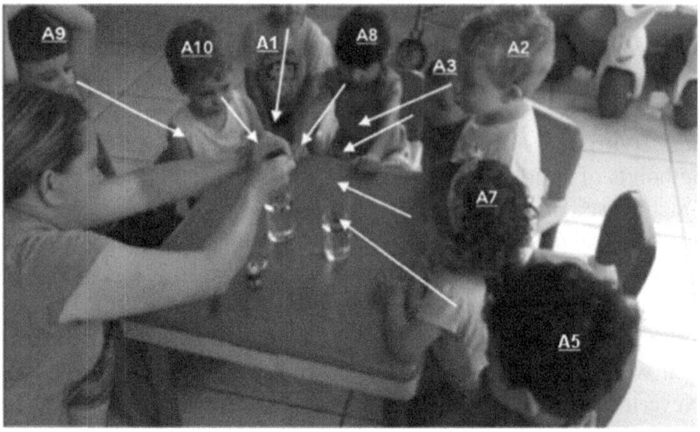

Fig. 20.2 Children pay attention to teacher assembled in a semi-circle, producing a barrier separating them from outsiders. A2 lifts rises from the chair and A8 tilts the body forward with the armrest, followed by A7 and A10

are fixed on the container, and some children point at him with their finger. A8 lifts his body and speaks loudly: "– It got stuck!" Then looks at the teacher and seems to seek confirmation in the teacher's eyes that she has heard his answer. We notice that the teacher's question and constant eye contact with the children stimulates their engagement and encourages them to move from the role of observers to participants in the process through the elaboration of causal explanations. Moreover, they vocalized shared observations.

The teacher moves the spoon into the cup. A9 moves his body forward and lowers his shoulders. A10 tilts his body to the side by touching the arm of A9, who displays mixed expressions of surprise and happiness. When tilting his body, A10 approaches the vessel to the center and shouts: "– Mix!". We advanced the recording at 1/24 speed and noticed that A10 almost touches his head to A9's chest. All children are staring at the experiment. A10 insists on mixing the ingredients, moves his arms, makes circular gestures (in frame-by-frame advance), he retreats towards A8, and A8 leans back to the table. The change in the color of the water provokes cries of collective vibration and the expression of happiness is recorded on the children's faces (Fig. 20.3).

A10, A9 and A1, with open mouths, experienced an emotional state of shared mood, which is also manifested by A2. A8 grabs her body, slouches in her chair, shrugs slightly and stretches her neck toward the experiment; A3 accompanies A8 and tilts his body forward. Mentor/teacher accompanies the group, her expresses happiness.

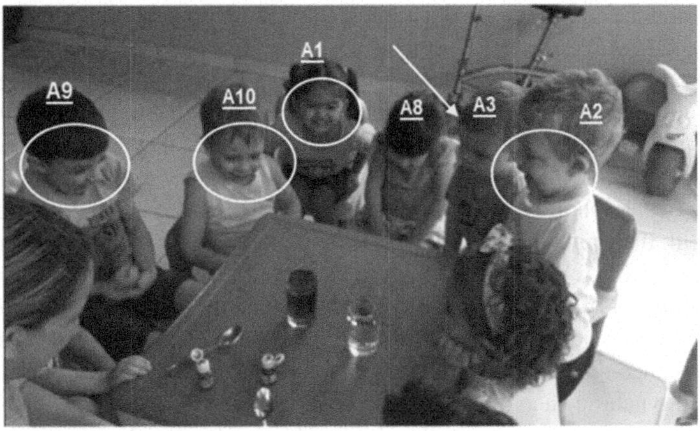

Fig. 20.3 Emotional state of shared mood. The change in the color of the water provokes cries of collective vibration and the expression of happiness is recorded on the children's faces

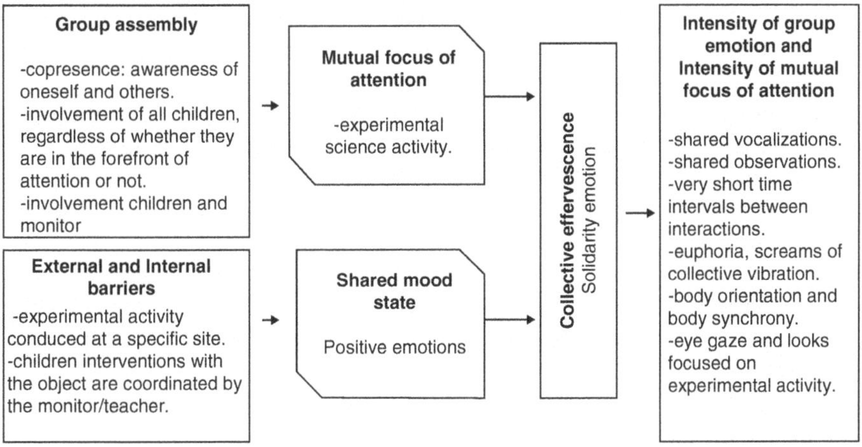

Fig. 20.4 Presence of ingredients and Ritual Interaction effects among participants. (Adapted from IRC theoretical approach (Collins, 2004))

20.4 Identifying Mutual Focus and Emotional Engagement

The body of evidence produced confirmed our initial impressions that we were facing a ritual of successful interactions. According to Collins (2004), for social interactions to be successful, they depend on people engaged in rituals and emotions, as rituals help to maintain the mutual focus of attention and produce emotional rhythmic feedback. We reached a significant advance in the interpretation of the event when we identified the fundamental *ingredients* and the *effects* produced by the interaction ritual (Fig. 20.4).

In this episode, the elements of an Interaction Ritual were all present in the activity and it is possible to state that it was a success ritual, with plenty of positive EE. When all the elements are present, there is synchrony between the participants and an emotional state of collective effervescence, according to Collins (2004).

The children shared emotions with such intensity that they have their attention completely dominated. As the ritual of interaction progresses, the group's enthusiasm increases. Euphoric by the change of color in the water, the children infect the teacher with their joy who participates in the collective expression of laughter. The EE experienced by each child during interactions with the teacher and group solidarity are effects of the rhythmic feedback between the mutual focus of attention and the shared emotional state that produces a positive collective experience. In a sense, the hands-on activity established an emotional microclimate where children manifested synchrony and fun, leading to positive expectations and increase in emotional engagement, generating in the children stimuli and likely interest to participate again in a similar experience. Collins (2004) states that intense ritual experiences such as those experienced by children generate EE and create symbolic objects that provide occasions for transformations. According to this author: "These moments of high degree of ritual intensity are high points of experience. They are high points of collective experience, […] the times when significant things happen. These are moments that […] shape new social structures." (Collins, 2004, p. 42).

This implies that positive emotional experiences are promising when they provoke interest to participate in similar moments in the future. Consequently, it is necessary to build positive affective memories in the child's earliest contacts with scientific knowledge. In their handbook chapter, Olitsky and Milne (2012) state the importance of research on collective emotions because it may have a powerful impact on the collective engagement of students in science classes, but also on individual identity, class perception and learning. As a final remark, the authors argue that emotions shared in collective sets are a precondition to the different dimensions of engagement.

In the activity reported and analyzed here, we saw a high degree of synchronization and the accumulation of positive emotions – happiness, joy, wonder, ecstasy – creating a common atmosphere of amusement among children. Is possible to state that the non-verbal stimuli at the beginning of the activity were fundamental for the success of student-student and student-teacher interactions, generating a progressive increase in the emotional engagement of the children. The progress of the activity offering some element of expectation and mystery also helped to keep the children attuned and sharing the same emotional states.

The non-verbal interactions between the children and the teacher serve as clear evidence of Collins' (2004) statements regarding moments of emotional climax and high solidarity, when: "[…] bodies tend to touch themselves, eyes fix in the same direction and the movements become rhythmic and synchronous" (p. 135). For teachers wishing to foster positive classroom changes, these studies suggest the need to provide a shared experience that is available to all within a context that has clear boundaries.

With respect to the problem of interest and engagement addressed earlier, we suggest that early participation in interaction rituals where positive EE is produced is likely to feed future participation in similar rituals. Experiences similar to these ones are essential to establish a positive relationship with scientific knowledge, which has been a constant concern in the last decades (e.g. Sjøberg & Schreiner, 2006; Schreiner & Sjøberg, 2007). We must remember that experimental scientific activities are not part of the daily routine of most young children who attend formal education in many countries, and wait until the final years of Basic Education for encouraging this kind of engagement may be too late.

20.5 Conclusions

In particular, the theory of interaction rituals and the method of analysis presented here offers a way forward to understand the study of science as an emotional and situated social practice. In scientific experimental activities with young children, this theoretical-methodological model enables investigators to understand the structure of interactions and face-to-face behaviors associated with the organization, establishment, and maintenance of emotional engagement.

Among the most evident clues, we highlight the potential for emotional engagement in scientific activities regarding (a) producing positive interactional environments and (b) potentially increasing children's interest in science. The joy and pleasure manifested in activities in early childhood education should not be seen as just a side effect of a learning process. It is also not just entertainment! Activities that become successful interaction rituals with the production of positive EE are preparation for participation in future rituals. Even if class activities provide bodily co-presence, the focus of attention, barriers to outsiders and the shared moods, it is necessary to have previously accumulated EE for these activities to have significance later on.

The evidence produced by this research and the analytical lens described, can inform new paths for early childhood teacher training, helping teachers create and maintain positive emotional climates in science classes. Teachers should be aware of the importance of developing activities where microclimates with positive EE are produced. Emotions produced throughout life fuel participation in future situations. If we want our students to engage in scientific and technological activities, we should be able to provide a repository of positive EE from the very first experiences with science.

Acknowledgement One of the authors (Pietrocola, M.) acknowledges *National Scientific Research Council* (CNPq) and the Institute for Advanced Studies – USP for the partial support for this research.

References

Alsop, S., & Watts, M. (2003). Science education and affect. *International Journal of Science Education, 25*, 1043–1047.

American Association for the Advancement of Science. (1989). *Science for all Americans: Project 2061* (Report on goals in science, mathematics and technology). Author.

Aultman, L., Williams-Johnson, M., & Schutz, P. (2009). Boundary dilemmas in teacher–student relationships: Struggling with "the line". *Teaching and Teacher Education, 25*, 636–646. https://doi.org/10.1016/j.tate.2008.10.002

Bellocchi, A. (2015). Methods for sociological inquiry on emotion in educational settings. *Emotion Review, 7*(2), 151–156.

Bellocchi, A. (2017). Interaction ritual approaches to emotion and cognition in science learning experiences. In A. Bellocchi, C. Quigley, & K. Otrel-Cass (Eds.), *Exploring emotions, aesthetics and wellbeing in science education research* (pp. 85–106). Springer.

Bellocchi, A., & Ritchie, S. (2015). "I was proud of myself that I didn't give up and I did it": Experiences of pride and triumph in learning science. *Science Education, 99*. https://doi.org/10.1002/sce.21159

Bellocchi, A., Quigley, C., & Otrel-Cass, K. (Eds.). (2017). *Exploring emotions, aesthetics and wellbeing in science education research*. Springer.

Brasil. (2018). Ministério da Educação (MEC). *Base Nacional Comum Curricular*: Educação é a Base. Brasília: MEC.

Brasil. (2019). Decreto Federal n° 9.765, de 11 de abril de 2019. Institui a Política Nacional de Alfabetização. *Diário Oficial da União*. Edição: 70-A, Seção 1 – Extra, p. 15.

Charpak, G. (1996). *La main à la Pâte*. As Ciências na escola primária. Trad. Ana Maria Moreira Cezar Fernandes.

Collins, R. (1981). On the microfoundations of macrosociology. *American Journal of Sociology, 86*(5), 984–1014.

Collins, R. (1987). Interaction Ritual Chains, power and property: The micro-macro connection as an empirically based theoretical problem. In J. C. Alexander, B. Giesen, R. Münch, & N. J. Smelser (Eds.), *The micro-macro link* (pp. 193–206). University of California Press.

Collins, R. (1990). Stratification, emotional energy, and the transient emotions. In T. D. Kemper (Ed.), *Research agendas in the Sociology of Emotions* (pp. 27–57). State University of New York Press.

Collins, R. (2004). *Interaction ritual chains*. Princeton University Press.

Dearden, R. F., Hirst, P. H., & Peters, R. S. (1972). The education of the emotions. In R. F. Dearden, P. H. Hirst, & R. S. Peters (Eds.), *Education and the development of reason* (pp. 466–483). Routledge & Kegan Paul.

Durkheim, E. (1912). *2001*. Oxford University Press.

Ekman, P., & Friesen, W. V. (2003). *Unmasking the face*. Malor.

Franks, D. (2006). The neuroscience of emotions. In J. E. Stets & J. H. L. Turner (Eds.), *Handbook of the sociology of emotions* (pp. 1–664). Springer. https://doi.org/10.1007/978-0-387-30715-2_3

Fredricks, J. A., Blumenfeld, P. C., & Paris, A. H. (2004). School engagement: Potential of the concept, state of the evidence. *Review of Educational Research, 74*(1), 59–109. https://doi.org/10.3102/00346543074001059

Goffman, E. (1967). *Interaction Ritual: Essays on face to face behavior*. Anchor books.

Goodwin, M. H. (2005). Interaction, language practice, and the construction of the social universe. *Calidoscópio, 3*(3), 184–195, set./dez. 2005.

Hampden-Thompson, G., & Bennett, J. (2013). Science teaching and learning activities and students' engagement in science. *International Journal of Science Education, 35*(8), 1325–1343. https://doi.org/10.1080/095006900750036253

Hargreaves, A. (2005). The emotions of teaching and educational change. In A. Hargreaves (Ed.), *Extending educational change* (pp. 278–295). Springer.

Harrigan, J. A. (2008). Proxemics, kinesics, and gaze. In J. A. Harrigan, R. Rosenthal, & K. S. Scherer (Eds.), *The new handbook of methods in nonverbal behavior research* (pp. 137–198).

Herrenkohl, L. R., & Guerra, M. R. (1998). Participant structures, scientific discourse, and student engagement in fourth grade. *Cognition and Instruction, 16*, 431–473.

Hidi, S. (1990). Interest and its contribution as a mental resource for learning. *Review of Educational Research, 60*(4), 549–571.

Juslin, P. N., & Scherer, K. R. (2008). Vocal expression of affect. In J. A. Harrigan, R. Rosenthal, & K. S. Scherer (Eds.), *The new handbook of methods in nonverbal behavior research* (pp. 65–136). Oxford University Press.

Milne, C. & Otieno, T. (2007). Understanding engagement: Science demonstrations and emotional energy. *Science Education, 91*(4), 523–553.

Olitsky, S. (2007). Promoting student engagement in science: Interaction rituals and the pursuit of a community of practice. *Journal of Research in Science Teaching, 44*(1), 33–56. https://doi.org/10.1002/tea.20128

Olitsky, S., & Milne, C. (2012). Understanding engagement in science education: The psychological and the social. In B. J. Fraser, K. Tobin, & C. McRobbie (Eds.), *Second international handbook of science education* (pp. 19–33). Springer. https://doi.org/10.1007/978-1-4020-9041-7_2

Piaget, J. (1965). *The moral judgment of the child (1932)*. Free Press.

Ritchie, S. M., & Tobin, K. (Eds.). (2018). *Eventful learning. Learner emotions*. Brill.

Ritchie, S. M., Tobin, K., Hudson, P., Roth, W.-M., & Mergard, V. (2011). Reproducing successful rituals in bad times: Exploring emotional interactions of a new science teacher. *Science Education, 95*(4), 745–765. https://doi.org/10.1002/sce.20440

Schreiner, C., & Sjøberg, S. (2007). Science education and youth's identity construction – Two incompatible projects? In D. Corrigan, J. Dillon, & R. Gunstone (Eds.), *The re-emergence of values in the science curriculum*. Sense Publishers.

Siry, C. (2012). Towards multidimensional approaches to early childhood science education. *Cultural Studies of Science Education, 9*. https://doi.org/10.1007/s11422-012-9445-8

Siry, C., & Brendel, M. (2016). The inseparable role of emotions in the teaching and learning of primary school science. *Cultural Studies of Science Education, 11*(3), 803–815. https://doi.org/10.1007/s11422-016-9781-1

Siry, C., Ziegler, G., & Max, C. (2012). "Doing science" through discourse-in-interaction: Young children's science investigations at the early childhood level. *Science Education, 96*, 311–326. https://doi.org/10.1002/sce.20481

Sjøberg, S., & Schreiner, C. (2006). How do learners in different cultures relate to science and technology? Results and perspectives from the project ROSE (the Relevance of Science Education) *APFSLT: Asia-Pacific Forum on Science Learning and Teaching*, *7*(1), Foreword.

Stifterverband für die Deutsche Wissenschaft. (1999). Wissenschaft im dialog. .

Summer-Effer, E. (2006). Ritual theory. In J. E. Stets & J. H. L. Turrner (Eds.), *Handbook of the sociology of emotions* (pp. 1–664). Springer. https://doi.org/10.1007/978-0-387-30715-2_7

Sutton, R. E., & Wheatley, K. F. (2003). Teachers' emotions and teaching: A review of the literature and directions for future research. *Educational Psychology Review, 15*(4), 327–358.

Tobin, K. (2010). Reproducir y transformar la didáctica de las Ciencias en un ambiente colaborativo. *Enseñanza de las Ciêncas, 28*(3), 301–303.

Tobin, K. G., & Capie, W. (1982). Relationships between classroom process variables and middle-school science achievement. *Journal of Educational Psychology, 74*(3), 441–454. https://doi.org/10.1037/0022-0663.74.3.441

Tobin, K., & Ritchie, S. M. (2012). Multi-method, multi-theoretic, multi-level research in the learning sciences. *The Asia-Pacific Education Researcher, 20*, 117–129.

Tomas, L., Rigano, D., & Ritchie, S. M. (2016, February). Students' regulation of their emotions in a science classroom. *Research in Science Teaching, 53*(2), 234–260.

Turner, J. H. (2007). *Human emotions: A sociological theory*. London, UK: Routledge.

Wilmes, S. E. D., & Siry, C. (2018). Interaction rituals and inquiry-based science instruction: Analysis of student participation in small-group investigations in a multilingual classroom. *Science Education, 102*(5), 1107–1128. https://doi.org/10.1007/978-3-319-43516-9_18
Zembylas, M. (2005). Discursive practices, genealogies, and emotional rules: A poststructuralist view on emotion and identity in teaching. *Teaching and Teacher Education, 21*, 935–948. https://doi.org/10.1016/j.tate.2005.06.005

Chapter 21
Crossing Boundaries – Examining and Problematizing Interdisciplinarity in Science Education

Shulamit Kapon and Sibel Erduran

21.1 Introduction

Recent visions of science education call for creating explicit connections between STEM disciplines in science education. These visions are motivated by realizations of the fundamental social, political, cultural, and economic changes likely to unfold over the course of the twenty-first century (Schwab, 2017). The widespread availability of digital technologies, as well as the ever-growing convergence of digital, biological, and physical innovations raise many concerns over the current disconnected nature of STEM education (European Commission, 2015; World Economic Forum, 2017). Different calls and curricular innovations have attempted to overcome this disconnection by (a) incorporating engineering challenges into the instruction of science and mathematics (Berland et al., 2014); (b) engaging students in mathematical (Lehrer & Schauble, 2012) and computational (Sengupta et al., 2013) modeling as a central component of their science learning; (c) devising integrated STEM curricula (Struyf et al., 2019), and (d) engaging students in scientific inquiry contextualized in real-life problems that inherently require the integration of STEM disciplines (NGSS Lead States, 2013).

This chapter discusses the talks presented in an invited symposium during ESERA 2019 entitled *'Crossing boundaries – Examining and problematizing interdisciplinarity in science education'*. Our goal in this chapter is to problematize disciplinary boundary crossings by examining the potential, affordances, challenges,

S. Kapon (✉)
Technion – Israel Institute of Technology, Haifa, Israel
e-mail: skapon@technion.ac.il

S. Erduran
University of Oxford, Oxford, UK
e-mail: sibel.erduran@education.ox.ac.uk

O. Levrini et al. (eds.), *Engaging with Contemporary Challenges through Science Education Research*, Contributions from Science Education Research 9,
https://doi.org/10.1007/978-3-030-74490-8_21

and impairments to science education they entail. The three presentations in the symposium provide rich terrain for this analysis since each took a different vantage point on the issue. Schvartzer et al. (2019) employed ethnography and discourse analysis to examine learners' engagement in detail when learning and using disciplinary knowledge in different interdisciplinary contexts. Levy et al. (2019) reflected on a series of design-based studies as a way to probe the explanatory potential of interdisciplinarity for disciplinary-based problems. Branchetti and Levrini (2019) took a historical and curriculum development perspective to examine the inherent interdisciplinarity of STEM disciplines in discipline-based educational systems. Taken together, this diversity of methodological approaches provides a dynamic platform to zoom in and gradually zoom out on the issues at hand.

In order to contextualize boundary crossing in interdisciplinarity, we begin with a broad overview of the role of interdisciplinarity in science education. This helps set the stage for the conceptualization of interdisciplinarity and illustrates its relevance for STEM education. We raise questions about the curriculum relevance of interdisciplinarity, present examples of interdisciplinary integrations in STEM education and tackle more fundamental questions about the nature of the constituent disciplines of STEM. The literature on STEM education on boundary crossing and integration suffers from a number of under-researched issues; specifically, issues such as what exactly is being integrated and how have not received the attention they deserve. As shown in this chapter, the integration of STEM disciplines often involves a complex and rich dialogical process of bringing together values, language, concepts, and practices from different STEM disciplines, which evolve as a result of this process. This chapter discusses a set of studies that have explicitly and directly addressed this issue from a range of perspectives.

21.2 Interdisciplinarity and STEM Education

An interdisciplinary approach in STEM education involves learning across the subject boundaries of science, technology, engineering, and mathematics for enhanced understanding. STEM education has been advocated in recent curriculum policy and research literature over the last 20 years (e.g. Eurydice, 2011; National Science and Technology Council, 2013). One of the key rationales for interdisciplinarity in STEM education is that many problems are complex and cannot be solved through a single and discrete disciplinary approach. Consider, for example, issues such as climate change and nuclear energy that draw not only on disciplines such as biology and physics respectively but also environmental science. An interdisciplinary approach also provides the opportunity to reflect on how STEM disciplines work and examine potential misconceptions about science and the scientific method (Nagle, 2013). For example, a typical misconception about NOS is that the scientific method is linear and unproblematic, whereas there is a diversity of scientific methods that operate in fairly complex ways (e.g. Woodcock, 2014).

An interdisciplinary approach encourages students to explore and integrate multiple perspectives from different subject disciplines, sub-disciplines, and areas of expertise (Golding, 2009). Interdisciplinary teaching can take various forms including integrated STEM courses, coordinated STEM courses and subject-focused courses (Nagle, 2013) and multidisciplinary approaches (Klein, 1990). Hurley (2001) reviewed empirical studies on mathematics-science integration and observed that there was a reasonable increase in science achievement resulting from integration. This effect increased significantly, with large effects on achievement in science associated with higher levels of integration. Hurley (2001) noted that integration is difficult to define given the complexities of timetabling, sequencing and the relative emphasis on the subjects integrated. The studies reviewed also lacked a careful conceptualization of the integration itself. In contrast, Redish and Kuo (2015) showed that the use of mathematics in physics education does not simply involve the transfer of mathematical skills from mathematics classes to physics classes, but rather a transformation of the transferred mathematical constructs themselves, since doing physics involves meaning-making with mathematical constructs in a different way than meaning-making with mathematics constructs employed by mathematicians.

Pang and Good (2000) argued for more sophisticated understanding and explicit discussions of the nature of science and mathematics. They stressed that science seeks to understand the world through empirical evidence external to the field itself whereas mathematics deals with internal, logical deduction. Park et al. (2020) went further than these broad characterizations and examined the disciplinary nuances between science and mathematics as represented in curriculum standards. They found that in the influential Next Generation Science Standards (NGSS Lead States, 2013), in particular, NGSS explicitly points to the similarities and differences between argumentation in science and mathematics:

> … [Like mathematics,] science too involves making arguments and critiquing them. However, there is a difference between mathematical arguments and scientific arguments—a difference so fundamental that it would be misleading to connect any of the standards to MP.3. here. The difference is that *scientific arguments are always based on evidence, whereas mathematical arguments never are*. It is this difference that renders the findings of science provisional and the findings of mathematics eternal. Blurring the distinction between mathematical and scientific arguments leads to a misunderstanding of what science is about. For more information about argumentation in science, see the NGSS science and engineering practice 'Engaging in argument from evidence'. (Appendix p. 140)

Hence, interdisciplinary integration raises some fundamental questions about the nature of the constituent disciplines. For example, what is the nature of knowledge in science, mathematics, engineering, and technology? Is knowledge in each discipline have similar characteristics or are there fundamental differences between knowledge from different constituent STEM domains? Even within domains of science there may be variations about the nature of knoweldge. Fore example, such questions were raised about how laws and explanations might compare in biology and chemistry (Dagher & Erduran, 2014). Park et al. (2020) addressed the issue of the epistemic nature of STEM by focusing on the epistemic components of each disciplinary system. They looked at the impact of the theoretical framework on aims and values, practices, methods and knowledge in science, technology, engineering, and mathematics drawing on the work of Erduran and Dagher (2014). They investigated several curriculum standards such as *Science for All Americans (SfAA)* (American Association for the Advancement of Science, 1990) and *Next Generation Science Standards (NGSS)* (NGSS Lead States, 2013), to examine their coverage of epistemic aspects of STEM.

The curriculum standards of the SfAA and the NGSS were published about 24 years apart and have been very influential in the USA and worldwide. The authors concluded that although there are numerous similarities between the SfAA and the NGSS (e.g., advocating the epistemic aim of "accuracy" in science), the SfAA seemed more detailed on some topics and NGSS in others. For example, while SfAA emphasizes the kinds of methodological approaches utilized in science (e.g., references to hypotheses as well as quantitative and qualitative methods), NGSS details kinds of scientific knowledge in more depth in terms of theories and laws. With respect to aims and values, practices, methods and knowledge in science, technology, engineering and mathematics, the two documents include references to all categories except for aims and values, and methods in the case of the framing of mathematics in NGSS. Whereas mathematics is considered to be critically important for addressing STEM problems, these disparities in curriculum standards will pose challenges to integration in STEM. These observations illustrate the basic assumptions embedded in curriculum standards on the ways in which knowledge operates in disciplines subsumed within STEM fields.

21.3 Boundary Crossing – Three Vantage Points

This section discusses each ESERA presentation separately. Each subsection starts with a brief summary of the main arguments presented by the authors, followed by an analysis of these arguments through the lens of boundary crossing as a dialogical enactment (Akkerman & Bakker, 2011). The discussion of the various boundary crossing in the three presentations can be framed by Akkerman and Bakker's (2011) conceptualization of boundaries as dialogical phenomena with four "dialogical learning mechanisms of boundaries" (p. 150) which represent a family of procedures that promote learning across boundaries. The key constructs in these authors' framework include the following:

- *Identification* – Identification has to do with the ways in which people find out about the diverse practices on each side of the boundary and how they relate to one another. Characteristic processes include *othering* and *legitimating coexistence*. One example is delineating how one practice differs from another.
- *Coordination* – Coordination involves the formation of cooperative and routinized exchanges between practices on each side of the boundary. Characteristic processes include *communicative connections*, *efforts at translation*, *increasing boundary permeability*, and *routinization*. Examples cover efforts at translating

between the worlds on each side of the boundary, or the process of automatizing and operationalizing these practices (i.e., routinization).

- *Reflection* – Reflection refers to the ways in which learners can expand their perspectives on other practices. Characteristic processes include *perspective making* and *perspective taking*. Examples involve making explicit one's understanding and knowledge of a particular issue, or deliberate attempts to take a different perspective than one's own.
- *Transformation* – Transformation encompasses the processes of collaboration and co-development of new practices. Characteristic processes include *confrontation, recognizing a shared problem space, hybridization, crystallization, maintaining the uniqueness of intersecting practices, continuous joint work at the boundary*. Examples include confronting discontinuities that are not easily surpassed, when creating a shared problem space (often in direct response to this confrontation), and creating a hybrid practice that is meaningful in both worlds and is somewhat different from the original practices from which it emerged.

21.3.1 Learning Physics Through Maker Projects – Between Disciplinary Authenticity and Personal Relevance (Schvartzer et al., 2019)

Schvartzer et al. (2019) discussed the boundary crossing between the Maker movement[1] and the formal instruction of science, as well as school science and personal relevance. They presented an ethnographic case study that followed a pair of students engaged in a long-term (15 month) engineering Maker-based inquiry that was an integral part of the students' formal matriculation in advanced level physics. The study (Kapon et al., 2021) provided a fine-grained examination of the evolving discourse between the students, their project mentor and other members of the educational staff, and revealed how students' forms of participation were socially constructed and evolved over time. The students' engagement was conceptualized as participating in a particular figured world (Holland et al., 1998). To illustrate the boundary crossing involved, the authors juxtaposed it with the figured worlds of authentic scientific inquiry in school (Kapon, 2016) and traditional school physics. Using fine-grained discourse analysis of student-student and student-educational staff interactions in authentic working sessions, complemented by interviews and other ethnographic accounts, the authors identified two legitimate forms of participation that contributed extensively to the engineering Maker-based inquiry goal of

[1] Making is an emerging contemporary "do it yourself" trend that capitalizes on the growing accessibility of digital fabrication tools and open source hardware and software (Dougherty, 2012). It has been argued that the Making movement has great promise for STEM education because it can lead to a democratization of knowledge in engineering and science (Blikstein, 2013), alternative pathways to engineering (Martin & Dixson, 2016), and be a venue for STEM learning that offers equitable opportunities to engage underrepresented youth (Calabrese Barton et al., 2016).

creating a working artefact: participating as an engineer and participating as a technician. The analysis articulates the social construction of these forms of participation and showed that participating as an engineer facilitated many foundational aspects of learning and doing physics. However, while participating as a technician fostered a sense of agency and efficacy with regard to physics in a student who did not find ways to express himself in the regular physics classroom (i.e., promoting personal relevance – Kapon et al., 2018, 2021), it did not facilitate the learning of scientific content and practices.

The Schvartzer et al. presentation is an interesting case to examine boundary crossing as a dialogical phenomenon (Akkerman & Bakker, 2011). The juxtapositions of the different figured worlds (engineering Maker-based inquiry, authentic scientific inquiry in school, and school science) is a manifestation of *identification*. Specifically, it involves a process of *othering*; namely, discussing one figured world in light of the other and delineating the differences. The focus of the study was the nature of the practices and the roles involved (i.e., participating as…). The findings highlighted participating as engineer as an important legitimate form of participation in the figured world of engineering Maker-based inquiry, while providing various and frequent opportunities to engage in meaningful acts that characterize legitimate participation in the figured world of authentic scientific inquiry. This observation marks participation as an engineer as a *shared problem space* between the two figured worlds, which is a hallmark of what Akkerman and Bakker termed the *transformation* of practices. The study showed that participating as a technician was an important form of participation in the figured world of engineering Maker-based inquiry, and contributed to its ultimate goal of creating a working artefact, although at the same time it constituted an insignificant form of participation in the figured worlds of authentic scientific inquiry and school physics. This incongruency points to one of the arenas of *confrontation* between Making and doing science. One of the school staff members who took part in the study articulated this confrontation in an interview as stemming from the different goals of the figured worlds. For him this insight resulted from the *reflection* prompted by the interview. Resolving the *confrontation* between Making and doing science in school thus may require some sort of *hybridization* of practices, which may most likely result in further *transformation* of both.

21.3.2 Slipping Between Disciplines: How Forming Causal Explanations May Compel Crossing Disciplinary Boundaries (Levy et al., 2019)

Levy et al. reflected on instances of boundary crossing in three design-based studies in their group. They argued for boundary crossing between STEM disciplines when practices and explanatory means in one discipline can significantly improve mechanistic explanations of phenomena in another discipline and thus enhance students' understanding. Their argument was supported by three design-based studies that examined students' learning in technological learning environments that deliberately incorporated representational change in chemistry and in biology. In the first study (Zohar & Levy, 2019) a force-based explanation, which characterizes explanations in physics (classical mechanics), was incorporated into the instruction of chemistry to support students' understanding of chemical bonding, a notoriously difficult concept for students to grasp. The pre-post interviews suggested a significant improvement in high school students' understanding of the chemical bond as a dynamic equilibrium between forces of attraction and repulsion. In the second and third studies (Dagan et al., 2019; Dubovi et al., 2018) ideas and representations from chemistry; i.e., conservation of matter at a molecular level, were implemented and adapted into a learning environment that aimed to support learners understanding of the biochemical process related to diabetes, by specifically helping learners to visually follow individual molecules throughout the system. One study examined nursing students studying the related pharmacology, and another study examined the learning of adolescent patients during routine visits to a diabetes clinic. The pre- and post- tests results highlighted the growth in the learners' conceptual understanding, and their ability to transfer the learned reasoning to other relevant problems. Levy et al. argued that "*the explanations and representations developed in these studies were particularly generative in supporting the understanding of difficult topics, transferring this knowledge to other topics, and supporting related behaviors.*" (Levy et al., 2019).

Levy et al.'s presentation highlights several facets of boundaries as dialogical phenomena (Akkerman & Bakker, 2011). In our view, the most striking learning mechanisms can be attributed to *coordination* and *transformation*. These authors explicitly worked to *enhance boundary permeability* in the digital learning environments they designed. Specifically, the representations of force diagrams in the case of chemical bonding, and the representations of the molecular dynamics in the case of diabetes, were an integral part of the learning environment, but the learners did not seem to experience any discontinuity in their forms of reasoning when "shifting" from chemistry to physics or from chemistry to biology. The "new" representations formed an integral part of the design, so that no explicit transitions were required. The reported transfer suggests that at least some level of *routinization* was achieved as well. The designers identified a potential *shared problem space*, and the new representations they introduced to this space generated a *hybridization*, since the original practices took on a new form. For example, the use of force diagrams in the chemistry learning environment was not identical to the use of force diagrams in classical mechanics.

21.3.3 Disciplines and Interdisciplinarity in STEM Education to Foster Scientific Authenticity and Develop Epistemic Skills (Branchetti & Levrini, 2019)

Branchetti and Levrini described the tension between the robust separation between disciplines in traditional schooling and the need to develop STEM interdisciplinary skills for the labor market. They argued that discipline-based instruction can and should continue to play an important educational role in current schooling, provided it is used as a platform to develop students' epistemic skills rather than knowledge per-se. By examining the structural role of mathematics in the development of physics, they further argued that throughout the history of science, interdisciplinarity has been an important authentic aspect of disciplinary-based science. This argument formed the basis of their claim that even from a disciplinary authenticity perspective, students should explicitly learn and experience the interdisciplinary aspects of the disciplined-based sciences they study in school. These arguments were illustrated by two case studies involving efforts at curriculum development. In the first case (Branchetti et al., 2019) the designers had to cross boundaries between physics and mathematics to effectively support college level students' understanding of the nature, meaning, and significance of quantum mechanics to the problem of black-body radiation, which puzzled scientists at the end of nineteenth century. In the second case the designers grappled with how to meaningfully introduce the complex and novel idea of artificial intelligence to secondary school students. The presentation showed that interdisciplinarity should not be confused with a–disciplinarity or multidisciplinarity, and that epistemic skills can be more effectively developed when different disciplines are compared and contrasted, and when both specific and transversal skills are made explicit.

Branchetti and Levrini's presentation constitutes an intriguing case of boundaries as dialogical phenomena (Akkerman & Bakker, 2011). They clearly acknowledged the importance and the unique features of the individual disciplines in students' education. "*The meaning of interdisciplinarity cannot ignore the meaning of 'discipline'. The term 'discipline' contains the Latin root 'discere', whose meaning is to learn. Disciplines can be seen as re-organizations of knowledge within the scope of teaching it.*" They claimed that disciplinary-based teaching is far more than a repository of knowledge since it must "*transform knowledge into rigorous and recognizable definitions and its practices into repeatable methods.*" (Branchetti & Levrini, 2019). This is an example of stressing the importance of *identification* (Akkerman & Bakker, 2011) of the unique epistemic practices of each discipline to students' learning. Branchetti and Levrini emphasized the process of *othering* each discipline as a crucial aspect of interdisciplinary learning. The case study of the curriculum development in quantum mechanics (boundary crossing between mathematics and physics) employed *reflection* as a central learning mechanism, in that the students were explicitly engaged in deliberate attempts to employ different historical perspectives to examine the problem at hand. *Reflection* and *coordination* (Akkerman & Bakker, 2011) were central learning mechanisms in the second case study of curriculum development as well (teaching artificial intelligence to high school students). The artificial intelligence example illustrated how disciplinary knowledge could foster the learning of new disciplines or when dealing with new problems that are not yet organized into a discipline. For example, the designers made an analogy between some of the epistemic differences between mathematics and physics reasoning to explain the epistemic differences between the logical approach and the machine learning approach to artificial intelligence.

21.4 Examining the Three Vantage Points on Boundary Crossing

The three presentations discussed above highlight the multidimensional nature of crossing boundaries between STEM disciplines. The first presentation (Schvartzer et al., 2019) demonstrated how the dialogical nature of crossing boundaries is socially constructed in discourse. The second presentation (Levy et al., 2019) demonstrated how specific boundary crossing in design (i.e., changes in representation) come to bear on students' learning. The third presentation (Branchetti & Levrini, 2019) demonstrated the historical and curricular considerations involved.

Whereas in Branchetti and Levrini's work, the design effort seemed to reside in carefully reconstructing the boundary through *identification* and *reflection*, in Levy et al.'s work the design effort seemed to reside in overcoming the boundary and facilitating effortless movement between the disciplines (i.e., *coordination* and *transformation*). All three studies highlighted the affordances for learning. Schvartzer et al. (2019) and Branchetti and Levrini (2019) stressed the potential contribution to students' sense of personal relevance, and the possibilities of connecting school science to modern societal and economic trends of interdisciplinarity; Branchetti and Levrini (2019) and particularly Levy et al. (2019) pointed to the different explanatory affordances entailed by boundary crossing. Nevertheless, Schvartzer et al. (2019) and Branchetti and Levrini (2019) also highlighted conflicts. Specifically, Branchetti and Levrini (2019) noted the importance of maintaining the identity of separate STEM disciplines as means of learning epistemic practices, and identified this effort as crucial for any meaningful boundary crossing. Schvartzer et al. (2019) underscored the discontinuities in practice that should be resolved to enable the integration of Making and engineering practices in the instruction of science. Taken together, these affordances and constraints reflect the complex dialogical nature of boundary crossing (Akkerman & Bakker, 2011) in STEM education and articulate it as an ongoing challenge for future research and development.

Taken together the presentations not only highlight why interdisciplinarity is important for science education but also raise questions about what counts as a "discipline" in the first place. Branchetti and Levrini's presentation traced the etymology of the word 'discipline' to the Latin root 'discere' meaning "to learn". In so doing, they emphasized the value of interdisciplinarity in forging new insights through boundary crossing as indicated in the following quote from their presentation:

> *Starting from a concrete problem, we showed the integration of S-T-E-M disciplines into a new STEM field of research and application, but we also used the traditional S-T-E-M disciplines epistemologies to shape and clarify the differences between the approaches, and we contributed indirectly to a better understanding of the traditional disciplines themselves* (Branchetti & Levrini, 2019)

A similar account of metaphor use emerged in Levy et al.'s work when they utilized terminology such as "slipping" and "sliding" to capture and conceptualize the features underpinning interdisciplinarity.

All three groups of researchers showed the relevance of interdisciplinarity for a range of stakeholders including science students (Schvartzer et al., 2019) and nursing students (Levy et al., 2019). A multiplicity of disciplines were represented including artificial intelligence (Branchetti & Levrini, 2019) as well as the boundaries between traditional disciplines such as physics-chemistry, chemistry-biology (Levy et al.) and physics-engineering (Schvartzer et al.). This set of studies utilized a range of methodological approaches including discourse analysis (Schvartzer et al., 2019), design-based research (Levy et al., 2019) and historical case studies (Branchetti & Levrini, 2019).

21.5 Discussion and Conclusion

This chapter illustrates the opportunities and challenges of boundary crossings in STEM education. The literature on interdisciplinarity in science education points to the curricular and instructional rationales as well as the relevance of interdisciplinarity for science education. The ESERA 2019 conference presentations provide a wealth of perspectives for characterizing and detailing how interdisciplinary boundary crossing can be situated in science education. Whereas Schvartzer et al. (2019) problematized learners' engagement when learning or using disciplinary knowledge in different interdisciplinary contexts and problems, Levy et al. (2019) drew attention to the explanatory potential of interdisciplinarity for disciplinary-based problems. Branchetti and Levrini's (2019) presentation problematized the inherent interdisciplinarity of STEM disciplines in discipline-based educational systems from a historical and curriculum development perspective. The discussion of the various boundary crossing in the three presentations was framed by Akkerman and Bakker's (2011) conceptualization of boundaries as dialogical phenomena which provides a distinct analytical lens for a discussion of the interdisciplinarity embedded in each project. The enactment of boundary crossing in these three projects provides concrete evidence on ways in which recent policy calls in STEM education can be materialized at the level of teaching and learning. As such, they highlight how higher order twenty-first century skills can be fostered meaningfully and constructively in education.

References

American Association for the Advancement of Science. (1990). *Science for all Americans: Project 2061*. Oxford University Press.

Akkerman, S. F., & Bakker, A. (2011). Boundary crossing and boundary objects. *Review of Educational Research, 81*(2), 132–169. https://doi.org/10.3102/0034654311404435

Berland, L., Steingut, R., & Ko, P. (2014). High school student perceptions of the utility of the engineering design process: Creating opportunities to engage in engineering practices and apply math and science content. *Journal of Science Education and Technology, 23*(6), 705–720. https://doi.org/10.1007/s10956-014-9498-4

Blikstein, P. (2013). Digital fabrication and "making" in education: The democratization of invention. In J. Walter-Herrmann & C. Büching (Eds.), *FabLabs: Of machines, makers and inventors*. Transcript Publishers.

Branchetti, L., Cattabriga, A., & Levrini, O. (2019). Interplay between mathematics and physics to catch the nature of a scientific breakthrough: The case of the blackbody. *Physical Review Physics Education Research, 15*(2), 020130. https://doi.org/10.1103/PhysRevPhysEducRes.15.020130

Branchetti, L. , & Levrini, O. (2019). Disciplines and interdisciplinarity in STEM education to foster scientific authenticity and develop epistemic skill. In S. Kapon (Chair) & S. Erduran (Discussant), Crossing boundaries – Examining and problematizing interdisciplinarity in science education, invited symposium held at the *European Science Education Research Association (ESERA) Biannual Conference*.

Calabrese Barton, A., Tan, E., & Greenberg, D. (2016). The makerspace movement: Sites of possibilities for equitable opportunities to engage underrepresented youth in STEM. *Teachers College Record, 119*(6), 11–44. Retrieved from http://invincibility.us/wp-content/uploads/2015/08/EquityMakerspaces.TCR_.pdf

Dagan, E., Dubovi, I., Levy, M., Zuckerman Levin, N., & Levy, S. T. (2019). Adherence to diabetes care: Knowledge of biochemical processes has a high impact on glycaemic control among adolescents with type 1 diabetes. *Journal of Advanced Nursing, 75*(11), 2701–2709. https://doi.org/10.1111/jan.14098

Dagher, Z., & Erduran, S. (2014). The role of disciplinary knowledge in science education: The case of Laws and Explanations in biology and chemistry. In M. Matthews (Ed.), *Handbook of research on history, Philosophy and Sociology of Science*. Springer.

Dougherty, D. (2012). The maker movement. *Innovations: Technology, Governance, Globalization, 7*(3), 11–14. https://doi.org/10.1162/INOV_a_00135

Dubovi, I., Dagan, E., Sader Mazbar, O., Nassar, L., & Levy, S. T. (2018). Nursing students learning the pharmacology of diabetes mellitus with complexity-based computerized models: A quasi-experimental study. *Nurse Education Today, 61*, 175–181. https://doi.org/10.1016/j.nedt.2017.11.022

Erduran, S., & Dagher, Z. (2014). *Reconceptualizing the nature of science for science education: Scientific knowledge, practices and other family categories*. Springer.

European Commission. (2015). Science education for responsible citizenship. Luxembourg. Retrieved from http://ec.europa.eu/research/swafs/pdf/pub_science_education/KI-NA-26-893-EN-N.pdf

Eurydice (2011). Science education in Europe. National Policies, practices and research. Education, Audiovisual and cultural executive agency. .

Golding, D. (2009). *Integrating the disciplines: Successful interdisciplinary subjects*. University of Melbourne.

Holland, D. C., Lachicotte, W., Skinner, D., & Cain, C. (1998). *Identity and agency in cultural worlds*. Harvard University Press.

Hurley, M. M. (2001). Reviewing integrated science and mathematics: The search for evidence and definitions from new perspectives. *School Science and Mathematics, 101*(5), 259–268.

Kapon, S. (2016). Doing research in school: Physics inquiry in the zone of proximal development. *Journal of Research in Science Teaching, 53*(8), 1172–1197. https://doi.org/10.1002/tea.21325

Kapon, S., Laherto, A., & Levrini, O. (2018). Disciplinary authenticity and personal relevance in school science. *Science Education, 102*(5), 1077–1106. https://doi.org/10.1002/sce.21458

Kapon, S., Schvartzer, M., & Peer, T. (2021). Forms of participation in an engineering maker-based inquiry in physics. *Journal of Research in Science Teaching,* 58(2), 249–281. https://doi.org/10.1002/tea.21654

Klein, J. T. (1990). *Interdisciplinarity: History, theory and practice*. Wayne State University Press.

Lehrer, R., & Schauble, L. (2012). Seeding evolutionary thinking by engaging children in modeling its foundations. *Science Education, 96*(4), 701–724. https://doi.org/10.1002/sce.20475

Levy, S. T., Zohar, A. R., & Dubovi, I. (2019). Slipping between disciplines: How forming causal explanations may compel crossing disciplinary boundaries. In S. Kapon (Chair) & S. Erduran (Discussant), Crossing boundaries – Examining and problematizing interdisciplinarity in science education, invited symposium held at the *European Science Education Research Association (ESERA) Biannual Conference*.

Martin, L., & Dixson, C. (2016). Making as a pathway to engineering. In K. Peppler, E. R. Halverson, & Y. B. Kafai (Eds.), *Makeology: Makers as learners* (pp. 183–195). Routledge.

Nagle, B. (2013). Preparing high school students for the interdisciplinary nature of modern biology. *Life Sciences Education, 12*, 144–147.

National Science and Technology Council. (2013). *A report from the committee on STEM education*. National Science and Technology Council.

NGSS Lead States. (2013). *Next generation science standards: For states, by states*. The National Academies Press.

Pang, J., & Good, R. (2000). A review of the integration of science and mathematics: Implications for further research. *School Science and Mathematics, 100*(2), 73–82.

Park, W., Wu, J., & Erduran, S. (2020). The Nature of STEM disciplines in the science education standards documents from the USA, Korea and Taiwan. *Science & Education*. https://doi.org/10.1007/s11191-020-00139-1

Redish, E. F., & Kuo, E. (2015). Language of physics, language of math: Disciplinary culture and dynamic epistemology. *Science & Education, 24*(5–6), 561–590.

Schvartzer, M., Peer, T., & Kapon, S. (2019). Learning physics through Maker projects – Between disciplinary authenticity and personal relevance. In S. Kapon (Chair) & S. Erduran (Discussant), Crossing boundaries – Examining and problematizing interdisciplinarity in science education, invited symposium held at the *European Science Education Research Association (ESERA) Biannual Conference*.

Schwab, K. (2017). The fourth industrial revolution. London: Portfolio Penguin.

Sengupta, P., Kinnebrew, J. S., Basu, S., Biswas, G., & Clark, D. (2013). Integrating computational thinking with K-12 science education using agent-based computation: A theoretical framework. *Education and Information Technologies, 18*(2), 351–380. https://doi.org/10.1007/s10639-012-9240-x

Struyf, A., De Loof, H., Boeve-de Pauw, J., & Van Petegem, P. (2019). Students' engagement in different STEM learning environments: Integrated STEM education as promising practice? *International Journal of Science Education, 41*(10), 1387–1407. https://doi.org/10.1080/09500693.2019.1607983

Woodcock, B. A. (2014). The scientific method as myth and ideal. *Science & Education, 23*, 2069–2093.

World Economic Forum. (2017). Realizing human potential in the fourth industrial revolution: An agenda for leaders to shape the future of education, gender and work. Geneva, Switzerland. Retrieved from http://www3.weforum.org/docs/WEF_EGW_Whitepaper.pdf

Zohar, A. R., & Levy, S. T. (2019). Students' reasoning about chemical bonding: The lacuna of repulsion. *Journal of Research in Science Teaching, 56*(7), 881–904. https://doi.org/10.1002/tea.21532

Part IV
Designing Research-Based Instruction

Chapter 22
Augmented Reality in Lower Secondary Science Teaching: Teachers and Students as Producers

Birgitte Lund Nielsen and Harald Brandt

22.1 Introduction

The research on the use of ICT in education shows that technology alone cannot be seen as a catalyst for change (Higgins et al., 2012). The question about the pedagogical approach and teaching and learning practices with technology should come before the question about effects from using a specific technology (Hennessy et al., 2007). In the field of science education, it is, in particular, discussed how students could learn from using digital artifacts in inquiry-based projects in real-life contexts (Krajcik & Mun, 2014). The importance of their high-level use of ICT in modeling, animating, and communicating about science phenomena are highlighted. Research and science curricula across national contexts refer to students' representational competence (Waldrip & Prain, 2012) and the importance of their meta-modeling knowledge (Schwarz et al., 2009; Oh & Oh, 2011). The digital artifacts must be included in the process of generating, testing, and revising explanatory models if students are expected to develop these competencies.

Furthermore, contemporary research emphasizes student-teacher and student-student exploratory dialogues when working with digital artifacts to help students make sense of science phenomena (Mercer et al., 2019). Hence, the teachers' role in scaffolding (Hammond & Gibbons, 2005) student dialogues and demonstrating strategies for handling the problems involved is crucial in inquiry-based approaches, i.e., model-based inquiry (Kind et al., 2011; Windschitl et al., 2008). The present paper discusses the scaffolding and mediation of students' inquiries and modeling activities in science with the use of Augmented Reality (AR).

B. L. Nielsen (✉) · H. Brandt
VIA University College, Aarhus, Denmark
e-mail: bln@via.dk; habr@via.dk

O. Levrini et al. (eds.), *Engaging with Contemporary Challenges through Science Education Research*, Contributions from Science Education Research 9,
https://doi.org/10.1007/978-3-030-74490-8_22

22.2 AR in Educational Settings

AR is defined by the combination of real and virtual objects in a real environment, running interactively, and in real-time (Radu, 2014). AR can enhance the user's sensory perception of the real world by adding virtual objects and a contextual layer of information (Ibáñez & Delgado-Kloos, 2018). Smartphones and tablets are equipped with sensors such as cameras, GPS, accelerometer, and gyroscope allowing the smartphone to form a virtual perception of the real world and use this to add a layer of augmented content. The AR content is often launched by letting the device read a visual marker placed in the real environment. Still, AR can also be markerless using location data, such as GPS, when overlaying information (Cheng & Tsai, 2013). Affordances by using AR in an educational setting are that students can see the content in a 3D perspective and get a sense of presence, immediacy, and immersion. AR can, therefore, help them visualize the invisible, bridging formal and informal learning. Challenges are mainly about usability difficulties and ineffective classroom integration (Radu, 2014; Wu et al., 2013). A systematic review of the literature on the use of AR in STEM learning highlight that similar design features are used across contexts with students mainly getting information through the interaction with AR. They emphasize the need for more than access to information for this to be a learning experience. Assistance in selecting and interpreting data from AR is crucial and recommended to be included in future initiatives (Ibáñez & Delgado-Kloos, 2018). So, though there is a growing knowledge base about affordances related to AR in education, the field is still in its infancy as emphasized by Radu (2014). Further research is asked for to expand the knowledge about how meaningful learning can be mediated with AR. This point was part of the rationale for initiating the European ARsci project running from 2012–2015 (https://ar-sci.csesga.es). This context, where AR-animated models were developed and tested in lower secondary science classrooms in three countries (Denmark, Norway, and Spain) provided an opportunity to examine affordances for supporting students' inquiries into micro-processes in science like photosynthesis and combustion and for students to model with AR themselves. Some findings from this longitudinal project have been published (Nielsen et al., 2016, 2018), but will be summed up to discuss how the findings can inform the pedagogical use of AR in a new and somewhat different context, namely students' work with processes related to the global Ocean, focusing on Ocean Literacy (Fauville, 2017). The rationale for pursuing this is that findings from the ARsci project (elaborated below) indicated a need to support students' model-based inquiry (Windschitl et al., 2008). The global Ocean provides an excellent context for focusing on models and macro-processes. This context can include, as it will be elaborated and exemplified in the literature study, what Tran et al. (2010) call a system-thinking approach supporting student capability to use models, and more specifically, create, manipulate, and revise models. Furthermore, Ocean Literacy accentuates socio-scientific perspectives and contemporary global environmental challenges related to the content students are inquiring into using AR. Ocean Literacy refers to the aim of students in particular and citizens, in general, building a civic relationship with the Ocean (Fauville, 2017, 2018; Tsai, 2018).

22.3 Research Questions

- What do students from lower secondary science emphasize as possibilities, challenges, and perceived outcomes from the testing activities in the ARsci project?
- How can the experiences from applying AR to support student learning about micro-processes in science in the ARsci project inform the continuing work with AR to promote students' Ocean Literacy, referring to macro-processes in science and including also socio-scientific perspectives on the science-content?

22.4 Methods

The ARsci project operated with an iterative approach to design, testing, re-design, and adaptation: design-based research (DBR) (Barab & Squire, 2004). The main focus was on teachers and students as producers. Still, in the first round, showcase material developed by the ARsci-team (using software like Blender, Unity, and Daqri) was tested. The project team aimed to develop the pedagogical framing for the later test-phases where first teachers and then students were producers of AR-animation models. The science content used in designing the first AR-animations (photosynthesis and combustion) was informed by analyzing science curricula across countries and by a survey at the initiation of the project (Nielsen et al., 2016). In the second round of testing, teachers were designing AR-animation models, and in the third round, students were producers using the same software as the teachers (Blippbuilder). The first two rounds of testing included two classes from each country (n = 73) and their teachers. The third round of testing included students and teachers from Denmark and Spain (n = 46). All participating students were from lower secondary, 7th, and 8th grade, aged 13–15 years old. Multiple types of data were collected, including student questionnaires, interviews with students and teachers, and classroom observations using observation schemes and video, both full class, and video following dialogues in groups of students. Data from questionnaires were analyzed by frequency analysis and cross-tabulations. Dialogues from the video were analyzed, looking into (1) elements of teacher scaffolding, (2) various kinds of discussion, e.g., exploratory talk (Mercer et al., 2019), and (3) indications of students' representational competence (Waldrip & Prain, 2012).

Following the DBR approach also across project contexts, the findings from the ARsci project are at present informing the development in an EU-ERASMUS+ project about Ocean Literacy (https://www.ocean-connections.net). The method to answer the second research question included a focused state-of-the-art literature study (Gough et al., 2017) to discuss the findings from the ARsci project with the particular perspectives on ICT, AR, and Ocean Literacy. The focused review included searches in the databases Eric and Teacher Reference Centre using the search strings, [(ocean literacy) AND (science education) AND (augmented Reality OR AR OR (virtual Reality) OR VR OR ICT] and [(ocean literacy) AND science AND (socioscientific)]. Research applying a wide range of quantitative and qualitative methods was included in the narrative synthesis (Gough et al., 2017; Popay et al., 2006). First, a summary of each study was made using a review template. In the next step of the analysis, themes related to pedagogical principles for teaching about Ocean Literacy with the ICT tools were identified. This approach to narrative synthesis was informed by the thematic analysis (Braun & Clarke, 2006). The full review will be publicly available: https://www.ocean-connections.net/project-results/.

22.5 Results

We start by addressing the first research question presenting findings from the ARsci project, before the conclusions of the literature study about Ocean Literacy. This order of sections mirrors the DBR approach, also followed across studies.

22.5.1 Findings from the ARsci Project

In the first round, the teachers and students from Norway, Spain, and Denmark were using two examples of pre-produced AR-materials “Lost in the woods” and “Catalytic converter” focusing on photosynthesis and combustion, e.g. adding an augmented layer to leaves from a real tree and exploring the chemical reactions in a car’s catalytic converter (Nielsen et al., 2018) (an example in Fig. 22.1, the full description can be found in the ARsci user guide).

Detailed analyses of the dialogues from students’ inquiries are presented in Nielsen et al. (2018), revealing the first phase with questions like: *What can this app do? What happens, if…?* When interviewed, the students emphasized the affordance of visualizing what is inside – the invisible – as immediately catching their interest. In later phases, the teachers were supporting students in using science concepts to communicate about their models and name the different molecules. The teachers were asking questions, e.g., to stimulate reflections about substance conservation in the processes in the catalytic converter. After their initial work in class with the models from ‘lost in the woods,’ the students moved outdoors, where they physically placed the AR-markers on different parts of a tree. Later they presented their results in class. Most of the students found this to be a motivating and engaging task. One student said: “*You see it more like it is for real – it is meaningful, and you get a sense of how it is happening.*” When asked to elaborate on the specific task of

Fig. 22.1 This example from the resource 'Lost in the woods' shows a student in the forest investigating an AR-animation with her smartphone. The second photo is a screenshot showing the digital layer on top of the real leaf, as seen through the smartphone. This case is part of a range of AR-animations illustrating models related to photosynthesis, water transport in roots and xylem, etc. The task for the students was to examine these animation models and connect them to a real, physical tree in the outside area

finding out what processes the models represented, another student said: "*I think it is fun [...] it would have been too boring and easy if she* [the teacher] *had explained about the models before we started. It is fun to examine it yourself*". But a few students experienced it to be a difficult task. One student used the term "*confusing*" about the task. These experiences from the first testing were used in adapting the materials, describing in the user-guide possibilities for using 'Lost in the woods' in differentiated ways with various degrees of teacher guiding.

In the second round, four AR-animation models were produced by the teachers using the software Blippbuilder. It was tried in their classes with students using the app Blippar. Teachers in all three countries produced AR models closely connected to the science content in the curricula, e.g., modeling a magnetic field, rocks and plate tectonics, and processes related to carbohydrates (more about these AR-animation models in the ARsci user guide). The students in all three countries reported a high level of perceived outcomes. For example, did around 80% of the students report that the inquiries with AR to a high or a very high degree helped them acquire new knowledge about the science content? A particular issue was that some of the students at this point were a little disappointed that the resources designed by the teachers were not of the same high technical quality as the ones from the first testing.

In the third round, students were working with the Blippbuilder software with teachers' scaffolding (Hammond & Gibbons, 2005) informed by the first findings, e.g., with a variation in the level of openness and teacher guiding. Danish 8th-grade students, for example, designed an augmented world map as a first task where they at the same time learned to use the software. Later they collaboratively developed AR-models showing various elements of the global water cycle. In the project with the global water cycle, students worked in groups to produce AR-animations modeling the various complex processes invisible to the naked eye. Those processes must be put together to understand how water is transported between the reservoirs. The aim was that students could explore, understand, and explain the processes and phenomena by creating their own AR-animation models. The 7th-grade students were involved in a more structured task of illustrating electric circuits in the home and connecting it to the ones from the science lab. The students reported about some challenges with the Blippbuilder software being slow, but they anyway across tasks reported a high level of perceived learning outcomes: "*It is worth the effort*." For example, 90% reported the activities to be very or mostly useful for the purpose. Compared to test 2, they referred to the possibility to be creative designing themselves. Many groups included a reference to environmental issues related to the global water cycle, e.g., chemical pollution. This particular focus was new compared to the micro-processes in science in most of the materials from the ARsci project until then. Another interesting issue was that the students emphasized a new respect for the teachers' design in test 2, after being AR-producers themselves, realizing that someone designs all the ICT they use in their everyday life. When interviewed about perceived outcomes, students used terms like "*meaningful*" and "*get a sense of what happens*," and they referred like in test 1 very positively to the possibility of seeing the invisible. The students furthermore valued the degrees of freedom, the possibility to make their own decisions, and creativity. Still, they also highlighted learning outcomes from the first more structured tasks, e.g., the task of working collaboratively by augmenting the world map.

Summing up, the students across the phases of testing referred positively to learning outcomes, and there was based on the multiple data evidence that many students over the testing period developed a level of representational competencies (Waldrip & Prain, 2012). They, for example, began to emphasize signifiers in the AR-animation models. Neither students nor teachers did, however, refer to AR-animations as a model, and there were some student utterances like: "*I did not think it looked like this*," indicating a naïve understanding of the nature of models (Nielsen et al., 2018). So this (pedagogical) approach did not seem to contribute sufficiently with what can be called meta-modeling knowledge: knowledge about the nature of models and the purpose of using models (Schwarz et al., 2009). Hence, throughout the project, the research team developed a renewed interest in how students are engaged more deeply with the content and epistemic characteristics of scientific knowledge. The students can realize that the ideas represented in the models are testable, revisable, explanatory, conjectural, and generative (Windschitl et al., 2008).

Together with inspiration from the students' choices to work with environmental issues in the third phase, this was the rationale for exploring further the affordances of AR to explore the macro-processes that can be exemplified in the Ocean globally.

22.5.2 State-of-the-Art Related to Ocean Literacy and Science Teaching with ICT/AR

While various concepts related to environmental literacy can be traced back to the sixties, Ocean Literacy emerged as a specific term at the start of the twenty-first century (Cava, 2002). Multiple goals and benefits of using the Ocean as a context are referred to, e.g., to help to teach complex topics in a way that captures students' imagination and to provide a portal for introduction of cutting-edge science and technology into the classroom (Cava, 2002). The focused searches in the literature revealed many examples of classroom trials from a U.S. context published in peer-reviewed practitioner journals (Plankis & Marrero, 2010). Nearly all studies refer to 7 key principles of Ocean Literacy described by UNESCO (Santoro et al., 2017). Many of these studies include inquiry-based activities for the students with, e.g., the 5E model (Eidietis & Rutherford, 2009; Gillan & Raja, 2016). There are several examples where students are generating data and instances where they are cooperating with and/or working with real-time data from scientists (Adams & Matsumoto, 2009). Finally, there are examples where students work on creating their own models (Weersing et al., 2010).

The searches also revealed publications from high ranking, international journals, and dissertations. In Fauville (2018), the teaching context was students' work with ocean acidification, using a virtual lab, and involving online discussions with a marine scientist. They conclude that students' interaction and dialogue with scientists allow them to explore and reason about a wider range of ideas beyond the range offered by the school setting. The affordances of the virtual lab of making the invisible visible are also emphasized (Fauville, 2018). About socio-scientific arguing that findings are illustrating how the 7th-grade students can apply ocean concepts about physical and biological processes to personal and societal decision making related to, e.g., pollution and food choice (Marrero & Mensah, 2010). So, though it is a new field and most of the literature is practitioner reports, there are more solid studies documenting how school science can be combined with citizen science, including also a literature review about research on learning and teaching ocean sciences (Tran et al., 2010). In the review, a system-thinking approach to critical concepts and processes, such as the water and carbon cycles, is emphasized as being a key to support students' Ocean Literacy. System-thinking means the cognitive ability to see and consider the whole system, the parts (sub-systems), the mutual interrelationships between them (the dynamics and change intra-impact), and the overall mode of operation.

An example is an interplay between understanding how the density of seawater is affected by the change in temperature and salinity in one location (micro-system) and the thermohaline circulation driving the global ocean currents (macro-system). It also involves understanding how ocean currents transport heat influencing regional climate patterns and how it can cause upwelling of nutrition at a location on the opposite side of the globe boosting biodiversity and production (micro-system). Tran et al. (2010) define students' system-thinking skills as their capability to use models, and more specifically, create, manipulate, and revise models using ICT and virtual environments. They emphasize that system-thinking has great explanatory and predictive power, but understanding global processes from a system perspective requires types of thinking that are challenging for students. Strategies that can support system-thinking include ensuring that teachers have advanced pedagogical knowledge to scaffold student thinking and for designing activities to give students control to create and manipulate models (virtual and physical). There is a need to provide opportunities for students to be involved in dialogue with peers to articulate and share their thinking.

Furthermore, Tran et al. (2010) suggest to include external learning environments like aquariums and science centers, as such sites can provide access to objects, organisms, and phenomena that create personal connections for learners. With AR, in particular, a study by Hsiao, Chang, Lin, and Wang (2016) showed that key characteristics of the manipulative AR, such as the simultaneity of virtual and real objects, high interactivity, and hands-on experience lead to a greater positive impact on the students' academic achievement and motivation. A study by Chen and Wang (2018) using a game-type AR to construct a mixed-reality environment facilitating conceptual learning among 5th and 6th graders report about a relationship between presence and learning achievement. Low presence in an AR-mediated learning environment is correlated with low learning achievement. The study suggests that enhancing interactive experience could increase learner presence in AR-mediated environments.

22.6 Discussion & Conclusions

The findings from the ARsci project highlighted the importance of students as producers. They exemplified the role of teacher questioning in scaffolding students' exploratory dialogues and learning activities with AR – a focus asked for by Ibáñez and Delgado-Kloos (2018). The students emphasized across the phases of testing, in particular the possibility for seeing the invisible and the experience of presence in phenomena, issues highlighted by Chen and Wang (2018), and Wu et al. (2013) as a determent for this to be a genuine learning experience. The possibilities for students to work collaboratively modeling the complex processes and phenomena in science were exploited in the third round. The students themselves were producing and modeling with AR. Students' work in this phase also revealed their interest in environmental issues related to macro-processes connected to the global water cycle. The new project Ocean Connections is exploring the affordances of using AR in modeling the large-scale partly invisible processes related to the Ocean. The positive experiences with students as producers and with teacher scaffolding of dialogues to support their representational competence (Waldrip & Prain, 2012) can also be used when working with the global Ocean. But the findings from the ARsci project also revealed some challenges about helping teachers and students to discuss models more profoundly as representations that are testable, revisable, explanatory, conjectural, and generative (Windschitl et al., 2008). The state-of-the-art review provided insights from both practitioner reports and more solid research about the development of Ocean Literacy through technology-supported, inquiry-based, and data-driven science – findings that can further inform how to approach this pedagogical challenge. The fact that the Earth has one big Ocean with many features, the first of seven principles of Ocean Literacy (Santoro et al., 2017), can provide an excellent context for challenging students to consider both micro and macro-processes and the interplay between them (Tran et al., 2010).

A systems approach to critical concepts and processes related to the global water cycle should include students' design of their own models with layered information. Building on the positive findings from the ARsci project, these models should also be used when students test their own ideas, but adding a more explicit focus on the whole system and the relation to the sub-systems. The affordances of AR in this context are in particular about illustrating the invisible phenomena and processes and linking students mediated and situated inquiries to live data streams, e.g. AR can be data-driven by add-on sensors as frequently explored, but also from open databases (Nielsen et al., 2016), and the literature provides examples where students are using real-time data and communication with scientists when working with ocean processes (Adams & Matsumoto, 2009). Tran et al. (2010) furthermore point to the need for support for teachers as their advanced pedagogical knowledge is a critical aspect in supporting students' system-thinking, and they also suggest the cooperation with external partners like aquaria. They emphasize aquaria as *a context* for working with the ocean processes. But it might be that working closely together across groups of stakeholders like educational researchers, educators at aquaria, natural scientists, and teachers can contribute to both supporting teachers and scaffolding students' modeling practices? System-thinking is known to be challenging for many students. Still, meeting stakeholders like scientists in the situated practice at aquaria might support the combination of perspectives and meta-thinking about AR-animation models. The research from the rather new field of Ocean Literacy suggests the potential of cross-sectorial cooperation. Hence, the findings from the ARsci project, where students highlight both the experiences from producing AR-animation models themselves, and the teacher-guided work to use the models to 'see' and feel a presence in the invisible science processes, can inform the ongoing work in the Ocean Connections project. This project can particularly add to knowledge about using technology like AR in mediating students' system-thinking and meta-modeling competencies.

References

Adams, L. G., & Matsumoto, G. (2009). Enhancing ocean literacy using real-time data. *Oceanography, 22*(2), 8–9.

ARsci user guide.:https://ar-sci.cesga.es

Barab, S., & Squire, K. (2004). Design-based research: Putting a stake in the ground. *The Journal of the Learning Sciences, 13*(1), 1–14.

Braun, V., & Clarke, V. (2006). Using thematic analysis in psychology. *Qualitative Research in Psychology, 3*(2), 77–101.

Cava, F. (2002). Promoting ocean literacy in the classroom. *California Journal of Science Education, 3*(1), 49–63.

Chen, Y. H., & Wang, C. H. (2018). Learner presence, perception, and learning achievements in augmented-reality-mediated learning environments. *Interactive Learning Environments, 26*(5), 695–708.

Cheng, K. H., & Tsai, C. C. (2013). Affordances of augmented reality in science learning: Suggestions for future research. *Journal of Science Education and Technology, 22*(4), 449–462.

Eidietis, L., & Rutherford, S. (2009). Sailing toward understanding surface currents. *Science Activities, 46*(3), 5–14.

Fauville, G. (2017). Questions as indicators of ocean literacy: Students' online asynchronous discussion with a marine scientist. *International Journal of Science Education, 39*(16), 2151–2170.

Fauville, G. (2018). *Digital technologies as support for learning about the marine environment: Steps toward ocean literacy*. Göteborg University.

Gillan, A., & Raja, S. (2016). Aquaponics: What a Way to Grow! *Science and Children, 53*(7), 48–56.

Gough, D., Sandy, O., & Thomas, J. (2017). *An introduction to systematic reviews* (2nd ed.). Sage.

Hammond, J., & Gibbons, P. (2005). Putting scaffolding to work: The contribution of scaffolding in articulating ESL education. *Prospects, 20*(1), 6–30.

Hennessy, S., Wishart, J., Whitelock, D., Deaney, R., Brawn, R., la Velle, L., et al. (2007). Pedagogical approaches for technology-integrated science teaching. *Computers & Education, 48*, 137–152.

Higgins, S., Xiao, Z., & Katsipataki, M. (2012). *The impact of digital technology on learning: A summary for the education endowment foundation*. Education Endowment Foundation.

Hsiao, H. S., Chang, C. S., Lin, C. Y., & Wang, Y. Z. (2016). Weather observers: A manipulative augmented reality system for weather simulations at home, in the classroom, and at a museum. *Interactive Learning Environments, 24*(1), 205–223.

Ibáñez, M. B., & Delgado-Kloos, C. (2018). Augmented reality for STEM learning: A systematic review. *Computers & Education, 123*, 109–123.

Kind, P. M., Kind, V., Hofstein, A., & Wilson, J. (2011). Peer-argumentation in the school science laboratory – Exploring effects of task features. *International Journal of Science Education, 33*(18), 2527–2558.

Krajcik, J. S., & Mun, K. (2014). Promises and challenges of using learning technology to promote student learning of science. In N. G. Lederman & S. K. Abell (Eds.), *Handbook of research in science education* (Vol. II, pp. 337–360). Routledge.

Marrero, M. E., & Mensah, F. M. M. (2010). Socioscientific decision making and the ocean: A case study of 7th grade life science students. *Electronic Journal of Science Education, 14*(1), 1–27.

Mercer, N., Hennessy, S., & Warwick, P. (2019). Dialogue, thinking together and digital technology in the classroom: Some educational implications of a continuing line of inquiry. *International Journal of Educational Research, 97*, 187–199.

Nielsen, B. L., Brandt, H., & Swensen, H. (2016). Augmented reality in science education – Affordances for student learning. *NorDiNa, 12*(2), 157–117.

Nielsen, B. L., Brandt, H., & Swensen, H. (2018). Students developing representational competence as producers with and of augmented reality in science. *NorDiNa, 14*(2), 138–153.

Oh, P. S., & Oh, S. J. (2011). What teachers of science need to know about models: An overview. *International Journal of Science Education, 33*(8), 1109–1130.

Plankis, B. J., & Marrero, M. E. (2010). Recent ocean literacy research in United States public schools: Results and implications. *International Electronic Journal of Environmental Education, 1*(1), 21–51.

Popay, J., Roberts, H., Sowden, A., Petticrew, M., Arai, L., Rodgers, M., et al. (2006). Guidance on the conduct of narrative synthesis in systematic reviews. *A Product from the ESRC Methods Programme Version, 1*, b92.

Radu, I. (2014). Augmented reality in education: A meta-review and cross-media analysis. *Personal and Ubiquitous Computing, 18*(6), 1533–1543.

Santoro, F., Selvaggia, S., Scowcroft, G., Fauville, G., & Tuddenham, P. (2017). *Ocean literacy for all: a toolkit* (Vol. 80). UNESCO Publishing.

Schwarz, C. V., Reiser, B. J., Davis, E. A., Kenyon, L., Achér, A., Fortus, D., et al. (2009). Developing a learning progression for scientific modeling: Making scientific modeling accessible and meaningful for learners. *Journal of Research in Science Teaching, 46*(6), 632–654.

Tran, L. U., Payne, D. L., & Whitley, L. (2010). Research on learning and teaching ocean and aquatic sciences. *NMEA Special Report, 3*(1), 22–26.

Tsai, C. Y. (2018). The effect of online argumentation of socioscientific issues on students' scientific competencies and sustainability attitudes. *Computers & Education, 116*, 14–27.

Waldrip, B., & Prain, V. (2012). Learning from and through representations in science. In B. J. Fraser, K. G. Tobin, & C. J. McRobbie (Eds.), *Second international handbook of science education* (pp. 145–155). Springer.

Weersing, K., Padilla-Gamino, J., & Bruno, B. (2010). What microbe are you! *The Science Teacher, 77*(6), 40–44.

Windschitl, M., Thompson, J., & Braaten, M. (2008). Beyond the scientific method: Model based inquiry as a new paradigm of preference for school science investigations. *Science Education, 92*(5), 941–967.

Wu, H., Lee, S. W., Chang, H., & Liang, J. (2013). Current status, opportunities and challenges of augmented reality in education. *Computers & Education, 62*, 41–19.

Chapter 23
Visualisation and Spatial Thinking in Primary Students' Understandings of Astronomy

Russell Tytler, Peta White, and Joanne Mulligan

23.1 Introduction

Astronomy is taught across school grade levels, from a focus in the primary school on night and day and planetary features and exploration, to more complex considerations of moon phases, celestial movement, and cosmology in secondary schools. Yet, research has consistently shown a variety of misconceptions of a diverse range of phenomena such as earth-sun relations (Vosniadou & Brewer, 1992), seasons, phases of the moon, and lunar eclipses etc., despite repeated exposure to teaching (Danaia & McKinnon 2008; Lelliot & Rollnick, 2010). A major problem at each level is the coordination of earth- and space-bound perspectives on these phenomena. This is essentially a problem of visualisation and spatial reasoning involving coordination of different earth-based and space-based representational systems (Hegarty & Waller, 2004; Padalker & Ramadas, 2008; Plummer, 2014).

Visualisation is increasingly recognised as central to conceptual learning across a range of sciences (Gilbert, 2005), from learning to coordinate visual-spatial sub-micro particle arrangements to explain macro material properties, to the abstracted visualisations of forces and fields governing the movements of objects in space in physics. Underpinning all of these are systems of representation through which we visualise, and coordinate visual and spatial relations, such as diagrams, models and simulations of earth movements, and apparent movement of celestial objects. In theorising the role of these representations in creating meaning, we draw

R. Tytler (✉) · P. White
School of Education, Deakin University, Melbourne, Australia
e-mail: russell.tytler@deakin.edu.au; peta.white@deakin.edu.au

J. Mulligan
School of Education, Macquarie University, Sydney, Australia
e-mail: joanne.mulligan@mq.edu.au

O. Levrini et al. (eds.), *Engaging with Contemporary Challenges through Science Education Research*, Contributions from Science Education Research 9,
https://doi.org/10.1007/978-3-030-74490-8_23

particularly on pragmatist semiotic perspectives of Peirce (1931/58). We argue that spatial reasoning and visualisation through these representations is particularly important for learning challenging astronomy concepts (Hubber & Tytler, 2017; Lelliot & Rollnick, 2010; Plummer, 2014; Testa et al., 2014; Tytler et al., 2017).

We are currently engaged in a three-year longitudinal project exploring interdisciplinary approaches to science and mathematics teaching and learning in primary schools (Interdisciplinary Mathematics and Science IMS: https://imslearning.org/), across Grades 1 through 6, across a range of science concepts. The principle underpinning the project is that robust learning involves the invention, evaluation, refinement, and coordination of representational systems in science and mathematics and that through a serious focus on how these science and mathematics systems interrelate more robust learning of foundational constructs will occur. Our past research indicates that deeper learning in both science and mathematics is achievable when students create and assess representations, make justified claims, and have first-hand experiences of the imaginative challenges and pleasures of these disciplinary processes (Hubber & Tytler, 2017; Lehrer & Schauble, 2012). Creative and critical reasoning can be developed when students are embedded in contexts of guided inquiry, where they see their overarching learning purpose is to generate, revise, and critique knowledge through making and critiquing representations (Lehrer & Schauble, 2012; Prain & Tytler, 2012; Tytler et al., 2013). In visualising, creating and analysing representations, data, and models, students learn about the varied purposes and specific affordances of different disciplinary representations for imagining, reasoning, and claim-making.

In this chapter we describe the planning and implementation of a primary school astronomy unit that focuses particularly on students constructing and coordinating representations to link the movement of shadows over the day, the sun's position in the sky, night and day, and the rotation of the earth. We link the spatial reasoning processes involved in this work with fundamental ideas of the mathematics curriculum, such as length, rotation and angle, time, location, direction, perspective taking and pattern recognition and representation.

In the learning sequence, students first tracked their shadows on the playground, linking this to the apparent movement of the sun in sky – rotation and angle. The changing length of shadows was tracked by a variety of means, including streamers, or counting the number of blocks as units of length. A record of the shadow movement of a garden gnome was used to model how shadow length and directions changed over the day, and this record used in the classroom to model the change in the sun's position using a torch. Students were challenged to represent the changing length and direction of their shadows, using the whole class results, and agreement was reached on the general patterns of shadow length and movement. In the fourth lesson a presentation of the earth's rotation using video animation was shown and students played with globes and torches, under teacher guidance, to consider the causes of day and night and link this to their shadow results.

The research questions were:

1. In what ways can students' construction of representations support productive learning in astronomy?
2. How can learning of astronomy concepts be productively linked with mathematical concepts and processes?
3. What teacher actions are effective in supporting students' learning of astronomy through constructing representations?

23.2 Methodology

Our research team collaborated with six teachers of Grade 1 students (150 of them, aged 6 years) across two schools to plan and implement a sequence of five lessons focused on students' understanding of the sun-earth relations with regard to shadows, daily solar movement, and night and day. Teachers had previously, with researcher support, used the approach in two prior topics. Teachers implemented the sequence flexibly and collaboratively according to students' grade level and needs. The pedagogy focused on students' guided observations and construction of diagrammatic and textual representations, with class discussion helping students evaluate and refine representational work.

Data consisted of students' representations over the sequence, field notes and video capture of teaching and learning interactions including students' collaborative discussion in selected classrooms, the pre/post test, and teacher and student interviews. The pre/post test included questions requiring drawn representations on the nature and cause of shadows, the movement of the sun in the sky, and the relation of the sun and earth to cause night and day. Students' interview questions focused on their changing understandings, their interpretation of the representations, and their response to aspects of the learning sequence. Data is mainly sourced from three classes, with an average of 25 students in each.

23.3 Findings

The findings relate, first, to the nature of the sequence and of the pedagogy, and students' response to the pedagogical moves made by the teacher. Second, we will describe the outcomes in terms of students' science and mathematical conceptual understandings, drawing on their responses in discussion, in their drawn representations and explanations, and on the pre/post tests.

23.3.1 The Sequence of Teaching and Learning Moves

In the first lesson the teacher, Colin (pseudonym), began with asking children what would happen to their shadow if they stood still, over the day. Students' suggestions were recorded, as a probe of prior ideas, and the task for the next lesson was described. The students were encouraged to invent their own methods for this task, focusing on: (1) what we might expect to notice; (2) what data to collect (length, direction, position of sun); and (3) how to measure and record the data.

In the following lesson students worked in pairs in an outside North-facing area. Compass points were established. One student in each pair stood in the same position a number of times during the day, with the other student tracing around the shadow. In between each measurement the class discussed how the shadow was changing, what might happen next, and what the data might show. Teachers used a variety of devices to relate the shadow to the position of the sun. Colin encouraged students to draw an image of the sun showing its direction (colour-coded) for each shadow tracing, in order to draw out the oppositional relation between the sun and shadow directions. When working with groups he encouraged them also to point to the sun and notice its position in relation to the shadow (Fig. 23.1).

While this data recording was going on, Colin gathered students around a garden gnome placed on an A3 sheet of paper, to model the tracing of shadows, and the positioning of the sun, as a basis for discussion of the spatial relations. Figure 23.2 shows Colin pointing out these relationships.

Colin's questioning during these sessions focused on establishing the sun's movement and relation to direction and length of the shadows. Students described

Fig. 23.1 Students were encouraged to point to the sun and notice the relation to the shadow direction. The girl is positioned in the sun's direction

Fig. 23.2 Grade 1 students discussing the relationship between sun position and shadow direction, shadow length and sun's altitude

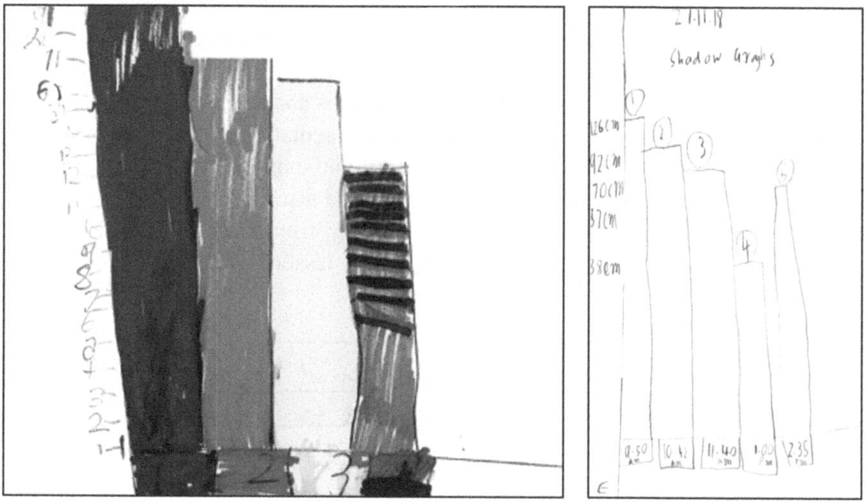

Fig. 23.3 Graphical representations of changes in shadow length, with differing attention to scale and time

how 'the shape of the shadow changed'. 'It got smaller and turned to the side'. Much of the discussion of angle and direction was modeled through gesture.

Having established the data on the courtyard, students were challenged to represent the data as a graph in their own way ("How are you going to show how your shadow has changed?", "How are you going to record how the time has changed?"). Figure 23.3 shows student-constructed graphs from Colin's and Ellie's classes where streamers and blocks were used. The graph in Fig. 23.3 (left-hand side) shows the decreasing height of the shadows at four intervals, labelled simply as 1 through 4, across the morning period from 9:30 am to midday. The scale is represented with equal units showing the number of blocks that were used to measure the height as an informal unit of measure. In contrast, the graph in Fig. 23.3 (right-hand side) is a more sophisticated representation with the vertical scale showing only the formal measures, using measuring tapes, of each height (cm) across five equal-sized intervals from 9:50 am to 2:35 pm. Students observed and measured the decrease in the height of the shadow leading up to midday and then the increase in shadow length in the mid afternoon. This enabled students to interpret the pattern of change in the shadows. In each class, students' graphs were compared and critiqued for features of data representation, in order to move students towards more powerful graphical representations.

Following the representation of shadow length, students were challenged to relate the direction to the position of the sun. Figure 23.4 shows one student's entry explaining the relation between the sun's movement and shadow size and directions.

In the subsequent lesson the gnome and shadow tracings were used to explore with a torch the movement of the sun over the day. The representation and modeling had the effect of stabilising the data in a way that allowed control over the spatial and temporal conditions of shadow formation. Figure 23.5 shows how, in Ellie's class, students were encouraged to role play the direction of the sun in relation to the shadow, established by the torch-gnome-shadow pattern modeling. Students discussed the similarities between their shadow patterns and that of the gnome.

In the next lesson, teachers showed a video representation of the earth rotating in space to demonstrate night and day (Fig. 23.6), and explored this further using a model globe and torch. This 3D modeling was used also to establish that from the earth's perspective the sun moved across the sky from East to West. Teachers included a variety of role plays in this night and day lesson, for instance of observers

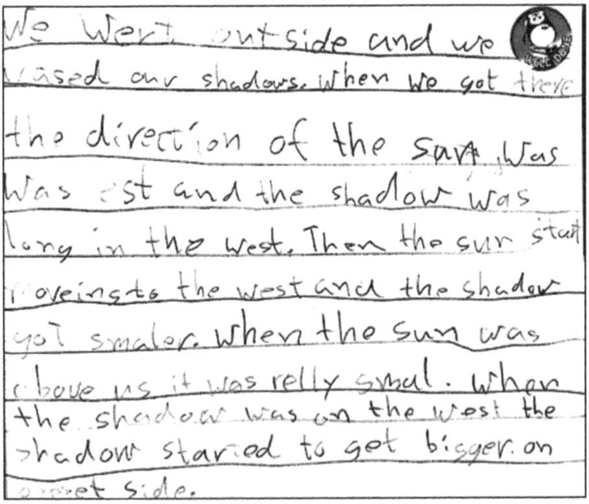
We Went outside and we
trased our shadows. When we got there
the direction of the sun Was
Was est and the shadow was
long in the west. Then the sun start
moveing to the west and the shadow
got smaler. When the sun was
above us it was relly smal. When
the shadow was on the west the
shadow started to get bigger on
oppset side.

Fig. 23.4 One student's entry describing sun direction in relation to the shadow

Fig. 23.5 Ellie moves the torch to duplicate the sun's movement to create the gnome shadow pattern, then has a student point to the sun and shadow tip to establish they are in the opposite direction

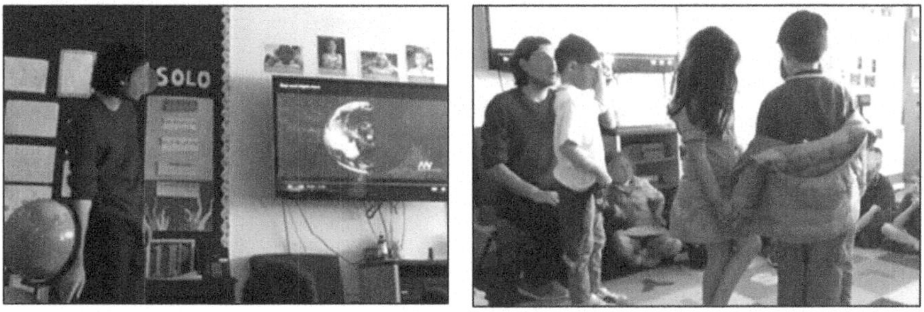

Fig. 23.6 Colin showing a day and night video with the globe visible, and a subsequent role play of students observing a stationary earth from a rotating platform

on a rotating earth perceiving the movement of a stationary light source. Fig. 23.6 shows students in Colin's class in such a role play.

23.3.2 Students' Learning Outcomes

The pre-test probed students' knowledge of the movement of the sun across the sky and the relation of this to shadow movement, and their perspectives on earth-sun relations and the relevance of this to day and night. The questions were mainly open-ended, and the post-test consisted of students being given their pre-test responses and asked to make changes to their original thinking and representations.

In the one multiple choice question, relating to the sun's movement, the results were compelling. Table 23.1 shows the shift from pre- to post-test in students' realisation of the sun moving from East to West, represented in diagrams, for two case classes.

Table 23.1 Pre- and post-test data showing students' perceptions of the sun's daily movement

	Incorrect (sun moves up and down, or from west to east)	Correct	Unclear
Pre test	36	7 (14%)	7
Post test	5	40 (80%)	5

Fig. 23.7 A student's changing view of the earth-sun relations, and shadow movement, pre- and post-test

In the pre-test there was little evidence of students articulating or understanding the movement of shadows. In the post-test, there was variation in the precision of students' representations, but many were able to represent movement during the day, with a subset of these showing an acceptable representation in relation to the sun. The questions were: 'Is a shadow different at different times during the day? How? Why?' and 'How do your shadows change during the day?' One student explained: "You can see better in the day as shadows and the sun can make it darker". This changed in the post-test to be explicit about "the earth spins and if it spins the sun rises, and the sun gets higher … and the shadow gets smaller". Figure 23.7 shows an example of a student's changing representations from "the sun moves" (right-hand side pre-test) to "they are longer and shorter because the earth moves" (left-hand side post-test).

Students' depictions of sun-earth relations improved from the pre- to the post-test. Most were clear that the earth spins to cause night and day. From the focus group interviews, it was evident that only a few students were not confident that night and day is caused by the earth spinning. Students often spontaneously asked to use the globe to illustrate, and some groups orchestrated role plays to convince the interviewer of their perspective. One student not only explained the changes between night and day, but shadow movement, using the globe:

> If America is daytime and the earth is spinning around, then people's shadow in America would be bigger. Rotate the earth please. Then America's night time now, so Australia would be day.

23.4 Discussion

This sequence for Grade 1 students followed the principles of the Representation-Construction -Approach (Tytler et al., 2013; Prain & Tytler, 2012) in which students are strategically challenged to construct representations in response to an established need, and the teacher guides a process of critique and refinement, moving towards scientifically valid representations. In the IMS project this approach, under the influence of Lehrer (Lehrer et al., 2006; Lehrer, 2009; Lehrer & Schauble, 2012), has been further codified to entail four stages: (1) Orienting/material engagement; (2) Posing representational challenges; (3) Comparative review/building consensus; and (4) Application/extension to new settings. This can be seen in the way, for instance, students' ideas were probed and stimulated regarding shadows, and their ideas about how one might observe, measure and chart movement over a day. After enacting this measurement process, students discussed and then represented the changes in their graphs and drawings. Their graphs were refined through direct comparison, emphasising clarity and communication of the pattern of change. These agreed patterns then formed the basis of the next phase of the sequence, focusing on space-centred views to explain shadow movement and night and day.

It was observed during the sequence, and in teacher interviews, that students were highly engaged with the tasks and with the discussion. Evidence from the post-tests, students' representational work, and the final interviews, demonstrated that there had been considerable learning with respect to the pattern of movement of the sun in the sky, and of the cause of night and day being the spinning of the earth. Most students were also able to offer a coherent account of time zones depending on longitude, using the model globe. Students were confident also about the broad features of shadow movement over the day. The greatest conceptual difficulties lay in the spatial relations between the sun's changing position and shadow length and angle.

23.4.1 Spatial Reasoning Challenges and the Role of Mathematics

Students' understanding of spatial relations between the sun's movement and shadows required strong framing of embodied reasoning, evident from field notes and teacher interviews. Teachers needed to guide/scaffold students' observations of the sun in the sky, and the movement of the shadow, by guiding students' directions. They asked students to describe what happened to the shadow since the last measure, and why, and asked for predictions of what will happen next. Colin invented a device for linking these directions by representing the direction of the sun by drawing a sun image in that direction. In that way it could become clear that the shadow tracing was always opposite to the sun's position.

From field observations, these 6-year-old students struggled with specifying the sun's position with respect to compass points, or the shadow. Having students point to the sun, and relate this to the shadow direction as an 180° relation, was a device used by Colin, by Ellie when investigating the gnome, and by other teachers. Students seemed to need these gestural prompts to position themselves and these objects in space. Part of the issue seemed to be that the sun, on a clear day, has no markers of position when high in the sky, and students needed help in imposing spatial reference points that could accommodate descriptions of its movement.

The sequence then became an opportunity to work with and establish mathematical ideas of spatial relations; angle and rotation and relative position. Using the arm to specify direction, and to treat the changing angle of the shadow as a rotation, provided opportunities to establish these ideas in the context of problem-solving shadow movement, through a guided-inquiry context. The data generation and refinement provided scope for envisaging shadow *length* related to *angle* of the sun in the sky, the *rotation* of the shadow in relation to earth rotation, and for envisaging the shadow position over time as a *continuous function* of the sun's tracking.

A key principle of the interdisciplinary IMS project is that mathematics and science should be positioned such that each is enriched by the other, and that this occurs by focusing on concepts common to both disciplines. This sequence provides a good example of this mutually reinforcing relation. The science provided a context for meaningful exploration of spatial concepts, of measurement, and of data modeling/graphical concepts. This IMS sequence was in fact the third for this cohort of students, with previous motion and ecology sequences involving graph invention, critique and revision.

The mathematics, on the other hand, fed back into the science understandings. The graphical competences developed by these students were quite advanced compared to curriculum and teacher expectations, and were further developed through the astronomy sequence through this same process. Figure 23.3 shows a range of competence, but all nevertheless represent well-developed graphical skills for the age group. Students were able to visualise and sketch their representation of data in an ordered way with coordination of vertical and horizontal axes, the development of scale informally or with formal units, and the structuring of time intervals.

Representations included use of models such as blocks and the use of streamers to reflect a column or bar chart.

Focusing on measure, and data display from the tracking of the shadows, provided a representation of the shadow-sun relations that raised questions and enabled interrogation of the data against ideas developed by the students. In response to a question about whether the students were able to make a connection between time of day and shadow position and length, Colin responded:

> Some did and some did not at the start and then when we went back and we modeled it again in class with the globe and we looked at the shadows and the sun with a torch, when we did that activity we then found that the children were going, "Oh, the shadow's getting a bit longer here," so, then we went back to our data, we had a look at the length of the shadows, what time of the day was it, we went back and had a look and then we said, "Can we see a pattern?" and they were able to then identify at the end that the morning and afternoon the shadows were longer and in the middle of the day it was shorter, "But why?" because the sun was higher in the sky.

23.4.2 Creating and Working with Representations

Colin's quote articulates an important feature of these graphical/diagrammatic representations – that they effectively freeze time by representing it spatially, in a way that allows interrogation based on students' ideas about changes over time. Patterns can be discerned in both the sun and shadow that can be retrospectively related. Similarly, the creation of the gnome shadow representation allowed for a readily accessible artefact that could form the basis of class discussion – How did the shadow move? When was it longest? What was happening at 3 pm? What would have happened after that? Further, the torch, made to track a duplication of the sun's movement, rendered the phenomenon inspectable, solving the temporal problem of slow changes in position, and allowing a back and forth discussion of the effects of both height and angle of the torch/ sun. Further, the video representation of the earth, the globe and torch model, and the student role play of an earth perspective, offer models that are flexible in their representation of time (can be slowed down and stopped) and of space (particular locations can be identified and discussed).

In interpreting the role of representation in learning we draw on Peirce (1931/58) who developed a semiotic system based on the relations between a sign or representation, the referent being signified (movement of the sun, and shadows), and the meaning made of this (See Fig. 23.8). We have argued (Tytler et al., 2020) that any representation or model necessarily involves a strategic selection of key elements of the referent phenomenon. In itself, it imposes a view on what is important to notice. Therefore, in the students' tracings, length of the shadow and position relative to the sun's direction were key aspects to draw attention to. Each representation has particular affordances (Prain & Tytler, 2012) that productively constrain attention on features of import. Further, any such representation needs to satisfy the requirements of correspondence (does it appropriately represent the phenomenon) and

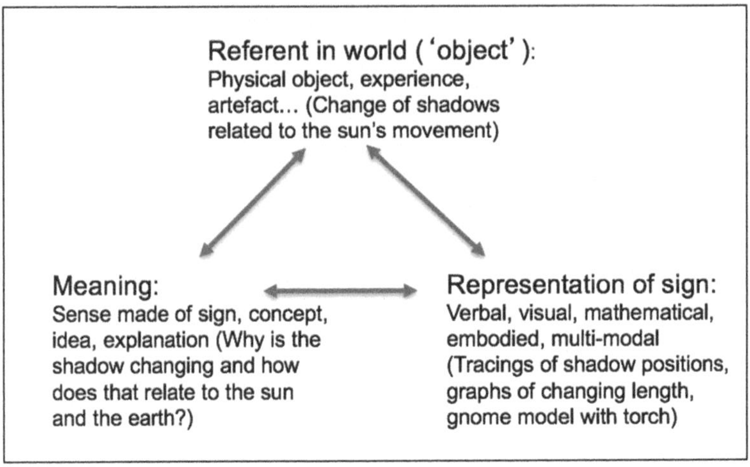

Fig. 23.8 Peirce's triadic model of meaning making

coherence (is it internally consistent, does it focus on productive features?). In these terms, we can note two features of the learning sequence: (1) The representations were deliberate in supporting productive noticing of features; and (2) Productive meaning is only established through the back and forth discussion and evaluation of representations (Lemke, 2004; Tytler et al., 2013).

23.4.3 Coordinating the Representational Sequence

The further feature of the sequence worth noting is the sequencing of activity through a series of representations, each of which builds on the previous to develop a multi-perspectival view of the night and day/shadows phenomena. A key challenge for learning astronomy is the need to coordinate earth and space perspectives (Hubber & Tytler, 2017), in order to interpret a space -centred model to account for observations from a rotating earth. From the current research, we can see that particularly for young students this involves a significant challenge for spatial perception and reasoning. The sequences of representations and the pedagogy associated with them were specifically designed to address these two problems. Previous research (Tytler et al., 2017) has shown how teachers of astronomy not only move through and across multiple representations, designed to knit together these perspectives, but that as they do this they pay attention to how to support students to make these links. In our sequence, teachers linked successive representations (a) sometimes by having both on display at the same time, and (b) often by making explicit reference to features in common. For instance, the students' shadow tracing, the gnome shadow, the gnome and torch model, and the gestural representations of sun and shadow directions, all were carried out in the same space, compared explicitly, and aligned spatially. We argue that only by explicitly modeling the translation process between representations and modes can students learn to productively coordinate them.

23.5 Conclusion

This research showed the fundamental importance of students constructing, evaluating, and coordinating multiple visual-spatial representations, including 2D drawings, 3D models and embodied gestures, to perceive the spatial relations underpinning earth sun relation that explain shadow movement, and the link between earth bound and space perspectives of day and night. It was clear that the representational challenge involved spatial reasoning practices/constructs associated with mathematics: measuring, representing, coordinating and interpreting data about shadow length, and spatial reasoning about the relation between the sun's angular movement and changes to shadows. Pedagogically, the sequence demonstrated the effectiveness of a guided inquiry, representation construction/modeling approach, with longer term planning involving careful sequencing of representational moves, and short term strategies involving teacher response to student learning needs through embodied strategies and model manipulation, and engagement of students in representational invention, critique, and revision. The research has provided fresh insight into the visual-spatial challenges for younger students in learning astronomy, and key strategies for constructing effective, representationally-rich learning sequences.

Acknowledgements This research was funded by the Australian Research Council Discovery Scheme DP180102333, as part of the project: *Enriching mathematics and science learning through an inter-disciplinary approach.* http://imslearning.org/

References

Danaia, L., & McKinnon, D. (2008). Common alternative astronomical conceptions encountered in junior secondary science classes: Why is this so? *The Astronomy Education Review, 2*(6), 32–53.

Gilbert, J. K. (2005). *Visualization in science education*. Springer.

Hegarty, M., & Waller, D. (2004). A dissociation between mental rotation and perspective-taking spatial abilities. *Intelligence, 32*, 175–191.

Hubber, P., & Tytler, R. (2017). Enacting a representation construction approach to teaching and learning astronomy. In D. Treagust, R. Duit, & H. Fischer (Eds.), *Multiple representations in physics education* (pp. 139–161). Springer.

Lehrer, R. (2009). Designing to develop disciplinary dispositions: Modeling natural systems. *American Psychologist, 64*(8), 759–771.

Lehrer, R., & Schauble, L. (2012). Seeding evolutionary thinking by engaging children in modeling its foundations. *Science Education, 96*(4), 701–724.

Lehrer, R., Konold, C., & Kim, M. J. (2006). Constructing data, modeling chance in the middle school. Paper presented at the annual meeting of the *American Educational Research Association*, San Francisco, CA.

Lelliott, A., & Rollnick, M. (2010). Big ideas: A review of astronomy education research 1974–2008. *International Journal of Science Education, 32*(13), 1771–1799. https://doi.org/10.1080/09500690903214546

Lemke, J. (2004). The literacies of science. In E. W. Saul (Ed.), *Crossing borders in literacy and science instruction: Perspectives on theory and practice*. International Reading Association and National Science Teachers Association.

Padalker, S., & Ramadas, J. (2008). Modelling the round earth through diagrams. *The Astronomy Education Review, 2*(6), 54–57.

Peirce, C. S. (1931–1958). *Collected papers of Charles Sanders Peirce*. 8 volumes (Eds. C. Hartshorne, P. Weiss & A. W. Burks, Vols 1–6), (Ed., A. W. Burks, vols 7–8). Cambridge, MA: Harvard University Press.

Plummer, J. D. (2014). Spatial thinking as the dimension of progress in an astronomy learning progression. *Studies in Science Education, 50*(1), 1–45. https://doi.org/10.1080/03057267.2013.869039

Prain, V., & Tytler, R. (2012). Learning through constructing representations in science: A framework of representational construction affordances. *International Journal of Science Education, 34*(17), 2751–2773.

Testa, I., Leccia, S., & Puddu, E. (2014). Astronomy textbook images: Do they really help students? *Physics Education, 49*(3), 332.

Tytler, R., Prain, V., Hubber, P., & Waldrip, B. (Eds.). (2013). *Constructing representations to learn in science*. Sense Publishers.

Tytler, R., Murcia, K., Hsiung, C.-T., & Ramseger, J. (2017). Reasoning through representations. In M. Hackling, J. Ramseger, & H.-L. S. Chen (Eds.), *Quality teaching in primary science education: Cross-cultural perspectives* (pp. 149–179). Springer. https://doi.org/10.1007/978-3-319-44383-6_7

Tytler, R., Prain, V., Aranda, G., Ferguson, J., & Gorur, R. (2020). Drawing to reason and learn in science. *Journal of Research in Science Teaching, 57*, 209–231. https://doi.org/10.1002/tea.21590

Vosniadou, S., & Brewer, W. F. (1992). Mental models of the earth: A study of conceptual change in childhood. *Cognitive Psychology, 24*(4), 535–585.

Chapter 24
Discipline-Based Educational Research to Improve Active Learning at University

Daniele Buongiorno, Robert Harry Evans, Sergej Faletič, Jenaro Guisasola, Paula Heron, Marisa Michelini, Gorazd Planinšič, Paulo Sarriugarte, Alberto Stefanel, and Kristina Zuza

24.1 Introduction

The competencies required for university students require global innovation that goes beyond the dissemination of good practices (Corbo et al., 2016). This multidimensional problem requires the active involvement of students in the learning process (Freeman et al., 2014) and teacher innovation opportunities. The goal is to produce students who not only have knowledge of content and methods, but awareness of the role of physics in different contexts and of physics methodologies (Hoskinson et al., 2014).

This paper offers research on active learning proposals in physics. Each contribution represents an aspect of discipline-based educational research through a coordinated and coherent research approach, in addition the proposals offer strategies and methods for flexible planning of active learning in different contexts. Active learning "tutorials" are designed to encourage evolution in students' thinking, starting with their prior ideas (Heron, 2018). A problem-solving strategy based on

D. Buongiorno · M. Michelini (✉) · A. Stefanel
Department of Mathematics, Informatics and Physics, University of Udine, Udine, Italy
e-mail: marisa.michelini@uniud.it

R. H. Evans
Institute for Science Education, University of Copenhagen, Copenhagen, Denmark

S. Faletič · G. Planinšič
Faculty of Mathematics and Physics, University of Ljubljana, Ljubljana, Slovenia

J. Guisasola · P. Sarriugarte · K. Zuza
Donostia Physics Education Research Group, Department of Applied Physics, University of the Basque Country, Leioa, Biscay, Spain

P. Heron
University of Washington, Seattle, WA, USA

O. Levrini et al. (eds.), *Engaging with Contemporary Challenges through Science Education Research*, Contributions from Science Education Research 9,
https://doi.org/10.1007/978-3-030-74490-8_24

developing guided research can help students improve their skills in scientific methodology (Guisasola et al., 2014). Competence in planning and carrying out experiments can be improved with the use of rubrics for student project work management and assessment. Developing awareness of the role of physics in different contexts is particularly relevant for introductory courses for students who are not pursuing physics degrees, such as those in biology. Design-based research aimed at stimulating a functional understanding of physical concepts can give students the opportunity to learn how physics methodologies are used in biological fields, integrating lab work, problem solving, and in-depth analysis of professional applications (Michelini & Stefanel, 2019).

24.2 "Intentional Teaching": Using Students' Ideas as the Basis for Teaching Physics

In typical physics courses, most students learn to apply rules and procedures sufficiently well enough to obtain passing marks. However, they often cannot reason qualitatively with concepts and principles. They often fail to distinguish closely related quantities (velocity and acceleration, potential and potential difference, *etc.*). They display uneven skills with representations (vectors, graphs, field-line diagrams, *etc.*). They frequently ignore one variable in reasoning with multivariable relationships, such as $PV = NkT$. Students often fail to grasp simplifications such as frictionless surfaces, ideal capacitors, *etc.* Above all, many students lack models they can use when formulas are inadequate.

These findings have motivated teachers and researchers to design instructional strategies intended to engage students at a deeper intellectual level than is typical of conventional teaching methods. Here we discuss some principles that underlie "tutorials," which are designed for small-group discussions in large-enrollment university physics courses (McDermott et al., 2002).

24.2.1 Perspective on Learning

The view presented here can be called ***pragmatic constructivism*** (Heron, 2018). It is *pragmatic* because we address problems that arise in authentic classroom situations. It is *constructivist* because we assume that learners construct new knowledge on the basis of their existing knowledge (whether canonical or not), which they constructed on the basis of prior formal instruction and everyday experience. Prior knowledge is viewed as both the foundation and building materials for new knowledge. No assumptions about the structure or stability of learners' initial ideas are needed. Specifically, not all errors are believed to stem from firmly held mistaken beliefs; some stem from the deployment of finer-grained cognitive resources in situations in which they do not apply.

The process of constructing understanding is not smooth, linear or trouble-free. However, there are some predictable pathways learners may follow. *Difficulties* can arise when learners interpret the content of their formal physics instruction in terms of what they already know. Thus difficulties are not necessarily beliefs that predate instruction. For example, some students interpret the wave properties of matter to mean that particles follow sinusoidal paths (Ambrose et al., 1999). While incorrect, this is a *rational* attempt to link prior knowledge with new ideas. Moreover, it is unlikely that students constructed this view by synthesizing their experience of the real world; the sinusoidal trajectory idea came into being at the moment it was needed.

24.2.2 "Intentional Teaching"

Tutorials are designed to help "bridge the gap" between what students are taught through conventional instruction, and what they actually learn (McDermott, 1993). The approach is an example of ***intentional teaching*** (Heron, 2018): each step in a tutorial is deliberately chosen with specific goals in mind and takes into account how students think about the topic. A tutorial does not provide a rigid pathway, nor are the step sizes so small they can be followed unerringly. Thus instructors must exercise judgment and creativity in responding as groups tackle thorny issues: When to ask an additional question to sharpen the issue? When to ensure that all viewpoints have been heard? When to step back and allow conversation to proceed organically? The result is that no group, or individual, navigates a tutorial in exactly the same way.

24.2.3 Incorporating Students' Prior Ideas

In many cases, tutorials build on what students know or expect to be true. For example, a tutorial on angular momentum leverages students' expectations that certain quantities will be conserved (Close & Heron, 2011). Specifically, many students make predictions about collisions between point particles and rigid objects that *would be correct* if linear momentum could be transformed into angular momentum. The tutorial relies on the fact that they prefer to accept that a point particle moving in a straight line can have angular momentum rather than to accept the appearance or disappearance of quantities of motion. In most tutorials, students' attention is drawn to situations in which research has revealed a tendency to falter. Thus, tutorials help students construct, interpret and apply fundamental concepts and principles, while at the same time addressing difficulties. However, developing a coherent conceptual framework that enables students to interpret physical scenarios, choose and apply appropriate analysis tools, and evaluate their conclusions, cannot be achieved by a piecemeal approach that treats difficulties in isolation.

To the extent that the tutorials use an over-arching instructional principle, it is to *teach by questioning*. Therefore, rather than merely telling students what we wish them to learn, we attempt to elicit from them the meaning of concepts and relationships. However, they do not discover the laws of physics on their own. In most cases, tutorials assume that an idea has been presented previously (i.e., that momentum is conserved) but that what it *means* to be conserved can be worked out by answering a series of questions. Thus students go through the relevant reasoning themselves, rather than attempting to follow while someone else (instructor, textbook author, *etc.,*) outlines that reasoning, which leaves many learners unaware of the steps that were skipped, the choices that were made, and the counter-arguments that may arise. Questioning also serves to elicit students' own intuitions, and to promote reflection on what they do and do not understand.

24.3 The Role of Exercises in Learning: Examples of Research in Engineering Studies

Science education research has repeatedly shown that many students in university Introductory Physics courses that use traditional teaching formats learn to solve the type of quantitative problems found at the end of each textbook chapter. These students, however, usually are incapable of explaining the meaning of their own numerical solutions to the problems (McDermott, 1991). Undergraduate students have to acquire not only conceptual understanding of physics principles but also the science process abilities that are needed to solve physics problems. It will be necessary to design teaching sequences that do not just explain the problems numerical resolutions but also use a teaching strategy giving students the opportunity to use scientific procedures. These procedures involve putting forward a hypothesis about the relevance of modelling for the problem proposed, suggesting a number of different strategies to explain it, and justifying the answer based on evidence.

We developed a programme of teaching-learning problem solving, which sets the student problems within techno-scientific references. Students are presented with "situations" and, by posing some questions, they are given opportunities to use evidence to solve problems and use epistemic practice to communicate their ideas. The traditional roles of authority and novice are blurred, as students work in cooperative teams solving problems that have already been solved (as opposed to novel investigations) under the direction of a teacher who knows the solution well. We call this teaching strategy "Teaching/learning as oriented research activity" (TELORA) (Zuza et al., 2014). Here we use the topic of "Work and Energy" in Mechanics to illustrate the teaching approach.

24.3.1 Context of the Study and Results

The students involved in this study were enrolled in the introductory physics course that is required for engineering degrees at the University of the Basque Country (Spain). All of them had passed an examination to enter the University and had taken at least 2 years of calculus-based physics that included topics in electromagnetism, during their pre-university education.

The effectiveness of the TELORA approach in problem solving was evaluated by different tools including pre- and post-tests, which were administered under exam conditions. The test, administered in exam conditions, was given to students from experimental groups who follow TELORA and control group students who followed traditional teaching in solving problems. The results were included as a part of the final marks for the subject unit. The scores for each group's category of answer were compared by chi-square statistic.

One example of the type of problems from the test is given below.

Problem P4. – A car moves at constant velocity on a horizontal highway. Is work done on the car? Is there variation of system energy?

The students have to define the system and apply the General Work and Energy Principle. Correct answers have to define a system that includes car and the surface of highway, because there is friction between both car and surface. This friction is an internal force of the system and contributes to the variation of the internal energy. Table 24.1 shows the results.

The results of question P4 show a significant improvement in the percentages of correct answers in the experimental group. In addition, the chi-squared statistic for two questions obtained results with p < <0.0001. It can be stated that the results depend on the teaching method being used. As we have already indicated, to show whether the difference between the experimental and control groups is relevant from an educational point of view, we calculate the statistic size of the effect. The learning objectives for Problem P4 the size of the effect in implementing TELORA lies greater than 0.5, which means a large size of the effect.

In relation to the use of scientific skills by experimental students, we passed a pre-test and post-test to experimental students (see results in Table 24.2).

Table 24.1 The frequency of correct answers for the Problem P4

	All courses	Post-2015–16		Post-2016–17	
Problem	Pre (N = 257)	Control (N = 115)	Exp. (N = 175)	Control (N = 115)	Exp. (N = 178)
P4	0.0	9.0	55.5	8.0	62.0

During the 2 years of the experiment, the percentages of correct answers in the pre-test did not have statistically significant differences, so we have presented the average percentages in the first column of the table

Table 24.2 Frequency of use of scientific procedures of experimental students in Problem 4

Scientific procedures in problem solving	Pre-test (N = 353)	Post-test (N = 175)	Post-test (N = 178)
A. Approach and modeling: It relates theoretical knowledge to the specific context of the problem. Define the chosen "system"	0,0%	77,0%	80,0%
B. Resolution strategies: Define stages in the resolution of the problem: Identify the initial and final situation, types of energy, work as energy transfer.	0,0%	65,0%	78,0%
C. Resolution: Apply the general principle of work and energy, $E_{begining} + W_{ext} = E_{final}$	89,0%	76,0%	80,0%
D. Analysis of the result	0,0%	71,0%	74,0%
D. Choosing inappropriate resolution ways, consider partial incorrect resolutions without considering the problem as a whole.	82,0%	23,0%	20,0%
No answer	10,0%	–	–

24.4 A Context-Independent Way of Guiding and Assessing Students' Work in Project Laboratory

An important part of university instruction is to produce competent experts, capable of designing and carrying out their own projects. The Project Laboratory course is designed to develop these competences early in university education. The projects given to groups of students are open ended and their work is guided by rubrics. The Rutgers Scientific abilities rubrics are clear and context-independent guidelines that provide clear goals, but are applicable in very different contexts and settings. As such, they transcend a specific intervention and instead become a general tool to provide guidelines across the curriculum. The assessment of the reports using the rubrics was considerably faster than writing comments, and the reports improved in quality.

24.4.1 Rubrics

The rubrics are tables, which list various assessed instructional goals and the criteria by which to determine, how well the goal has been achieved. The "Scientific abilities rubrics" developed at Rutgers University, NJ, USA (Etkina et al., 2006; Rutgers, 2016) assess scientific abilities or scientific process skills. An example of a row is listed in Table 24.3. In Project laboratory, we use five rubrics with a total of 35 rows. For each project students use three of the rubrics with approximately 19 rows total. The rubrics do not change for each specific project.

Table 24.3 One row of a rubric. The first cell states the process skill being evaluated and the rest contain criteria by which it is determined to which level the skill has been developed

	Missing (0)	Inadequate (1)	Needs improvement (2)	Adequate (3)
Is able to record and represent data in a meaningful way.	Data are either absent or incomprehensible.	Some important data are absent or incomprehensible. They are not organized in tables or the tables are not labelled properly.	All important data are present, but recorded in a way that requires some effort to comprehend. The tables are labelled but labels are confusing.	All important data are present, organized, and recorded clearly. The tables are labelled and placed in a logical order.

24.4.2 Setting

Project Laboratory is a course for first and second year physics students. Groups of four to five students receive an experimental problem to solve or a physical phenomenon to investigate. The experimental part guided by the rubrics lasts for nine contact hours over 3 weeks and the writing of the report, also guided by the rubrics for another 2 weeks. Instructors then evaluate the report and give feedback to students. Before the rubrics, the feedback was given in the form of comments/annotations to the report. With the rubrics, the feedback consists mostly of scores on the rubrics.

The students have the opportunity to hand in an improved report and the process can be repeated multiple times until the end of the semester. The course is graded pass/fail.

24.4.3 Methods

We selected an instructor who has been assessing the students' reports without rubrics (2008–2014) and with rubrics (2015–2018). We collected all the feedback sent to the students by this instructor in the entire period. As a measure of the instructor's workload we took the number of words in the total feedback to each.

In the period 2008–2014, the judgement of acceptability was left to the instructor. In the period 2015–2018, a perfect score was required for a passing grade. To compare the quality of the reports, we scored a random sample of six reports from the period 2008–2014 with the rubrics. Failure to achieve a perfect score implies that now the reports are of better quality.

24.4.4 Findings

In the period 2008–2014, the average number of words for all the feedback on one report was 1400. In the period 2015–2018, the average number of words was 730 (Fig. 24.1). The ratio is 0.52 and the difference is statistically significant ($p < 0.01$). The quality of the reports increased (Fig. 24.2). Only one of the sample of six reports scored an almost perfect score.

Feedback using almost only scores on the rubrics requires that the students actively engage in reflection about their report to identify necessary improvements. By increasing the active engagement of students in their learning and providing clear learning goals, the rubrics reduced the instructor's workload and improved the quality of the reports.

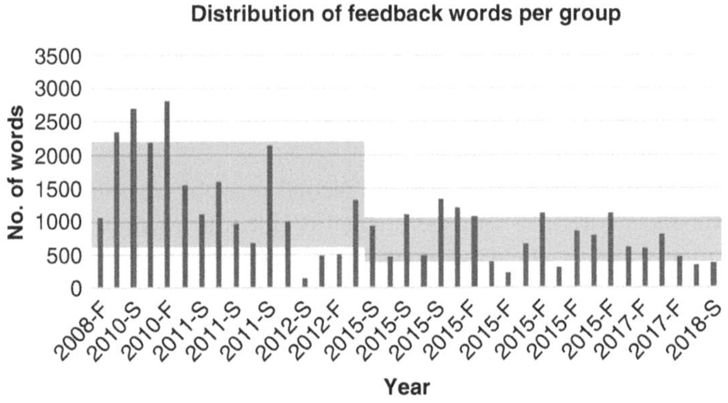

Fig. 24.1 The average number of words in a feedback for all examined reports. The labels on the horizontal axes are of the form year-semester. The shaded area represents the average number of words with the standard deviation for the period 2008–2014 and the period 2015–2018

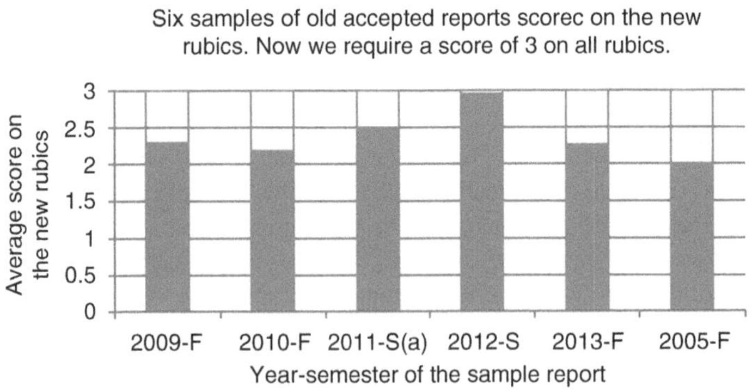

Fig. 24.2 The average scores received on the rubrics by the pre-rubrics reports. Only one report is very close to receiving a perfect score of 3

24.5 Curricular Innovation in Physics for Bio-Area

Building an introductory physics course for the bio-area degrees is a research curricular challenge involving selection and contextualization of contents and methods to produce the awareness of the role of physics in the specific bio-area contexts and the ownership of the applied methodologies (Hoskinson et al., 2014 AAAS 2011; Donovan et al. 2013). We focus on the integration of classroom activities and experimental laboratories, by means of active learning modalities (Laws, 2004; Watkins et al., 2012). Pragmatic constructivism methodologies (Heron et al. 2004; Heron, 2018), problem solving activities (Zuza et al., 2014) and responsibility based laboratory work (Etkina et al., 2006; Holmes and Wieman 2018) became the last 5 years' strategies for the approach in the introductory physics courses for 500 students per year in biotechnology, agronomy, and science of food degrees at University of Udine. The research framework for the teaching/learning proposal was the Model of Educational Reconstruction (Duit et al., 2012) in the perspective of Redish group (Meredith & Redish, 2013; Redish et al., 2014). Design Based Research (Anderson & Shattuck, 2012) guided the core proposal implementation and conceptual model-based links between apparently different fields (as for instance fluid flow in a river or in a body and electrical current flow in electric circuits) by means of student outcomes step by step.

The laboratory activities, integrated in each topic, offer to the students an operative responsibility and intellectual challenge: only the goals, the available instruments and employment suggestions were provided to students, allowing them to point out the problematic aspects, to choose the best data collection procedures, the conduction of the experiment, the relative data analysis and reporting. Student problem solving included the interpretation of experimental data, in order to obtain the value of physical quantities, as for example the thermal conduction coefficient of a solid.

The dynamics of real fluids represent here one of the main contents contextualized. Fluids in equilibrium are the module premise (6 hours overall). Examples of the problem posed to the students by means of a flipped approach were *Why does the water flow faster in the middle of the river? Why does the pressure of the blood decrease as it increases its path? The different solutions obtained by paper and pencil or computer modelling are usually discussed in the class group to structure conceptual and theoretical aspects.*

Research questions are: (RQ1) How the teaching/learning approach contributes to learning outcomes, taking into account fluids contents and lab work on thermal conduction, (RQ2) How does students' engagement contribute in learning outcomes on the contents of fluids? (RQ3) How do the chosen strategy for lab activities contribute to improving student competence?

Both qualitative (Denzin & Lincoln, 2011) and quantitative data analyses were carried out on tutorials, lab reports and final exam questionnaires and multiple-choice tests.

The active involvement of students, measured by means of their participation in teaching activities, passed from 37% before this research based approach (2013) to

98% in the last year. The learning outcomes in terms of percentage of students passing the final examinations (RQ1) increased from 40% before 2013 to a 3 year consolidated 63% for all cohort students, being 75% taking into account only the active students. In the case of selected students in biotechnology degree reached 95%. The gain index (Hake, 1998) in the final examination was 0.5.

Table 24.4 shows the percentage of correct answers (homogenous for cohort and degree – $p < 0.001$) to the items on fluids included in the final examination (cohorts 2014/18). The items (Michelini & Stefanel, 2019) regard the concept of pressure (Q1–Q2) the Pascal principle (Q3), non-viscous flow and continuity equation (Q4, Q6–Q7), viscous flow (Q6–Q7–Q8).

The highest score (RQ2) is obtained for the more critical questions according with the students difficulties in literature (Heron, 2018): concept of pressure (Q2), quantities dependence and pressure conditions in complex contexts (Q5) and dynamic behavior of bodies in viscous fluids (Q8), but from qualitative analysis shows that about 25% of students do not overcome their learning problem concerning Pascal principle. The 35 ± 5% of students showed more difficulties with not-viscous flow, than with real flow proposed in contextualized items and in passing from static to dynamic situations. The formative success gain for thermal conduction is from 40% (N = 54) in 2016/17 to 66% (N = 61) in 2017/18, when the new approach to lab was implemented.

The positive effect of problem solving based on lab work emerges also in the final examination with a 3 year consolidated percentage of learning success of 78% in items regarding thermal conduction (RQ3).

24.6 Concluding Remarks

To improve learning outcomes in introductory physics courses, producing students' ownership and competence in physics is the main goal.

One way to achieve this is student active engagement in developing conceptual understanding, reasoning competences and process skills, as well as understanding the role of physics in other scientific areas. This involves the planning of content, instruments, methods and strategies for each different competence that physics teaching/learning process in introductory course addresses. Each of these four studies is based in constructivism in the form of problem-based instruction to activate

Table 24.4 Percentage of correct answers to the items on fluids in the examination (cohorts 2014/18)

Item	Q1	Q2	Q3	Q4	Q5	Q6	Q7	Q8
Sample number	N = 475	N = 475	N = 583	N = 1474	N = 979	N = 303	N = 412	N = 258
Correct (%)	44	68	60	58	68	67	62	70
Uncorrect (%)	40	30	35	36	30	25	30	21
NA (%)	16	3	5	7	2	8	8	9

student cognitive engagement in learning physics. In each case students worked cooperatively to construct their solutions to problems, sometimes with structured dialogic questioning, sometimes with rubrics to guide them, and without being given answers. Clear and consistent feedback is provided for the students to give them the motivation to continue their problem-solving, while simultaneously providing instructors with feedback on lesson progress and therefore encouragement to also continue with this learning style. The context for each of these studies was authentic physics problems which required active consideration of scientific processes for resolution.

Traditional based physics teaching where teachers and texts provide algorithms and answers is altered in these different examples of student activation. Relinquishing the ease and comfort of teaching in an authoritarian method, these innovative methods suggest alternative ways to achieving success for students who are guided to construct the concepts and knowledge of physics.

The introductory university physics course designs require discipline-based educational research on different aspects in a coherent way. In this paper, we offer evidence-based results on how tutorials, the TELORA method for problem solving, and specific rubrics in ISLE labs contribute to produce active learning for conceptual understanding, and how the integration of these strategies in curricular design for introductory physics courses for bio-area students produces gains in student ownership and significantly increased learning outcomes.

References

AAAS-American Academy for the Advancement of Science. (2011). *Vision and change in undergraduate biology education*. AAAS.

Ambrose, B. S., Shaffer, P. S., Steinberg, R. N., & McDermott, L. C. (1999). An investigation of student understanding of single-slit diffraction and double-slit interference. *American Journal of Physics, 67*(2), 146–155. https://doi.org/10.1119/1.19210

Anderson, T., & Shattuck, J. (2012). Design-based research: A decade of progress in education research? *Educational Researcher, 41*(1), 16–25.

Close, H. G., & Heron, P. R. L. (2011). Student understanding of the angular momentum of classical particles. *American Journal of Physics, 79*(10), 1068–1078. https://doi.org/10.1119/1.3579141

Corbo, J. C., Reinholtz, D., Daney, M. H., Deetz, S., & Finkelstein, N. (2016). Framework for transforming departmental culture to support educational innovation. *Physical Review Physics Education Research, 12*, 010113.

Denzin, N. K., & Lincoln, Y. S. (2011). *Handbook of qualitative research*. Sage.

Donovan, D. A., Kratz, R. F., & Nelson, G. D. (2013). Advantages and challenges of using physics curricula as a model for reforming an undergraduate biology course. *CBE-Life Sciences Education, 12*, 215–229.

Duit, R., Gropengießer, H., Kattmann, U., Komorek, M., & Parchmann, I. (2012). The model of educational reconstruction. In D. Jorde & J. Dillon (Eds.), *Science education research* (pp. 13–37). Sense Publishers.

Etkina, E., Van Heuvelen, A., White-Brahmia, S., Brookes, D. T., Gentile, M., & Murthy, S. (2006). Developing and assessing student scientific abilities. *Physical Review. Special Topics, Physics Education Research, 2*, 020103.

Freeman, S., et al. (2014). *Active learning increases student performance in science, engineering and mathematics*, loc cit by Corbo J. C.

Guisasola, J., Ceberio, M., & Zuza, K. (2014). Problem solving by developing guided research in introductory university physics courses. In D. W. Sunal, C. S. Sunal, E. L. Wright, & D. Zollman (Eds.), *Research based undergraduate science teaching*. Information Age Publishing.

Hake, R. (1998). Interactive-engagement versus traditional methods. *American Journal of Physics, 66*(1), 64–74.

Heron, P. R. L. (2018). Identifying and addressing difficulties: Reflections on the empirical and theoretical basis of an influential approach to improving physics education. In C. Henderson & K. A. Harper (Eds.), *Getting started in PER, Reviews in PER Vol. 2*. American Association of Physics Teachers.

Heron, P. R. L., Shaffer, P. S., & McDermott, L. C. (2004). Research as a guide to improving student learning. In *Invention and impact: Building excellence in undergraduate STEM education* (pp. 33–38). AAAS.

Holmes, N. G., & Wieman, C. E. (2018). Introductory physics labs: We can do better. *Physics Today, 71*(1), 38.

Hoskinson, A. M., Couch, B. A., Zwickl, B. M., Hinko, K., & Caballero M. D. (2014). Bridging physics and biology teaching through modeling. *American Journal of Physics 82*(5): 434–441.

Laws, P. W. (2004). *Promoting the diffusion of undergraduate science curriculum reform: The activity-based physics suite*. AAAS. https://www.aaas.org/sites/default/files/08_Cre_App_Laws.pdf

McDermott, L. (1991). Millikan lecture 1990: What we teach and what is learned-closing the gap. *American Journal of Physics, 59*, 301–315.

McDermott, L. C. (1993). Guest comment: How we teach and how students learn – A mismatch? *American Journal of Physics, 61*(4), 295–298. https://doi.org/10.1119/1.17258

McDermott, L. C., Shaffer, P. S., & Physics Education Group at the University of Washington. (2002). *Tutorials in introductory physics* (1st ed.). Pearson.

Meredith, D. C., & Redish, E. F. (2013). Reinventing physics for life-science majors. *Physics Today, 66*(7), 38–43.

Michelini, M., & Stefanel, A. (2019). Chapter 20: Innovation in physics teaching/learning for the formative success. In M. Pietrocola (Ed.), *Perspectives in physics teaching and learning from physics education research*. Springer.

Redish, E. F., Bauer, C., & Carleton, et al. (2014). NEXUS/physics: An interdisciplinary repurposing of physics for biologists. *American Journal of Physics, 82*(5), 368–377.

Rutgers Physics and Astronomy Education Research Group. (2016). *Scientific abilities*. Acquired on May 22, 2016, https://sites.google.com/site/scientificabilities/

Watkins, J., Coffey, J. E., Redish, E. F., & Cooke, T. J. (2012). Disciplinary authenticity: Enriching the reforms of introductory physics courses for life-science students. *Physical Review Special Topics: Physics Education Research, 8*, 010112.

Zuza, K., Almudí, J. M., Leniz, A., & Guisasola, J. (2014). Addressing students' difficulties with Faraday's law: A guided problem solving approach. *Physical Review Special Topics: Physics Education Research, 10*, 010122.

Chapter 25
Instructional Activities Predicting Epistemic Emotions in Finnish Upper Secondary School Science Lessons: Combining Experience Sampling and Video Observations

Elisa Vilhunen, Xin Tang, Kalle Juuti, Jari Lavonen, and Katariina Salmela-Aro

25.1 Introduction

Academic emotions are emotions that people experience in educational settings or that relate to learning, studying or other academic activities (Pekrun et al., 2018). The importance of academic emotions in achievement, engagement and motivation has been widely acknowledged (Pekrun et al., 2018; Schneider et al., 2020; Sinatra et al., 2014). Despite the large body of research on academic emotions and their known effects on learning and student well-being, there remains a paucity of knowledge about the enhancement of beneficial affects in practice, and so far, little is known about the relations between emotions and learning activities in classrooms (Meyer, 2014). Furthermore, the research on academic emotions in the field of science education is particularly scarce (Sinatra et al., 2014). This study sheds light on how students experience different instructional activities in science classrooms by combining students' self-reported, real-time experience sampling data on epistemic emotions with qualitative video observation data on instructional activities.

E. Vilhunen (✉) · X. Tang · K. Juuti · K. Salmela-Aro
Faculty of Educational Sciences, University of Helsinki, Helsinki, Finland
e-mail: elisa.vilhunen@helsinki.fi; xin.tang@helsinki.fi; kalle.juuti@helsinki.fi; katariina.salmela-aro@helsinki.fi

J. Lavonen
Faculty of Educational Sciences, University of Helsinki, Helsinki, Finland

Department of Childhood Education, University of Johannesburg, Soweto, South Africa
e-mail: jari.lavonen@helsinki.fi

O. Levrini et al. (eds.), *Engaging with Contemporary Challenges through Science Education Research*, Contributions from Science Education Research 9,
https://doi.org/10.1007/978-3-030-74490-8_25

25.1.1 Epistemic Emotions

Pekrun et al. (2018) categorise academic emotions into four groups based on their antecedents or objects of focus. *Achievement emotions* relate to success or achievement in academic tasks. *Topic emotions* relate to the topics students study. *Social emotions* relate to social relationships in the class, such as those between students and teachers or among peers. *Epistemic emotions* relate to learning itself and have an object focus in knowledge or knowledge construction. Since epistemic emotions are theorised to relate directly to learning, they are especially interesting in terms of conducting, developing and evaluating educational practices. Epistemic emotions typically occur in situations of contradictory or incongruous information where new comprehensions are developed (Pekrun et al., 2018).

Based on previous research, some epistemic emotions can be considered as positive in terms of their effects on educational outcomes. For example, enjoyment and curiosity have been found to positively relate to learning and academic achievement (Ainley & Ainley, 2011; Gruber et al., 2014). Curiosity also predicts positive learning strategies, such as elaboration and rehearsal, knowledge exploring behaviour and metacognitive self-regulation (Chevrier et al., 2019; Vogl et al., 2020); it is also often linked to engagement (Bosch & D'Mello, 2017). By contrast, some epistemic emotions can be considered negative in terms of their effects on educational outcomes. For example, boredom, anxiety and frustration have been found to negatively relate to learning (Bosch & D'Mello, 2017; Eysenck, 1979; Pekrun et al., 2014). Boredom can also impede beneficial learning strategies, such as rehearsal and critical thinking (Chevrier et al., 2019). However, the role of some epistemic emotions in learning is more complex. For example, confusion, at appropriate levels, can be beneficial for learning (D'Mello et al., 2014), but when it goes unresolved, it can detract from learning (D'Mello & Graesser, 2012). In addition, the role of surprise is as yet unclear. It has been found that confusion and surprise can generate knowledge exploration (Vogl et al., 2020), but can also predict some learning strategies that are more negative (Bosch & D'Mello, 2017). Epistemic emotions also often co-occur, correlate and dynamically interact (D'Mello & Graesser, 2012). When encountering novel or even contradictory information, surprise is usually the primary emotion and is often followed by curiosity or confusion. If confusion is resolved, enjoyment follows. However, the persistence of confusion may lead to frustration and, eventually, boredom (D'Mello & Graesser, 2012).

In the context of science learning, it has been suggested that negatively valenced emotions are integral to learning and engagement and inherent in science (e.g. Engle & Conant, 2002; Jaber & Hammer, 2016; Radoff et al., 2019). Through emotion-aware teacher scaffolding, students can learn to experience negative epistemic emotions as pleasurable and as gateways to new realisations or achievements (Radoff et al., 2019). According to DeBellis and Goldin (2006), these meta-affects, or 'affect about affect' (p. 136), enable the enjoyment of learning situations that would otherwise feel undesirable. Thus, some researchers suggest that teachers should not try to minimise these negative emotions in the classroom, but rather support students in their experience of them as part of the learning process (DeBellis & Goldin, 2006; Jaber & Hammer, 2016). It is further argued that the self-regulation of affects, or meta-affects, should be learned as part of the substance in science education (Jaber & Hammer, 2016).

25.1.2 Instructional Activities in Science Teaching

Science lessons typically consist of multiple instructionally independent lesson segments, such as reviewing homework, teaching new content or doing independent classwork. According to Burns and Anderson (1987), these lesson segments can be seen as the context in which teaching and learning take place, thus providing a useful framework for examining and analysing instructional activities. Lesson segments can be distinguished by their different purposes, activity formats and topics or assignments (Burns & Anderson, 1987). In this paper, we use the term *instructional activities* to describe different classroom activities in order to emphasise that lesson segments, and their sequencing, are planned and managed by teachers. Previous research has indicated that activities emphasising students' own active participation and knowledge-construction, can promote interest, engagement, and positive emotions (Inkinen et al., 2019; Juuti et al., 2010, 2019). However, complex learning with high cognitive demands often involves emotions that are more negative, such as confusion, boredom, and frustration (D'Mello & Graesser, 2012).

According to Krajcik and Shin (2014), one attempt to increase students' interest and engagement in science and to tackle boredom is a method called project-based learning (PBL). In PBL, students make sense of phenomena and find solutions to problems by employing disciplinary core ideas and scientific and engineering practices. The PBL teaching and learning unit typically consists of multiple science lessons, and begins by introducing the driving question and engaging with anchoring events. Krajcik and Shin noted that other key PBL features include participating in scientific practices, engaging in collaborative activities, working with learning technologies and creating an artefact, such as an explanatory model. PBL can be regarded as a non-linear pedagogical method characterised by iterative and open-ended practices, in which students or groups of students can proceed at their own pace. Thus, PBL differs from the more traditional linear pedagogy in which educational practices are scripted and aim for task completion and knowledge reproduction. Though inquiry and experiments are fundamental parts of science education, science lessons also typically include teacher talk. According to Mortimer and Scott (2003), teacher talk is central to meaning making in science, and it develops 'the scientific story being taught' (p. 1). Teacher talk can be authoritative or dialogic, and interactive or non-interactive (Mortimer & Scott, 2003). Previous studies have shown the positive effect of PBL on students' interest, attitudes (Tseng et al., 2013), emotions (Hugerat, 2016) and motivation (Hung et al., 2012). Regardless of the extensive amount of research on PBL, far too little attention has been paid to how its particular features relate to students' epistemic emotions.

25.1.3 Research Questions

In this study, we aim to identify ways of promoting beneficial epistemic emotions and discourage adverse ones in educational settings by exploring the effect of instructional activities on epistemic emotions. Hence, we pose the research questions (RQs) below:

RQ1: What instructional activities can be found in PBL environments?
RQ2: How do these instructional activities predict the levels of epistemic emotions that students experience and report during science lessons?

25.2 Methods

25.2.1 Context and Participants

Altogether, 100 students from five upper secondary school physics classes participated in a six-lesson (× 75 min) PBL unit that was implemented by four teachers in two schools. The PBL unit was co-designed by the teachers and researchers and focused on the basics of Newtonian mechanics. Thus, the unit lessons shared many similarities, but were not identical. The units were implemented during the first physics courses that students take upon entering upper secondary school. The Finnish core curriculum for general upper secondary schools emphasises instructional methods similar to PBL, such as problem solving, scientific modelling and self-regulation (Finnish National Board of Education, 2016). Thus, the unit followed the principles of the national curriculum.

25.2.2 Data Collection and Measures

An experience sampling method (ESM) delivered via smartphone, was used to capture students' epistemic emotions (Goetz et al., 2016). Students responded to the ESM questionnaire on the basis of beeps coming to their smartphones during science lessons. The smartphones were for research use only and thus collected no personal data outside the questionnaire. The smartphones were preprogrammed to beep at random times three times per lesson and each student received the beep at the same time. Thus, each student had 18 opportunities to answer the questionnaire (only 17 in one group, due to a programming error), resulting in 1553 completed

ESM questionnaires from 89 measurement situations. Three questionnaires per lesson were considered to be enough to capture versatile learning situations, while distracting the teaching as little as possible. Depending on the teacher and the current activity in the classroom, the instruction or activity either continued or was paused while students answered the questionnaire. Each ESM questionnaire included identical items on social, emotional and contextual aspects. Students' experiences of situational epistemic emotions were examined using a modified seven-item short version of The Epistemically-Related Emotions Scales (Pekrun et al., 2017) in which students were asked: 'Were you feeling excited/anxious/bored/confused/surprised/frustrated/curious?'. A four-point Likert scale with the response categories from 1 = 'not at all' to 4 = 'very much' was used. The data collection design used in this study was described in more detail by Schneider et al. (2016).

Instructional activities were studied through classroom video observations. The video data were supplemented by corresponding lesson plans and stimulated recall sessions with the teachers of the learning units (Calderhead, 1981). Each teacher was asked to watch the selected excerpts and comment on them on the basis of the following questions: (1) What happens in this situation? (2) What is the instruction given to the students? (3) What is the goal of this activity? (4) How is this activity helping to achieve the learning goals? (5) What is the activity's role in the context of PBL? These questions guided teachers in their elaborations on the situations. The stimulated recall sessions were audio recorded and transcripts of these recordings were used in interpreting instructional activities.

25.2.3 Analyses

To answer RQ1, the study's qualitative data (video recordings, lesson plans and stimulated recall transcripts) were analysed. First, 89 video excerpts were combined with the corresponding parts of the lesson plans and stimulated recall transcripts. Second, a thematic analysis was conducted to systematically identify different instructional activities across the data set (Braun & Clarke, 2006). According to Braun and Clark, in thematic analyses, interesting features of the data are typically identified as initial codes that are then further collated into meaningful themes. In this study, we analysed all 89 classroom situations that took place prior to the ESM beeps. The situations were organised according to their similar instructional activities. The similarities of the activities were considered based on their purposes, activity formats and topics or assignments. These subordinate activities were further collated into representative superordinate activities, to enable the use of statistical analyses.

From the final data, we excluded five classroom situations (resulting in 84 situations) that did not focus on science learning or in which students were engaged in different phases of the inquiry task, which made it impossible to assign the situations to a single activity category. The data were initially analysed by one researcher and then checked when ambiguous by another researcher. Overall, the categorisation was an iterative process in which the research group discussed the subcategories and main categories together.

To answer RQ2, statistical analyses were conducted with Mplus 8.2 (Muthén & Muthén, 2017). First, we studied the hierarchical nature of our data by calculating intraclass correlations. Second, we used a multilevel regression analysis (random intercept, fixed slope) to explore whether the main instructional activity categories found in the thematic analysis predicted epistemic emotions (Snijders & Bosker, 2012). To examine the pairwise difference between the five instructional activities (the predictors), the activity categories were dummy coded and four different models, each with different reference predictor, were used. The model fit was excellent (RMSEA = 0.00; CFI = 1.00; TLI = 1.00) in each case.

25.3 Results

25.3.1 Instructional Activities in PBL Environment

The thematic analysis showed that the activities undertaken during PBL lessons could be categorised into five qualitatively distinctive main categories that illustrate the different instructional aims and activities (Table 25.1).

Table 25.1 Instructional activities identified in video-recorded classroom situations

Main category activities	Subcategory activities	n_{video}	n_{esm}
Orienting and engaging	Watching introductory videos Introductory, qualitative demonstrations Asking scientific questions	9	202
Conducting investigations	Collecting quantitative data Building experimental setup Cleaning up experimental setup	6	75
Analysing data and developing models	Analysing and interpreting data Developing models Constructing explanations	22	383
Teacher talk	Teaching new content Giving instructions for tasks Reviewing, elaborating on and wrapping up studied contents	24	443
Tasks for skills and content	Teacher assigned tasks for practicing the subject matter Teacher assigned tasks for practicing the use of digital analysis tools	23	397
TOTAL		84	1500

Note. 'n_{video}' refers to number of the main category instructional activities observed in the video data, i.e. class-level; 'n_{esm}' refers to number of ESM observations, i.e. student-level

25.3.2 Epistemic Emotions in Instructional Activities

Our data are hierarchical in nature, meaning the ESM observations are nested within students and students are nested within teachers and classes. To examine the nestedness of the data, we first calculated the intraclass correlations (ICCs) for students within teachers and ESM observations within students. The ICCs at the teacher level were mostly small (ranging from .01 to .03) and non-significant. Only in the case of excitement and curiosity were there significant differences between teachers (ICC = .03, p = .003, and ICC = .02, p = .003, respectively). By contrast, the ICCs of emotions at the student level were all high (ranging from .27 to .50) and highly significant. Thus, the nestedness of ESM observations within students was taken into consideration in subsequent two-level analyses.

The multilevel regression analysis showed that different instructional activities were significantly associated with different levels of epistemic emotions (Table 25.2). Students reported more excitement and curiosity during orienting and engaging activities than during all other activities. Anxiety was reported more when analysing data and developing models or working with tasks for skills and content than when conducting investigations or during teacher talk. Teacher talk induced more boredom in students than did orienting and engaging activities or when working with tasks for skills and content. Teacher talk was also experienced as less surprising when compared to orienting and engaging activities, working with tasks for skills and content, and analysing data and developing models. Confusion was reported least when conducting investigations. In addition, teacher talk activities were experienced as less confusing than analysing data and developing models and working with tasks for skills and content. Students reported being more frustrated when analysing data and developing models and when working with tasks for skills and content than during orienting and engaging activities, or when conducting investigations. Furthermore, analysing data and developing models induced more frustration than did teacher talk.

25.4 Discussion and Conclusion

25.4.1 Instructional Activities Reflect the Principles of PBL and National Curriculum

A thematic analysis of the video recordings, lesson plans and stimulated recall transcripts revealed that five distinctive instructional activities occurred during the science lessons. These activity categories strongly reflected the principles behind the implementation of the teaching, namely, PBL (Krajcik & Shin, 2014) and the Finnish core curriculum (Finnish National Board of Education, 2016). PBL emphasises a driving question and anchoring events as starting points for learning processes and as consistent threads throughout the PBL unit (Krajcik & Shin, 2014). In

Table 25.2 Multilevel regression analysis predicting epistemic emotions in different instructional activities, regression coefficients (β) and standard errors (S.E.). Each instructional activity is compared pairwise to other activities (rows), in terms of different epistemic emotions (columns)

	β (S.E.)						
Predictors/Instructional activities	Excited	Anxious	Bored	Confused	Surprised	Frustrated	Curious
Versus: Orienting and engaging							
Conducting investigations	−.43(.11)***	−.09(.07)	.05(.11)	−.20(.09)*	−.14(.09)	.04(.10)	−.41(.10)***
Analysing data and developing models	−.30(.06)***	.06(.07)	.11(.07)	.07(.07)	−.08(.06)	.30(.08)***	−.40(.06)***
Teacher talk	−.30(.06)***	−.05(.06)	.19(.07)**	−.05(.06)	−.16(.06)*	.13(.08)	−.34(.06)***
Tasks for skills and content	−.25(.06)***	.02(.06)	.02(.06)	.05(.07)	−.05(.07)	.20(.08)**	−.35(.06)***
Versus: Conducting investigations							
Analysing data and developing models	.13(.10)	.16(.06)*	−.06(.11)	.27(.07)***	.07(.08)	.24(.10)*	.02(.08)
Teacher talk	.12(.11)	.04(.06)	.14(.11)	.16(.07)*	−.02(.09)	.10(.10)	.08(.09)
Tasks for skills and content	.17(.11)	.12(.05)*	−.02(.10)	.26(.07)***	.10(.09)	.17(.08)*	.06(.09)
Versus: Analysing data and developing models							
Teacher talk	−.00(.06)	−.12(.06)*	.08(.06)	−.12(.06)*	−.08(.04)*	−.15(.07)*	.06(.04)
Tasks for skills and content	.05(.05)	−.04(.06)	−.09(.06)	−.02(.05)	.03(.04)	−.08(.07)	.05(.04)
Versus: Teacher talk							
Tasks for skills and content	.05(.05)	.08(.04)*	−.16(.05)**	.10(.05)*	.11(.05)*	.07(.06)	−.01(.04)

Note. *** $p \leq .001$, ** $p \leq .01$, * $p \leq .05$

our data, orienting and engaging activities are closely related to these themes in PBL. Introductory videos and qualitative demonstrations are anchoring events that help students see the value and different perspectives in the learning unit's driving question. This is followed by elaborations of scientific questions in order to study the driving question. Furthermore, our thematic analysis revealed categories that reflect the scientific practices emphasised in PBL, as well as in Finnish core curriculum, namely, conducting investigations, analysing data and developing models. Teacher talk and tasks for skills and content were also pertinent parts of science lessons (e.g. Mortimer & Scott, 2003).

25.4.2 Epistemic Emotions Can Be Aroused by Instructional Activities

The multilevel regression analysis indicated that the level of epistemic emotions varied significantly during science lessons with regard to different instructional activities. Orienting and engaging activities appeared to be particularly relevant in terms of inducing positive epistemic emotions such as excitement and curiosity. In addition, students reported being least bored during these activities. When the teacher aims to orient or engage students in relation to learning and see the value of the content being studied, excitement and curiosity can be expected to occur in these situations. Previous studies have shown the positive relation of curiosity to engagement (Bosch & D'Mello, 2017) and knowledge-exploration behaviour (Vogl et al., 2020).

Conducting investigations and analysing data and developing models most closely resemble scientific practices (Krajcik & Shin, 2014). They could even be regarded as subcategories of a parent category: *doing laboratory work*. However, in terms of the epistemic emotions they induce, they differ significantly from each other. The practical part of laboratory work (conducting investigations) induced relatively low levels of anxiety, confusion and frustration. By contrast, the computational part of the laboratory work (analysing data and developing models) induced significantly more anxiety, confusion and frustration. This implies that cognitive demands during the analysis and model development phase of the investigation were much higher than during the practical data collection phase. A similar difference could be seen among the more traditional instructional activities: teacher talk and tasks for skills and content. The most prevalent epistemic emotion students reported during teacher talk was boredom, which is in line with previous research (Mann & Robinson, 2009). By contrast, when students worked with tasks for skills and content, they reported relatively high levels of anxiety, confusion and frustration. These activities may both include different levels of cognitive demand, from low to high. However, students' roles in these activities differs significantly. During teacher talk, even when interactive, students can choose to play a passive role and thus eventually become bored. Tasks, by contrast, require students to be active learners, which may induce negatively valenced epistemic emotions, especially if the tasks are too cognitively demanding. Emotion-aware scaffolding is thus important in situations that induce anxiety, confusion or frustration.

Analysing data and developing models and working with tasks for skills and content can be seen as the most demanding classroom activities from the students' perspective: These activities were related to higher levels of confusion, frustration and anxiety in students than were some of the other instructional activities. It should be noted, however, that the essence of PBL is that students are not given the right answers; rather, they are scaffolded in open inquiry activities in order to construct knowledge by themselves. In such situations, confusion can seem very natural, and even fundamental as suggested in prior studies (DeBellis & Goldin, 2006; Radoff et al., 2019). Furthermore, it is obvious that not all students react the same way during the same activity. This underlines the importance of teachers' expertise in recognising and scaffolding those students who need support, whether emotional or cognitive.

25.4.3 Limitations

By definition, epistemic emotions have an object focus on knowledge or knowledge construction. In this study, we chose to investigate seven epistemic emotions identified in prior studies (Pekrun et al., 2017). However, in our questionnaire, students were only asked to indicate the extent to which they felt excited, anxious, bored, confused, surprised, frustrated or curious, and not what the object of their emotion was. Thus, the emotions studied were not necessarily *epistemic* in nature. For example, emotions such as excitement or anxiety can easily be related to some issue other than the knowledge processed in a given situation. Some emotions, such as confusion and curiosity, can be regarded as more specifically relate to knowledge.

Thematic analysis always requires the reduction of information. In this study, too, different kinds of activities were eventually included in one category. This was the case especially in the category of orienting and engaging activities in which we had activities with low cognitive demand and passive student participation (i.e. watching introductory videos) and activities with high cognitive demand and active student participation (i.e. asking and elaborating on scientific questions). A more detailed study design should be implemented to shed light on the features that actually make students excited and curious during these orienting and engaging activities.

25.4.4 Implications for Practice and Research

The findings of this study can help identify ways of promoting beneficial epistemic emotions and discouraging adverse ones in educational settings, thus helping teachers to design emotion-aware learning environments. The results offer two important lessons. First, orienting and engaging students in relation to learning by contextualising the subject matter and making it meaningful to them should be emphasised in teaching. Furthermore, future studies should focus on identifying the specific factors in these engaging activities that make students excited and curious and on determining whether these same factors could be incorporated into other activities. Second, all the phases of laboratory work are relevant. If investigations are conducted in a 'cookbook' manner, with a focus on completing the data collection task, a large amount of cognitive work and many epistemic emotions that are pertinent to science are missed. Instead, students should be active in asking scientific questions, planning investigations and collecting, analysing and interpreting data. To implement emotion-aware learning environments in practice, the importance of emotions in educational settings should be taken into consideration in teacher training and educational policy. Finally, we see the need for further studies to clarify the temporal fluctuations of epistemic emotions and their actual effects on, for example, learning, engagement, and well-being.

Acknowledgement This work was supported by the Ministry of Education and Culture, Teacher Education Forum (PI Kalle Juuti) and the Academy of Finland (#298323, PI Katariina Salmela-Aro).

References

Ainley, M., & Ainley, J. (2011). Student engagement with science in early adolescence: The contribution of enjoyment to students' continuing interest in learning about science. *Contemporary Educational Psychology, 36*(1), 4–12. https://doi.org/10.1016/j.cedpsych.2010.08.001

Bosch, N., & D'Mello, S. (2017). The affective experience of novice computer programmers. *International Journal of Artificial Intelligence in Education, 27*(1), 181–206.

Braun, V., & Clarke, V. (2006). Using thematic analysis in psychology. *Qualitative Research in Psychology, 3*(2), 77–101.

Burns, R. B., & Anderson, L. W. (1987). The activity structure of lesson segments. *Curriculum Inquiry, 17*(1), 31–53. https://doi.org/10.1080/03626784.1987.11075276

Calderhead, J. (1981). Stimulated recall: A method for research on teaching. *British Journal of Educational Psychology, 51*(2), 211–217.

Chevrier, M., Muis, K. R., Trevors, G. J., Pekrun, R., & Sinatra, G. M. (2019). Exploring the antecedents and consequences of epistemic emotions. *Learning and Instruction, 63*, 101209. https://doi.org/10.1016/j.learninstruc.2019.05.006

D'Mello, S., & Graesser, A. (2012). Dynamics of affective states during complex learning. *Learning and Instruction, 22*(2), 145–157. https://doi.org/10.1016/j.learninstruc.2011.10.001

D'Mello, S., Lehman, B., Pekrun, R., & Graesser, A. (2014). Confusion can be beneficial for learning. *Learning and Instruction, 29*, 153–170. https://doi.org/10.1016/j.learninstruc.2012.05.003

DeBellis, V. A., & Goldin, G. A. (2006). Affect and meta-affect in mathematical problem solving: A representational perspective. *Educational Studies in Mathematics, 63*(2), 131–147.

Engle, R. A., & Conant, F. R. (2002). Guiding principles for fostering productive disciplinary engagement: Explaining an emergent argument in a community of learners classroom. *Cognition and Instruction, 20*(4), 399–483. https://doi.org/10.1207/S1532690XCI2004_1

Eysenck, M. W. (1979). Anxiety, learning, and memory: A reconceptualization. *Journal of Research in Personality, 13*(4), 363–385.

Finnish National Board of Education. (2016). *National core curriculum for general upper secondary schools 2015*. Helsinki.

Goetz, T., Bieg, M., & Hall, N. C. (2016). Assessing academic emotions via the experience sampling method. In M. Zembylas & P. A. Schutz (Eds.), *Methodological advances in research on emotion and education* (pp. 245–258). Springer.

Gruber, M. J., Gelman, B. D., & Ranganath, C. (2014). States of curiosity modulate hippocampus-dependent learning via the dopaminergic circuit. *Neuron, 84*(2), 486–496.

Hugerat, M. (2016). How teaching science using project-based learning strategies affects the classroom learning environment. *Learning Environments Research, 19*(3), 383–395. https://doi.org/10.1007/s10984-016-9212-y

Hung, C., Hwang, G., & Huang, I. (2012). A project-based digital storytelling approach for improving students' learning motivation, problem-solving competence and learning achievement. *Journal of Educational Technology & Society, 15*(4), 368–379.

Inkinen, J., Klager, C., Schneider, B., Juuti, K., Krajcik, J., Lavonen, J., & Salmela-Aro, K. (2019). Science classroom activities and student situational engagement. *International Journal of Science Education, 41*(3), 316–329. https://doi.org/10.1080/09500693.2018.1549372

Jaber, L. Z., & Hammer, D. (2016). Learning to feel like a scientist. *Science Education, 100*(2), 189–220. https://doi.org/10.1002/sce.21202

Juuti, K., Lavonen, J., Uitto, A., Byman, R., & Meisalo, V. (2010). Science teaching methods preferred by grade 9 students in Finland. *International Journal of Science and Mathematics Education, 8*(4), 611–632.

Juuti, K., Loukomies, A., & Lavonen, J. (2019). Interest in dialogic and non-dialogic teacher talk situations in middle school science classroom. *International Journal of Science and Mathematics Education*. https://doi.org/10.1007/s10763-019-10031-2

Krajcik, J. S., & Shin, N. (2014). Project-based learning. In R. K. Sawyer (Ed.), *The Cambridge handbook of the learning sciences* (pp. 275–297). Cambridge University Press. https://doi.org/10.1017/CBO9781139519526.018

Mann, S., & Robinson, A. (2009). Boredom in the lecture theatre: An investigation into the contributors, moderators and outcomes of boredom amongst university students. *British Educational Research Journal, 35*(2), 243–258.

Meyer, D. K. (2014). Situating emotions in classroom practices. In R. Pekrun & L. Linnenbrink-Garcia (Eds.), *International handbook of emotions in education* (pp. 458–472). Routledge. https://doi.org/10.4324/9780203148211.ch23

Mortimer, E. F., & Scott, P. (2003). *Meaning making in secondary science classrooms*. Open University Press.

Muthén, L. K., & Muthén, B. O. (2017). *Mplus. User's guide*. Muthén & Muthén.

Pekrun, R., Hall, N. C., Goetz, T., & Perry, R. P. (2014). Boredom and academic achievement: Testing a model of reciprocal causation. *Journal of Educational Psychology, 106*(3), 696.

Pekrun, R., Muis, K. R., Frenzel, A. C., & Götz, T. (2018). *Emotions at school*. Routledge, Taylor & Francis Group. https://doi.org/10.4324/9781315187822

Pekrun, R., Vogl, E., Muis, K. R., & Sinatra, G. M. (2017). Measuring emotions during epistemic activities: The epistemically-related emotion scales. *Cognition and Emotion, 31*(6), 1268–1276. https://doi.org/10.1080/02699931.2016.1204989

Radoff, J., Jaber, L. Z., & Hammer, D. (2019). "It's scary but it's also exciting": Evidence of meta-affective learning in science. *Cognition and Instruction, 37*(1), 73–92. https://doi.org/10.1080/07370008.2018.1539737

Schneider, B., Krajcik, J., Lavonen, J., & Salmela-Aro, K. (2020). *Learning science: The value of crafting engagement in science environments*. Yale University Press.

Schneider, B., Krajcik, J., Lavonen, J., Salmela-Aro, K., Broda, M., Spicer, J., ... Viljaranta, J. (2016). Investigating optimal learning moments in U.S. and Finnish science classes. *Journal of Research in Science Teaching, 53*(3), 400–421. https://doi.org/10.1002/tea.21306

Sinatra, G. M., Broughton, S. H., & Lombardi, D. (2014). Emotions in science education. In R. Pekrun & L. Linnenbrink-Garcia (Eds.), *International handbook of emotions in education* (pp. 415–436). Routledge.

Snijders, T. A. B., & Bosker, R. J. (2012). *Multilevel analysis: An introduction to basic and advanced multilevel modeling* (2nd ed.). Sage.

Tseng, K., Chang, C., Lou, S., & Chen, W. (2013). Attitudes towards science, technology, engineering and mathematics (STEM) in a project-based learning (PjBL) environment. *International Journal of Technology and Design Education, 23*(1), 87–102. https://doi.org/10.1007/s10798-011-9160-x

Vogl, E., Pekrun, R., Murayama, K., & Loderer, K. (2020). Surprised–curious–confused: Epistemic emotions and knowledge exploration. *Emotion, 20*(4), 625–641. https://doi.org/10.1037/emo0000578

The manufacturer's authorised representative in the EU is Springer Nature Customer Service Centre GmbH, Europaplatz 3, 69115 Heidelberg, Germany. If you have any concerns regarding our products, please contact ProductSafety@springernature.com

Printed and bound by CPI Group (UK) Ltd, Croydon, CR0 4YY
30/09/2026
02225696-0019